I. Elishakoff · Y. K. Lin (Eds.)

Stochastic Structural Dynamics 2

New Practical Applications

Second International Conference on
Stochastic Structural Dynamics
May 9–11, 1990, Boca Raton, Florida, USA

Springer-Verlag

Berlin Heidelberg New York
London Paris Tokyo
Hong Kong Barcelona Budapest

Prof. I. Elishakoff
Center for Applied Stochastic Research and
Department of Mechanical Engineering
Florida Atlantic University
Boca Raton, FL 33431-0991
USA

Prof. Y. K. Lin
Center for Applied Stochastics Research
Florida Atlantic University
Boca Raton, FL 33431-0991
USA

ISBN-13:978-3-642-84536-9 e-ISBN-13:978-3-642-84534-5
DOI: 10.1007/978-3-642-84534-5

Library of Congress Cataloging-in-Publication Data
International Conference on Stochastic Structural Dynamics (2nd : 1990 : Boca Raton, Fla.)
Stochastic structural dynamics / Y. K. Lin, I. Elishakoff.
Contents: v. 2. New practical applications.
ISBN-13:978-3-642-84536-9
1. Structural dynamics. 2. Stochastic processes.
I. Elishakoff, Isaac. II. Lin, Y. K. (Yu-Kweng). III. Title.
TA654.I567 1990
624.1'7--dc 20 91-18956

Softcover reprint of the hardcover 1st edition 1991

Typesetting: Camera ready by authors

61/3020-543210 – Printed on acid-free paper

Dedicated
to the 25th Anniversary
of Florida Atlantic University

I. E., Y. K. L.

Preface

This volume contains eighteen selected papers presented at the Second International Conference on Stochastic Structural Dynamics, which are related to new practical applications in the field. This and a companion volume, related to new theoretical developments, constitute the proceedings of the conference, and reflect the state of the art of the rapidly developing subject.

The conference was held in Boca Raton, Florida during May 9-11, 1990 hosted by the Center for Applied Stochastic Research of Florida Atlantic University. A total of 20 technical sessions were organized, and attended by eighty participants from 12 countries.

Special emphases of the conference were placed on two areas: applications to earthquake engineering and stochastic stability of nonlinear systems. Two sessions were dedicated to the memory of late Professor Frank Kozin, one of the founders and most active contributors to the stochastic stability theory.

We are indebted to the National Center for Earthquake Engineering Research (NCEER) for financial support. Most credit belongs to each of the authors whose contributions were the very basis for the undoubted success of the conference. We are grateful to the reviewers who carefully refereed the contributions for these two volumes. Our special thanks are due to Mrs. Christine Mikulski, who carried out all the necessary secretarial tasks associated with the conference with dedication.

I. Elishakoff, Y. K. Lin
Editors
Boca Raton, February 1991

List of Participants

(The chairmen of the sessions are identified by asterisks)

* Ang, A.H-S., Department of Civil Engineering, University of California, Irvine, CA 92717, USA

* Ariaratnam, S.T., Solid Mechanics Division, University of Waterloo, Waterlo Ontario, CANADA N2L 3G1

* Benaroya, H., Department of Mechanical and Aerospace Engineering, Rutgers University, Piscataway, NJ 08855-0909, USA

Billah, K. Y. R., Department of Civil Engineering and Operations Research, Princeton University, E-209 Engineering Quadrangle, Princeton, NJ 08544 USA

Cai, G., Center for Applied Stochastics Research, Florida Atlantic University, Boca Raton, FL 33431-0991, USA

* Casciati, F., Dipartamento di Meccanica Strutturale, Dell'Universita Di Pavia, Via Abbiategrasso 211, 27100 Pavia, ITALY

Choi, S. T., Institute of Aeronautics and Astronautics, National Cheng Kung University, Tainan, TAIWAN 70101

Davoori, H., Department of Mechanical Engineering, University of Puerto Ric Mayaguez, PUERTO RICO 00708

Debchaudhury, A., Rockwell International, Rocketdyn, Department 545, 6633 Canoga Ave., Canoga Park, California 91303, USA

Di Paola, M., Dipartimento di Ingegneria Strutturale e Geotecnica, Universita' degli Studi di Palermo, Viale delle Scienze, I 90128, Palerm ITALY

Donley, M. G., Structural Dynamics Research Corporation, 2000 Eastman Drive, Milford, OH 45150-2789, USA

Elishakoff, I., Center for Applied Stochastics Research and Department of Mechanical Engineering, Florida Atlantic University, Boca Raton, FL 33431-0991, USA

Falsone, G., Dipartimento di Ingegneria Strutturale e Geotecnica, Universita' degli Studi di Palermo, Viale delle Scienze, I 90128, Palerm ITALY

Faravelli, L., Dipartamento di Meccanica Strutturale, Dell'Universita Di Pavia, Via Abbiategrasso 211, 27100 Pavia, ITALY

Gaonkar, G.H., Department of Mechanical Engineering, Florida Atlantic University, Boca Raton, FL, 33431-0991, USA

* Hilton, H. H., Department of Aeronautical and Astronautical Engineering, University of Illinois at Urbana-Champaign, 104 S. Mathews, Urbana, IL 61801-2997, USA

* Hoshiya, M., Department of Civil Engineering, Musashi Institue of Technology, 1-28-1 Tamazutsumi Setagaya-ku, Tokyo 157, JAPAN

Hou, Z.K., California Institute of Technology, 104-44, Pasadena, CA 91125, USA

* Irschik, H., Department of Civil Engineering, Institut für Allgemeine Mechanik, Technische Universität Wien, Karlsplatz 13/201 A-1040 Wien AUSTRIA

Kareem, A., Department of Civil Engineering, University of Notre Dame, Notre Dame, IN 46556-0767 USA

Larsen, C. E., Mail Code ES42, Johnson Space Center, NASA, Houston, TX 77058 USA

Leng, G., Department of Aeronautical and Astronautical Engineering, University of Illinois at Urbana-Champaign, 104 South Mathews, Urbana, IL 61801, USA

Li, Z., Department of Civil, Mechanical and Environmental Engineering, The George Washington University, Washington, D.C. 20052, USA

Lin, Y. K., Center for Applied Stochastics Research, College of Engineering, Florida Atlantic University, Boca Raton, Fl, 33431-0991, USA

Liu, W. K., Department of Mechanical Engineering, Northwestern University, Evanston, IL 60201, USA

* Lutes, L. D., Department of Civil Engineering, Texas A&M University, College Station, TX 77843, USA

Maymon, G., Ministry of Defence, RAFAEL - Div. 46, P.O. Box 2250, Stop 46 Haifa, 31 021 ISRAEL

* Minai, R., Disaster Prevention Research Institute, Kyoto University, Gokasho, Uji, Kyoto 611, JAPAN

Muscolino, G., Dipartimento di Ingegneria Strutturale e Geotecnica, Universita' degli Studi di Palermo, Viale delle Science, Palermo, ITALY

* Namachchivaya, N. Sri, Department of Aeronautical and Astronautical Engineering, University of Illinois at Urbana-Champaign, 104 South Mathews, Urbana, IL 61801, USA

Noori, M., Department of Mechanical Engineering, Worcester Polytechnic Institute, 100 Inst. Rd., Worcester, MA 01609, USA

Norwood, M., 5670 Spring Garden Road, Halifax, CANADA B3J 1H6

Orisamolu, I. R., Martec Limited, 5670 Spring Garden Rd., Halifax, Nova Scotia, CANADA B3J 1H6

Palo, P., Code L44, Naval Civil Engineering Laboratories, Ft. Hueneme, CA 93043, USA

Pradlwarter, H.J., Institut für Mechanik, Universität Innsbruck, Technikerstrasse 13, A-6020, Innsbruck, AUSTRIA

Quek, S.T., Department of Civil Engineering, National University of Singapore, 10 Kentridge Crescent, 0511, SINGAPORE

Rahman, M.S., Department of Civil Engineering, North Carolina State University, Box 7908, Raleigh, N.C. 27695-7908, USA

Roy, R.V., R. L. Spencer Laboratories, Department of Mechanical Engineering University of Delaware, Newark, DE 19716, USA

* Scanlan, R. H., Department of Civil Engineering, The Johns Hopkins University, Baltimore, MD 21218, USA

* Schuëller, G.I., Institut für Mechanik, Universität Innsbruck, Technikerstrasse 13, A-6020, Innsbruck, AUSTRIA

Shiao, M. C., Sverdrup Technology, Inc., NASA Lewis Research Center Group, 2001 Aerospace Parkway, Brook Park, OH 44142, USA

* Shinozuka, M., Department of Civil Engineering & Operations Research, Princeton University, E-209 Engineering Quadrangle, Princeton, NJ 08544 USA

* Singh, M. P., Dept. of Engineering Science and Mechanics, Virginia Polytechnic Institute and State University, Blacksburg, VA 24061-0219, USA

Socha, L., Department of Civil Engineering, 212 Ketter Hall, State Universi of NY, Buffalo, NY 14260, USA

* Spanos, P.D., Department of Mechanical Engineering, Rice University, P.O. B 1892, Houston, TX 77251, USA

Sternberg, A., RAPHAEL, Kiriat Mozkin, P.O.B. 2250, Haifa, 31021, ISRAEL

Su, H., Department of Mechanical Engineering, S-B 201, Concordia University 1455 de Maisonneuve Blvd. W., Montreal, Quebec, CANADA H3G 1M8

Su, T-C., Center for Applied Stochastics Research and Department of Ocean Engineering, College of Engineering, Florida Atlantic University, Boca Raton, FL, 33431-0991, USA

* Timashev, S., Science and Engineering Center, USSR Academy of Sciences, Ural Dept., 91 Pervomaiskaya St., 620219, Sverdlovsk, GSP-207, U.S.S.R

* To, C.W.S., Department of Mechanical Engineering, University of Western Ontario, London, Ontario, CANADA N6A 5B9

* Wedig, W.V., Institut für Technische Mechanik, Universität of Karlsruhe, Kaiserstrasse 12, D-7500 Karlsruhe 1, FRG

Xu, K., Dept. of Civil Engineering, Northwestern University, 2145 Sheridan Rd., Evanston, IL 60208, USA

* Yang, C.Y., Department of Civil Engineering, 137 DuPont Hall, University of Delaware, Newark, DE 19716, USA

Yim, S., Department of Civil Engineering, Apperson Hall 206, Oregon State University, Corvallis, OR 97331, USA

Yong, Y., Center for Applied Stochastics Research and Department of Ocean Engineering, Florida Atlantic University, Boca Raton, FL 33431-0991, USA

Zerva, A., Department of Civil Engineering, Drexel University, Philadelphia, PA 19104, USA

Zhang, R., Center for Applied Stochastics Research, Florida Atlantic University, Boca Raton, FL 33431-0991, USA

Zhang, X.T., Department of Engineering Mechanics, Tongji University, Shanghai, 200092 PEOPLE'S REPUBLIC OF CHINA

Zhang, Z.Y., Department of Electrical Engineering and Computer Science, Polytechnic University, 333 Jay Street, Brooklyn, NY 11201, USA

Zhu, W.Q., Department of Mechanics, Zhejiang University, Hangzhou, PEOPLE'S REPUBLIC OF CHINA

Contents

Stochastic Dynamics of Soil Liquefaction

A. H-S. Ang and J. A. Pires

Department of Civil Engineering
University of California, Irvine, California 92717, USA

A probabilistic analysis of sand liquefaction under random seismic loads is presented. The principal features of the methodology are: (i) considers the effects of underlying soft cohesive layers on the severity of shaking; (ii) represents the earthquake ground motions as random processes and calculates the response of the soil deposit in terms of its statistics and probabilities; (iii) uses methods for the dynamic analysis of nonlinear and hysteretic systems to calculate the stress conditions necessary to induce liquefaction; (iv) makes use of the historical data on liquefaction to define the resistance of the soil against liquefaction; and (v) computes the probability that liquefaction in the layer will spread over a specified area if liquefaction occurs anywhere in the layer.

INTRODUCTION

A method to compute the probability of liquefaction of horizontally layered soil deposits under random seismic loads is described. With the method the probability of the onset of liquefaction for any point at a specified depth can be evaluated. This probability of liquefaction for the critical depth plotted as a function of the intensity of the ground motion is the liquefaction fragility for the soil deposit. It is well known, however, that the adverse consequences of liquefaction on buried lifelines will depend on the lateral extent of liquefaction within the critical layer. Even in a seemingly homogeneous soil layer the contiguous lateral extent of liquefaction will be of finite dimensions as a result of random variability of soil properties.

The second part of the paper presents a methodology for computing the probability that liquefaction in the layer will spread over a specified area. The principal assumptions underlying this extension are: (i) the ground motions in the horizontal direction are

I. Elishakoff · Y. K. Lin (Eds.)
Stochastic Structural Dynamics 2
New Practical Applications
2nd International Conference on Stochastic Structural Dynamics
May 9–11, 1990, Boca Raton, Florida

assumed to be perfectly correlated; and (ii) the random spatial variability of the soil resistance against liquefaction can be described by a homogeneous random field, that is, the soil resistance against liquefaction for a layer varies about a constant mean in a manner which is independent of the location in the layer.

LIQUEFACTION IN THE LAYER

Ground Response Statistics

Since the ground acceleration is generally amplified as the seismic wave propagates towards the surface, and the soil properties may vary greatly with depth (layering), the response statistics must be determined at several depths in the deposit. The required statistics are calculated on the basis of a one-dimensional ground response analysis as follows:

(i) a lumped mass model of the soil deposit which includes a transmitting boundary [1,2] is constructed (see Fig. 1);

(ii) the nonlinear and hysteretic behavior of the soil is represented by a smooth hysteretic model [1,3]; and,

(iii) the response statistics of interest are evaluated through random vibration analysis using an equivalent linearization technique[3].

The effect of soft soil layers on the severity of shaking in the liquefiable soil layers is evaluated with the one-dimensional ground response analysis which idealizes the site amplification as the result of vertically propagating shear waves. This approach has been shown to offer a reasonably accurate means for estimating site amplification effects for horizontal ground accelerations [4] which are the accelerations of interest in this study.

Accordingly, a lumped mass model for the soil deposit (see Fig. 1) is constructed as follows. The soil deposit is divided into a number of elements (layers). The lumped masses, m_i, $i=1, 2, ..., n+1$, are obtained by lumping one-half the mass of each layer at the layer boundaries. At the interface between the lowest element and the base only one-half of the mass of the lowest layer is lumped. The motion of the system is described by the total displacements x_i, $i=1, 2, ..., n+1$, of the layer boundaries. Nonlinear springs with stress-strain properties representing the nonlinear, strain-dependent and hysteretic behavior of the soil connect the masses as shown in Fig. 1. The total shear strain at each layer is defined by $\gamma_i = (x_{i+1} - x_i)/\Delta h_i$.

With the smooth hysteretic model the hysteretic component, z, of the shear strain is described by

$$\frac{\partial z}{\partial \gamma} = A - \beta \frac{|\gamma|}{\dot{\gamma}} |z|^{r-1} - \delta |z|^r \tag{1}$$

where γ is the total shear strain, A, β, δ and r are parameters that describe the shape of the hysteresis loops. The shear stress is given by

$$\tau = \alpha G_m \gamma + (1 - \alpha) G_m z \tag{2}$$

where G_m is the (small strains) shear modulus and αG_m is the residual stiffness. The maximum hysteretic shear stress is given by

$$\tau_m = (1 - \alpha) G_m [A/(\beta + \delta)]^{1/r} \tag{3}$$

In this study, the following values are used for the parameters of the smooth hysteretic model: $A = 1.0$, $\delta = \beta$, $r = 0.5$ and $\gamma_r = 4.0 \times 10^{-4}$ (where γ_r is a reference strain defined as τ_m/G_m).

Dynamic equilibrium of each lumped mass requires (for $i = 1, 2, ..., n$),

$$\begin{aligned} m_i\ddot{x}_i + (1-\delta_{i1})[\alpha_{i-1}K_{i-1}(x_i - x_{i-1}) + (1-\alpha_{i-1}) K_{i-1}(\zeta_i - \zeta_{i-1})] \\ - [\alpha_i K_i(x_{i+1} - x_i) + (1-\alpha_i)K_i(\zeta_{i+1} - \zeta_i)] = 0 \end{aligned} \tag{4}$$

where $\delta_{i1}=0$ if $i \neq 1$ and $\delta_{i1}=1$ if $i=1$, $K_i=G_{mi}/\Delta h_i$, $i=1,2,...,n$ and $(\zeta_{i+1}-\zeta_i)/\Delta h_i = z_i$ for $i=1,2,...,n$.

For the n+1st mass the equation of equilibrium is

$$m_{n+1}\ddot{x}_{n+1} + \alpha_n k_n (x_{n+1} - x_n) + (1-\alpha_n) k_n (\zeta_{n+1} - \zeta_n) = \rho_R V_R (2U - \dot{x}_{n+1}) \tag{5}$$

The term $\rho_R V_R (2U - x_{n+1})$ represents the shear stress that is applied to the bottom of the soft soil deposit from the underlying stiffer soil or rock base, where ρ_R and V_R are the unit mass and the velocity of the shear wave propagation in the underlying base. The term U denotes the input motion in terms of particle velocity from the incident wave [2].

In the analysis, the acceleration of the strong ground motion phase of the incident wave is modeled by a stationary Gaussian random process. In this regard, it can be characterized by a power spectral density function, such as the Clough-Penzien spectrum

$$S(\omega) = S_0 \frac{1 + 4\zeta_B^2(\omega/\omega_B)^2}{[1 - (\omega/\omega_B)^2]^2 + 4\zeta_B^2(\omega/\omega_B)^2} \times \frac{(\omega/\omega_G)^4}{[1 - (\omega/\omega_G)^2]^2 + 4\zeta_B^2(\omega/\omega_G)^2} \tag{6}$$

To solve for the response statistics the equations of motion are written as a system of first order differential equations

$$\{\dot{y}\} + [G]\{y\} = \{f\} \tag{7}$$

where the state vector {y} contains the relative displacement and velocities between each two consecutive lumped masses, plus the hysteretic component of the shear-strain z_i for each spring. In addition, the state vector {y} also contains the relative displacements and velocities of the two second-order filters that define the Clough-Penzien power spectrum [5]. The solution for the response requires the determination of the covariance matrix [S], of the response variables satisfying the matrix differential equation

$$[S] + [G][S] + [S][G]^T = [B] \tag{8}$$

where [B] is the matrix of the ground motion parameters[3].

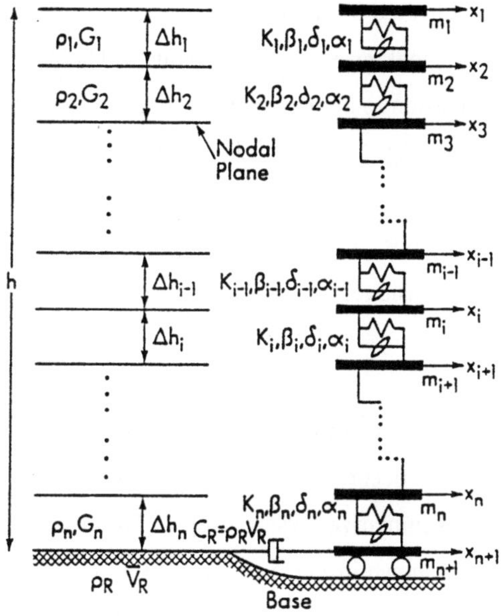

Figure 1. Lumped Mass Model

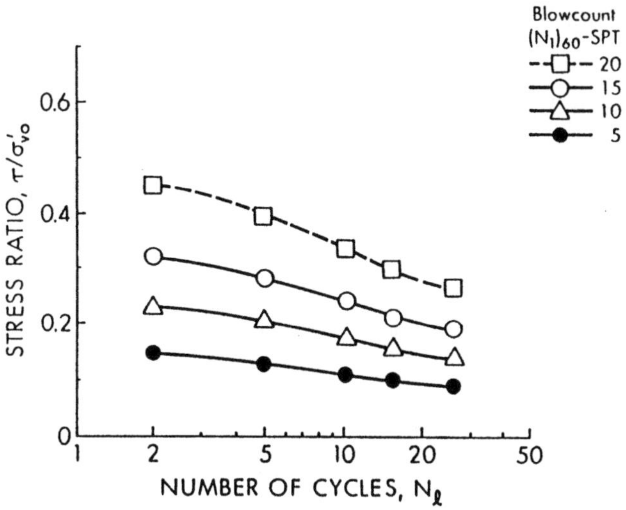

Figure 2. Cyclic Resistance Curves For Liquefaction

Example Application

Probabilities of the onset of liquefaction under various earthquake load intensities were calculated for the soil deposit shown in Fig. 3. The soil deposit was idealized with the lumped-mass model as indicated in Table 1 below and subjected to the respective levels of earthquake intensities (peak ground accelerations). The parameters of the Clough-Penzien spectrum were chosen to be ω_B = 16.9 rad/sec, ζ_B = 0.94, ω_G = 1.25 rad/sec and ζ_G = 0.7. The curve in Fig. 4 labeled "Liquefaction at a point" shows the calculated probabilities of liquefaction as a function of the earthquake intensity and for a strong motion duration of 8.0 seconds. It has been previously shown that the probabilities of liquefaction are far less sensitive to the strong motion duration than to the ground motion intensity[1].

To assess the reliability of the method probabilities of liquefaction were computed for some past case histories of occurrence or non-occurrence of liquefaction. The case histories investigated are summarized elsewhere[1]. The results of the investigation are shown in Fig. 5. The line shown in Fig. 5 separates the data for which liquefaction has occurred from those for which liquefaction has not occurred for 7.5-magnitude earthquakes. The further a data point plots below the line the smaller is the probabilitiy of liquefaction, whereas the further the data point plots above the line the greater is the probability of liquefaction. Predictions made with the method appear to be consistent with the observed data.

Table 1. Lumped-Mass Model for the Soil Deposit Analyzed

Element Number	σ'_v (ton f/m^2)	mass (ton)	Shear Modulus G_m (tonf/m^2)
1	1.065	2.13	3,710
2	3.195	2.13	3,710
3	4.680	2.84	3,710
4	5.520	2.84	3,710
5	6.640	3.90	4,760
6	8.040	3.90	4,760
7	9.765	4.55	9,210
8	11.815	4.55	9,210
9	13.680	5.04	11,858
10	15.90	5.04	11,858
11	(boundary element: C_R = 90 tonf - sec/m and a mass of 2.52 ton)		

G.L. (m)	Name of Layer	Width of Layer (m)	N-Value SPT	Shear Wave Velocity (m/sec)	Unit Weight (ton/m**3)	Fc(%)
3.0 GWT 7.00	Kanto loam	7.0	4	160	1.42	
12.0	Clay with loam	5.0	5	173	1.56	
17.0	Tokyo formation clayed sand	5.0	22	225	1.82	32
23.0	Tokyo formation silt	6.0	18	263	1.68	
				450		

Fc is the percent of fines

Figure 3. Soil Deposit Analyzed

Probability of Liquefaction

The resistance of the sand against liquefaction is defined by the number of cycles, $N_\ell(\bar{\tau})$, of constant amplitude shear stress ratio, $\bar{\tau}$, necessary to induce liquefaction; this is known as the "cyclic resistance curve". The shear stress ratio is defined as τ/σ'_o, where τ is the shear stress in the soil and σ'_o the effective overburden stress. Based on observations of the occurrence of liquefaction during past earthquakes, values of the cyclic shear stress ratio known to have caused liquefaction under a given intensity of shaking have been correlated with a readily available measure of the in-situ condition of the sand, the Standard Penetration Test blowcount normalized to an energy ratio of 60 percent of the theoretical maximum energy and to an effective overburden stress of 1.0 ton per sq. ft., i.e., $(N_1)_{60}$-SPT[6]. Based on those correlations it is possible to determine "cyclic resistance curves" for the sand for various values of the $(N_1)_{60}$-SPT, such as those shown in Fig. 2[1].

Any of the "cyclic resistance curves" can be interpreted by

$$N(\bar{\tau})[E_c(\bar{\tau})h(\bar{\tau})] = W_u \tag{9}$$

where W_u is an arbitrary constant, $E_c(\bar{\tau})$ is the energy dissipated through hysteresis in one cycle of shear stress amplitude $\bar{\tau}$, and $h(\bar{\tau})$ is a weighing function. The concept of the number of equivalent uniform loading cycles [6] implies that under N loading cycles with different shear stress, $\bar{\tau}_i$, liquefaction would occur when

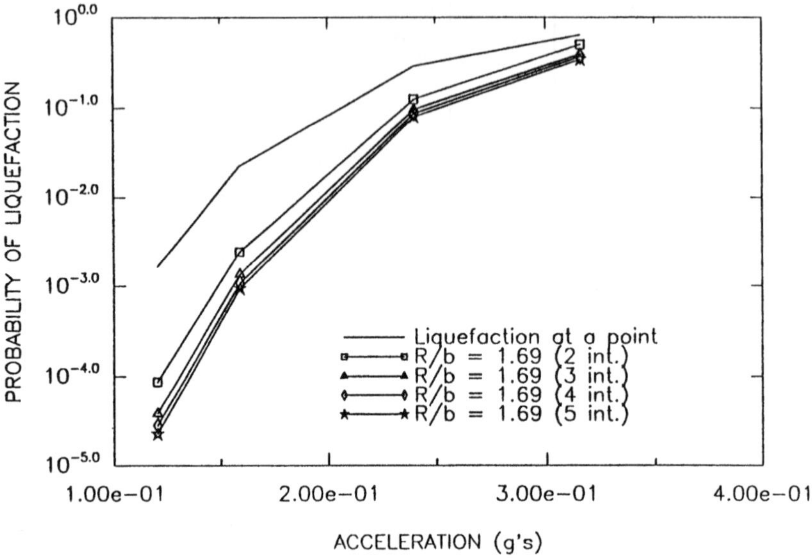

Figure 4. Liquefaction Fragilities

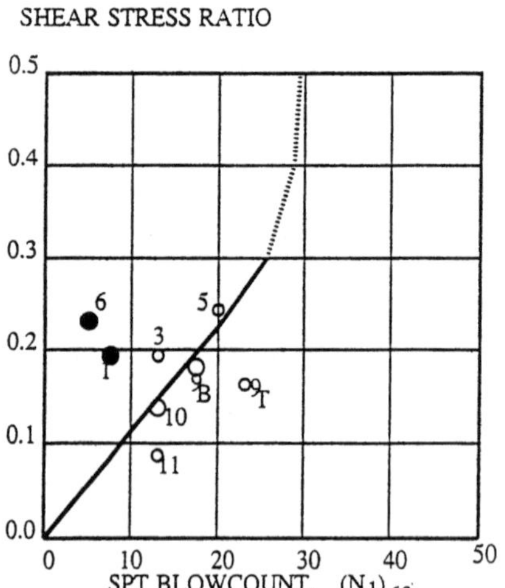

DATA CASE	PROBABILITY OF LIQUEFACTION
1	1.000
3	0.954
5	0.460
6	1.000
9_B	0.233
9_T	2.6×10^{-5}
10	0.142
11	5.8×10^{-5}

Figure 5. Probabilities of Liquefaction (Data Cases)

$$W = \sum_{i=1}^{N} E_c(\bar{\tau}_i) h(\bar{\tau}_i) \tag{10}$$

reaches W_u.

For an earthquake with a given intensity (expected maximum ground acceleration or rms ground acceleration), A = a, and a strong motion duration T_E = t, the energy dissipated by hysteresis is random and the quantity W becomes a random variable denoted W(a,t). Liquefaction occurs when

$$Z = W_u - W(a,t) < 0.0 \tag{11}$$

where W_u denotes the hysteretic energy dissipation capacity of the soil layer. The quantity W(a,t) is given by

$$W(a,t) = \int_0^t X(r) \dot{e}_T(r) dr \tag{12}$$

where $\dot{e}_T(r)$ is the rate of hysteretic energy dissipated at time r and X(r) is an equivalent weighing function to include the effect of random stress amplitude[1] given by

$$X(r) = \left[\int_0^{\bar{\tau}_m} h(\bar{\tau}) E_c(\bar{\tau}) f_T(\bar{\tau},r) d\bar{\tau}\right] / \left[\int_0^{\bar{\tau}_m} E_c(\bar{\tau}) f_{\bar{T}}(\bar{\tau},r) d\bar{\tau}\right] \tag{13}$$

where $\bar{\tau}_m$ is the normalized static yield shear stress of the sand, and $f_T(\bar{\tau},r)$ is the probability density function of the peaks of $\bar{\tau}$ at time r.

The mean and variance of W(a,t) are obtained from the random vibration analysis, in particular from the mean and variance of the hysteretic energy dissipated[7]. The statistics of W_u are obtained from the uncertainty analysis of the cyclic resistance against liquefaction. On this basis the "cyclic resistance curves" shown in Fig. 2 are regarded as mean "cyclic resistance curves", and the coefficient of variation of the number of loading cycles that will cause liquefaction at a given stress ratio is independent of the stress ratio and equal to 0.57[1].

LIQUEFACTION SPREAD

Methodology

Let E_0 denote the event that liquefaction occurs at any point in the layer of interest (see Fig. 5), independently of what happens elsewhere in the layer. With the previous notations (see Eq. 11) the probability of E_0 is given by

$$P[E_0] = \int_0^{\infty} F_{W_u}(q) f_{W(a,t)}(q) dq \tag{14}$$

where F_{W_u} denotes the CDF of W_u and $f_{W(a,t)}$ is the probability density function of W(a,t). To simplify the notation introduce $S = W_u$ and $R = W(a,t)$. To compute the probability that liquefaction will extend over a circle of radius R (axisymmetric random field), the circle is divided into segments of length D, as shown in Fig. 5 for 5 segments numbered 0, 1, ..., 4. The soil resistances against liquefaction at the center of each segment are random variables denoted by S_i, i = 0, 1, ..., 4. The random variables S_i are identically distributed, whereas the correlation coefficient between any pair of random variables (S_i, S_j) separated by a distance kD where k is a positive integer is

$$\rho_{S,S}(k,D) = \frac{R_{S,S}(kD)}{\sigma_{S_i} \sigma_{S_j}} \tag{15}$$

in which $R_{S,S}(x)$ is the autocarrelation function of the random field S(x) and σ_{Si} is the standard deviation of S_i.

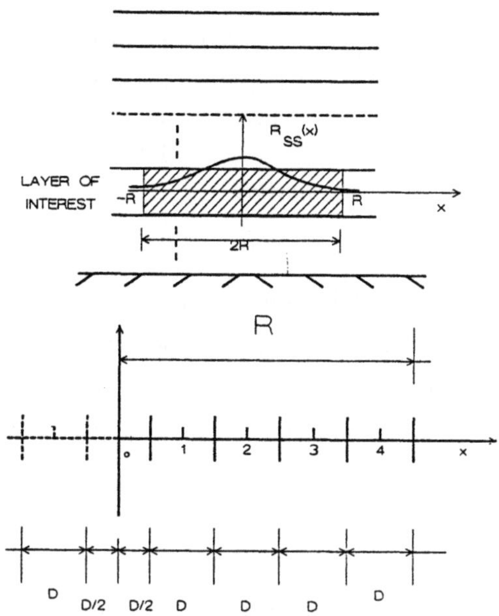

Figure 6. Liquefaction Spread in Layer

The probability that liquefaction will extend over the entire circle of radius R = 4.5D is the probability that all segments in the circle will liquefy. Therefore,

$$P_L(R = 4.5D) = \int_0^{\infty} F_{S_0 S_1 \ldots S_4}(q, \ldots, q) f_Q(q) dq \tag{16}$$

where

$$F_{S_0 S_1 \ldots S_4}(q, \ldots, q) = \int_0^{\infty} \ldots \int_0^{\infty} f_{S_0 S_1 \ldots S_4}(s_0, s_1, \ldots, s_4) \times dS_0 dS_1 \ldots dS_4 \tag{17}$$

The random variables S_i may be assumed to follow a lognormal distribution with median S_m and standard deviation σ_{Si}[1]. The matrix of the correlation coefficients for the random variables S_i is given by

$$[p] = \begin{bmatrix} 1.0 & \rho_{SS}(D) & \rho_{SS}(2D) & \rho_{SS}(3D) & \rho_{SS}(4D) \\ & 1.0 & \rho_{SS}(D) & \rho_{SS}(2D) & \rho_{SS}(3D) \\ & & 1.0 & \rho_{SS}(D) & \rho_{SS}(2D) \\ & & & 1.0 & \rho_{SS}(D) \\ & & symmetric & & 1.0 \end{bmatrix} \tag{18}$$

Evaluation of the multiple integral in Eq. 17 requires numerical integration. All such integrations were performed using an adaptive algorithm for numerical integration over an N-dimensional rectangular region[4]. With this algorithm, the numerical integrations can be performed without major difficulties except when the correlation matrix becomes nearly singular.

Example Application

The method is used to compute the probability of liquefaction spreading over a circle with a specified radius as a function of the expected peak ground acceleration. The soil deposit and loading conditions for the previous example (see Table 1 and Fig. 2) are chosen.

The correlation function of the soil resistance against liquefaction is assumed to be of the form

$$\rho_{S,S}(x) = \exp[-(x/b)^2] \tag{19}$$

where b is a positive parameter. This form of the autocorrelation function is extensively used in random fields applications in Civil Engineering[9] and has been suggested for some soil properties (e.g., relative density, grain-size distribution and shear strength[10,11]). Other possible forms for the correlation function have also been used[10,11]; e.g.,

$$\rho_{S,S}(x) = \exp(-|x|/a) \tag{20}$$

with a > 0.0. Here, b is treated as a parameter and the results are presented as a function of the dimensionless parameter R/b.

The radius of the circle is fixed at R = 1.69b and the sensitivity of the results to the number of segments in which the circle is subdivided is investigated; up to five segments are considered, which correspond to incremental lengths of D equal to 1.127b, 0.676b, 0.483b, and 0.376b, respectively. The computed probabilities of liquefaction over the radius R are shown in Fig. 4 as a function of the peak ground acceleration. The probabilities (over a circle of radius R = 1.69b) decrease as the number of segments increase and appear to converge for D = 0.376b. For a smaller D, D = 0.307b, the correlation matrix (Eq. 9) becomes almost singular and the integral in Eq. 17 cannot be evaluated as accurately as before. Nevertheless, the approximate values were found to be almost identical to those for D = 0.376b. As expected, the probability of liquefaction for a circle of radius R is significantly lower than that for a single point in the layer.

The probabilites of liquefaction for a peak ground acceleration of 0.24g for three different radii, namely R = 0.94b, R = 1.69b and R = 2.44b, as given by Eq. 17, are shown in Table 2 (second column).

Table 2. Probabilities of Liquefaction for Various Radii

Radius	Equation 17	Approximate Expression
17.8m	0.1583	0.1478
32.0m	0.07868	0.07868
50.3m	0.03867	0.04094

To compute the probability of liquefaction over a circle with radius $R_n = (n+0.5)D$, where n is a positive integer, and (n + 1)-fold integral needs to be evaluated numerically (see Eq. 17). As the radius of the circle increases to $R_{m+n} = (m+n+0.5)D$, a numerical integration of an (m+n+1)-fold integral is involved. If many different radii need to be considered, the numerical integrations can become very time consuming. Therefore, it is desirable to find a simplified method to compute the probability of liquefaction for a circle of a given radius, say $R_k = (k+0.5)D$, given that the probability of liquefaction for a circle with radius R_n is known. The following simplified method is proposed:

(i) the probability of liquefaction for a circle with radius $R_n = (n+0.5)D$ and for a load effect with intensity q may be written as

$$P_L(R_n,q) = F_{S_b}(q)\exp[-n\alpha(q)] \tag{21}$$

which can be solved for $\alpha(q)$, obtaining

$$\alpha(q) = \ln[P_L(R_n,q)/F_{S_b}(q)]/(-n) \tag{22}$$

and

(ii) for other radii, $R_k = (k+0.5)D$, where k is a positive integer, the value of $P_L(R_k,q)$ is

$$P_L(R_k,q) = F_{S_b}(q)\exp[-\alpha(q)k] \tag{23}$$

The probabilities of liquefaction over circles with radii, R = 0.94b, R = 1.69b and R = 2.44b, as computed by Eqs. 21-23, are shown in the third column of Table 2. It appears, that for this load intensity, sufficient accuracy is achieved with the approximate procedure. .

Figure 7 shows liquefaction fragility curves for circles with radii, R = 0.94b, R = 2.44b (solid lines) computed by Eqs. 21-23 and taking the results for R = 1.69b as the reference values. The same probabilities computed by Eq. 17 are shown in Fig. 7 by dashed lines. It thus appear that the approximate method (Eqs. 21-23) is sufficiently accurate. The results in Fig. 6 also show that the probabilities of liquefaction decrease significantly as the radius of the circle increases.

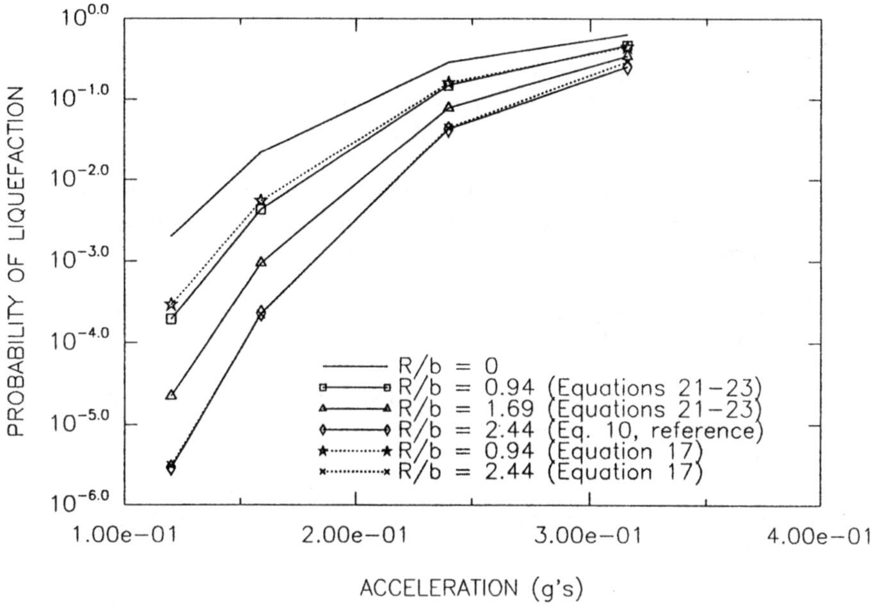

Figure 7. Liquefaction Fragilities (varying radii)

CONCLUSIONS

A method to calculate the probability of liquefaction of soil deposits under random seismic loads is presented. The method includes the effect of soft layers on the severity of ground shaking at a site. An important feature of the method is that it makes use of field observations from past earthquakes to define the sand resistance against liquefaction. Finally, the method also computes the probability that liquefaction will spread over a specified area on the condition that liquefaction occurs in some layer at the site. Prediction of the contiguous lateral extent of liquefaction is important for the reliability evaluation of buried lifelines.

References

[1] Pires, J.A., and Ang, A.H-S. and Katayama, I., (1989), "Probabilistic Analysis of Liquefaction", in Structural Dynamics and Soil Structure Interaction, A.S. Cakmak and I. Herrera, Eds., Computational Mechanics Publications, England.

[2] Ishihara, K. and Towhata, I., (1980), "One-Dimensional Soil Response Analysis During Earthquakes Based On Effective Stress Method", Journal of Faculty of Engineering, University of Tokyo, Vol. 35, No. 4, 1980.

[3] Wen, Y.K., (1980), "Equivalent Linearization for Hysteretic Systems Under Random Excitation", J. Appl. Mech., Trans. ASME, 47(1), 150-154.

[4] Idriss, I.M., (1990), "Response of Soft Soil Sites During Earthquakes", presented at the Memorial Symposium to Honor Professor Harry Bolton Seed, Berekeley, CA, May 10-11.

[5] Dynamics of Structures by Ray W. Clough and Joseph Penzien, McGraw Hill, New York, 1975.

[6] Seed, et al., (1984), "The Influence of SPT Procedures in Soil Liquefaction Evaluation", Report No. UCB/EERC-84/15, University of California, Berkeley, California.

[7] Pires, J.A. and Tang, M., (1990), "Statistics of Hysteretic Energy Dissipated Under Random Dynamic Loading," J. Engng. Mech., Div., ASCE, Vol. 116, No. 8, pp. 1706-1721.

[8] Genz, A.C. and Malik, A.A., (1980), "Remarks on algorithm 006: An adaptive algorithm for numerical integration over an N-dimensional rectangular region", Journal of Computational and Applied Mathematics, Vol. 6, No. 4.

[9] Yamazaki, F. and Shinozuka, M., (1989), "Statistical Preconditioning in Simulation of Stochastic Vectors", Proceedings, ICOSSAR '89, A.H-S. Ang, M. Shinozuka and G.I. Shueller, Editors, ASCE, New York, USA.

[10] Vanmarcke, E., (1977), "Reliability of Earth Slopes", Journal of the Geotechnical Engineering division, ASCE, Vol. 103, No. GT11, November, pp. 1247-1265.

[11] Fardis, M.N. and Veneziano, D., (1981), "Estimation of SPT-N and Relative Density", Journal of the Geotechnical Engineering Division, ASCE, Vol. 107, No. GT10, October, pp. 1245-1359.

Extreme Peak Distribution and Its Applications in Stochastic Structural Dynamics Problems

Siu-Tong Choi

Institute of Aeronautics and Astronautics
National Cheng Kung University
Tainan, Taiwan 70101, R.O.C.

Wen-Fang Wu

Department of Mechanical Engineering
National Taiwan University
Taipei, Taiwan 10764, R.O.C.

Abstract

The extreme peak is defined as the largest value of all peaks which one may encounter in a random process in stochastic structural dynamics analysis. The probability distribution and its statistical quantities of the extreme peak are very important to a structural engineer. In the present paper, order statistics and asymptotic theory of statistical extremes are applied directly to the peak distribution function of a stationary random process in order to find the probability distribution of the extreme peak within a time interval. The extreme peak distribution of a random process obtained from this approach is compared with results from some classical papers. It is concluded that the present approach can be applied not only to a narrow-band process but also to a wide-band process for which the classical papers used several approximations to find approximate results. After the extreme peak distribution is obtained, the present paper will also illustrate its possible applications in aerospace engineering, ocean engineering, wind engineering, and many other structural engineering fields.

Introduction

In modeling the structural loading and response as random processes, the extreme value problem is closely related to the reliability of the structure. It is a common interest among all probabilistic engineering mechanics fields and has been studied extensively by many researchers for the past three decades. The major concern of the extreme value problem is to determine the probability distribution function, or its statistical quantities, of the peaks and/or troughs of a random process at a certain time instant or within a certain time interval. It is of particular interest in ocean, wind and earthquake engineering applications.

Among many publications related to the extreme value problem, one cannot neglect Rice's classical paper where the peak distribution of a stationary zero-mean random process was first determined [1]. Following Rice's basic derivation, Longuet-Higgins [2]

I. Elishakoff · Y. K. Lin (Eds.)
Stochastic Structural Dynamics 2
New Practical Applications
2nd International Conference on Stochastic Structural Dynamics
May 9–11, 1990, Boca Raton, Florida

and Cartwright and Longuet-Higgins [3] derived various peak wave statistics in the ocean engineering field, first for a narrow-band Gaussian process, then extended the results to a Gaussian process of any spectral width. The probability density and cumulative distribution functions of the maxima of such random processes were derived in analytical forms. The same results can also be found in the aerospace engineering literature at almost the same time [4]. It is noted that in Longuet-Higgins' and Cartwright and Longuet-Higgins' papers, if the number of the ocean wave peaks can be counted by observation, then many statistical quantities related to the maximum peak can also be obtained based on the assumption of peak independence and on Gumbel's statistical theory of extreme values [5,6]. In wind engineering application, Davenport used some results derived by Rice and some from Cartwright and Longuet-Higgins to obtain several statistical quantities of the peak gust within a given period [7,8]. He concluded that the peak response of most structures to gusts may be expected to lie in the range of 3.5 to 4.5 standard deviations of the response in excess of the mean response. This conclusion has been widely used in the design of wind-loaded structures.

In the present paper, Gumbel's order statistics and asymptotic theory of statistical extremes are applied directly to the peak distribution function of a stationary random process in order to find the probability distribution of the extreme peak within a time interval. The extreme peak distribution of a random process obtained from this approach is then compared with results from some classical papers. It is shown that the present approach can be applied not only to a narrow-band process but also to a wide-band process for which the classical papers used several approximations to find approximate results. After the extreme peak distribution is obtained, the present paper will also illustrate its possible applications in aerospace engineering, ocean engineering, wind engineering, and many other structural engineering fields.

Extreme Value Distribution

Consider a stationary Gaussian random process, $X(t)$, having a zero mean and a standard deviation σ_x. The probability density function of its peak (including both positive and negative maxima) at any time instant is given by [3,4]

$$p(\eta) = \frac{\epsilon}{\sqrt{2\pi}\sigma_x} \exp\left(-\frac{\eta^2}{2\epsilon^2\sigma_x^2}\right) + \frac{\eta\sqrt{1-\epsilon^2}}{\sigma_x^2}\Phi\left(\frac{\eta\sqrt{1-\epsilon^2}}{\epsilon\sigma_x}\right)\exp\left(-\frac{\eta^2}{2\sigma_x^2}\right)$$

$$-\infty < \eta < \infty \tag{1}$$

in which

$$\Phi(u) = \frac{1}{\sqrt{2\pi}}\int_{-\infty}^{u} e^{-x^2/2}dx \tag{2}$$

$$\epsilon^2 = 1 - \frac{m_2^2}{m_0 m_4} \tag{3}$$

$$m_n = \int_{-\infty}^{\infty} \omega^n S(\omega) d\omega, \qquad n = 0, 2, 4 \tag{4}$$

and $S(\omega)$ is the spectral density function of the random process. It is well known that the parameter ϵ is a measure of the band width of the random process which has a value between zero and one. For a narrow-band process, ϵ is zero and Eq. (1) reduces to a Rayleigh probability density function given by

$$p(\eta) = \frac{\eta}{\sigma_x^2} \exp\left(-\frac{\eta^2}{2\sigma_x^2}\right) \qquad 0 \le \eta < \infty \tag{5}$$

For a relative wide-band process, ϵ reaches one and Eq. (1) reverts to a Gaussian probability density,

$$p(\eta) = \frac{1}{\sqrt{2\pi}\sigma_x} \exp\left(-\frac{\eta^2}{2\sigma_x^2}\right) \qquad -\infty < \eta < \infty \tag{6}$$

The cumulative distribution function of the peaks corresponding to Eq. (1) can also be derived in the following closed form [3]:

$$F(\eta) = \Phi\left(\frac{\eta}{\epsilon\sigma_x}\right) - \sqrt{1-\epsilon^2}\,\Phi\left(\frac{\eta\sqrt{1-\epsilon^2}}{\epsilon\sigma_x}\right) \exp\left(-\frac{\eta^2}{2\sigma_x^2}\right) \tag{7}$$

As before, as ϵ becomes zero, Eq. (7) reduces to a Rayleigh distribution function,

$$F(\eta) = 1 - \exp\left(-\frac{\eta^2}{2\sigma_x^2}\right) \qquad 0 \le \eta < \infty \tag{8}$$

and as ϵ reaches one, it reduces to a Gaussian distribution function,

$$F(\eta) = \Phi\left(\frac{\eta}{\sigma_x}\right) \qquad -\infty < \eta < \infty \tag{9}$$

In Refs. 2 and 3, the distribution of the largest peak in a sequence of N peaks was determined using Eqs. (1), (7) and the standard order statistics [5]. For a sequence of random variables $\eta_1, \eta_2, \cdots, \eta_N$ representing the peaks, which are assumed to be identically distributed and statistically independent, the cumulative distribution of the largest peak is given by

$$\begin{aligned} F_{\eta_{max}}(\eta) &= P\left[\max(\eta_1, \eta_2, \cdots, \eta_N) \le \eta\right] \\ &= \left[F(\eta)\right]^N \end{aligned} \tag{10}$$

where a "*max*" is added to η to denote that it is the maximum peak in a sequence of N peaks. The assumption that the random variables are identically distributed is valid since all of them follow Rice's peak distribution given by Eq. (1) or (7). The assumption of independence, however, may lead to slight overestimation if the random variables are dependent, which is likely to be the case for a narrow-band process of general interest.

According to the asymptotic theory of statistical extremes [9], as N becomes considerably large, Eq. (10) approaches to the following Type I asymptotic distribution of the largest value if the original peak distribution $F(\eta)$ is of the exponential type to the right [6]:

$$F_{\eta_{max}}(\eta) = \exp\{-\exp[-\alpha_N(\eta - u_N)]\} \tag{11}$$

in which the characteristic parameter u_N and the inverse measure of dispersion α_N are dependent on the original peak distribution, i.e., Eq. (7) in the present case. Once these parameters are determined, many statistical quantities related to the largest peak can be calculated. For example, the mean value and the standard deviation of the largest peak are, respectively,

$$\bar{\eta}_{max} = u_N + \frac{\gamma}{\alpha_N} \tag{12}$$

$$\sigma_{\eta_{max}} = \frac{\pi}{\sqrt{6}\alpha_N} \tag{13}$$

where $\gamma = 0.577216\cdots$ is the Euler constant.

For a peak distribution such as Eq. (7), the asymptotic maximum peak distribution in the form of Eq. (11) cannot be found analytically. Fortunately, for the special case that ϵ is zero, using the Rayleigh distribution, Eq. (8), and Eq. (10), one can find that

$$\frac{\bar{\eta}_{max}}{\sigma_x} = \sqrt{2\ln N} + \frac{\gamma}{\sqrt{2\ln N}} \tag{14}$$

and

$$\frac{\sigma_{\eta_{max}}}{\sigma_x} = \frac{\pi}{\sqrt{6}}\frac{1}{\sqrt{2\ln N}} \tag{15}$$

and for the case that ϵ reaches one, using the Gaussian distribution, Eq. (9), together with Eq. (10), one finds

$$\frac{\bar{\eta}_{max}}{\sigma_x} = \sqrt{2\ln N} + \frac{\gamma}{\sqrt{2\ln N}} - \frac{\ln\ln N + \ln 4\pi}{2\sqrt{2\ln N}} \tag{16}$$

and

$$\frac{\sigma_{\eta_{max}}}{\sigma_x} = \frac{\pi}{\sqrt{6}}\frac{1}{\sqrt{2\ln N}} \tag{17}$$

It is of interest to note that in Ref. 2, the following asymptotic expression was derived for the maximum wave amplitude of a stationary narrow-band Gaussian process which has N sample peaks chosen at random:

$$\frac{\bar{\eta}_{max}}{\sqrt{\mu_2}} = \sqrt{\ln N} + \frac{\gamma}{2}\frac{1}{\sqrt{\ln N}} \tag{18}$$

where μ_2 is the second moment (about the origin) of the amplitude of the process. Since the peak amplitude of a narrow-band process has a Rayleigh distribution, it can be shown that $\mu_2 = 2\sigma_x^2$, and Eq. (18) is exactly the same as Eq. (14). While in Ref. 3, the band-width was taken into consideration for the statistics of the maximum wave peak, and the expected maximum peak in a sequence of N independent peaks was derived as

$$\frac{\bar{\eta}_{max}}{\sqrt{\mu_2}} = \frac{\left[\ln\left\{\left(1-\epsilon^2\right)^{1/2} N\right\}\right]^{1/2} + \frac{1}{2}\gamma\left[\ln\left\{\left(1-\epsilon^2\right)^{1/2} N\right\}\right]^{-1/2}}{\left(1-\epsilon^2/2\right)^{1/2}} \tag{19}$$

where $\mu_2 = (2-\epsilon^2)\sigma_x^2$ was also derived. It is interesting to see that for a narrow-band process, ϵ becomes zero and Eq. (19) reduces to Eq. (14) again. On the contrary, for a wide-band process, as ϵ approaches one, Eq. (19) is no longer valid. In Cartwright and Longuet-Higgins' paper, the following expression was proposed for this special case, i.e., as ϵ approaches one,

$$\frac{\bar{\eta}_{max}}{\sigma_x} = \sqrt{2\ln\left(\frac{N}{\sqrt{2\pi}}\right)} \tag{20}$$

In fact, if the asymptotic theory of statistical extremes is used, Eq. (16) is also a proper expression for the case $\epsilon = 1$ just as Eq. (14) is for the special case $\epsilon = 0$. In order to check the accuracy of this proposition, numerical calculations are performed based on Eqs. (16) and (20), and results are shown in Table 1. The exact mean maximum peaks from Gumbel's book [6] are also included in Table 1. By comparing the exact and approximate solutions, it is found that Eq. (16) gives better approximations than Eq. (20) does. The differences between the exact values and those obtained from Eq. (16) are found almost negligible for the entire range of N shown in Table 1. It is also found that the asymptotic value from Eq. (16) becomes more and more accurate as N becomes larger and larger.

For an ϵ other than zero and one, Eqs. (7), (11), (12) and (13) can still be used to find the mean value and standard deviation of the maximum peak. However, some numerical techniques are needed. Some results are shown in Figs. 1 to 5. Figs. 1 and 2 plot the mean values and standard deviations of the maximum peaks with respect to the number of peaks for different values of ϵ. Figs. 3 and 4 show the probability density functions and distribution functions of the maximum peak under different values of ϵ for the case $N = 100$ and $N = 1000$, respectively. Fig. 5 shows the distribution functions of the maximum peak for the case that $\epsilon = 0.6$ and different values of N. In addition, for a comparison between the present proposition and that based on Eq. (19) proposed by Cartwright and Longuet-Higgins, the expected maximum peaks are calculated for $\epsilon = 0.99$ and different values of N. The results are tabulated in Table 2. It can be seen from Table 2 as well as from Table 1 that the present analysis does give us a reasonable

$\epsilon = 1.00$			
N	Eq. (20)	Eq. (16)	Ref. 6
10	1.6635	1.6310	1.6105
20	2.0380	1.9424	1.9247
25	2.1447	2.0356	2.0186
33	2.2749	2.1514	2.1353
50	2.4467	2.3073	2.2922
67	2.5616	2.4131	2.3987
100	2.7152	2.5565	2.5430
125	2.7962	2.6327	2.6196
167	2.8973	2.7285	2.7158
200	2.9595	2.7877	2.7754
250	3.0340	2.8589	2.8470
333	3.1273	2.9485	2.9371
500	3.2544	3.0712	3.0603
1000	3.4609	3.2718	3.2616

Table 1. Comparison of the expected maximum peaks for the case $\epsilon = 1$ obtained from Eq. (20), Eq. (16), and Ref. 6

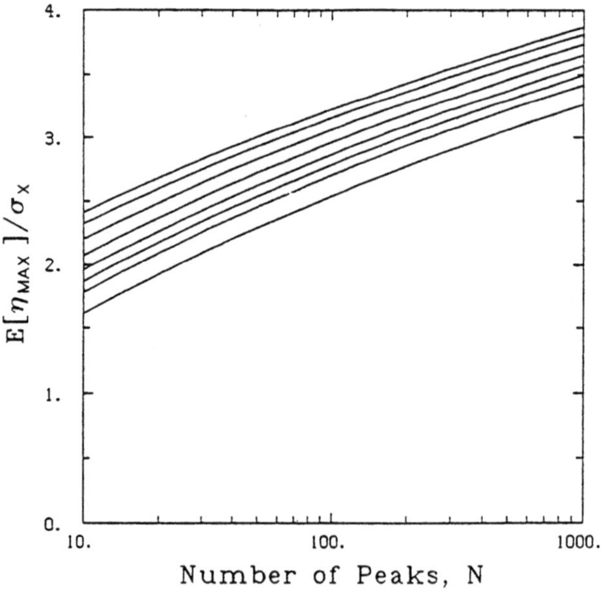

Fig. 1. Expected maximum peaks for cases $\epsilon = 0$ (the highest curve), 0.6, 0.8, 0.9, 0.95, 0.975, 0.99 and 1.00 (the lowest curve).

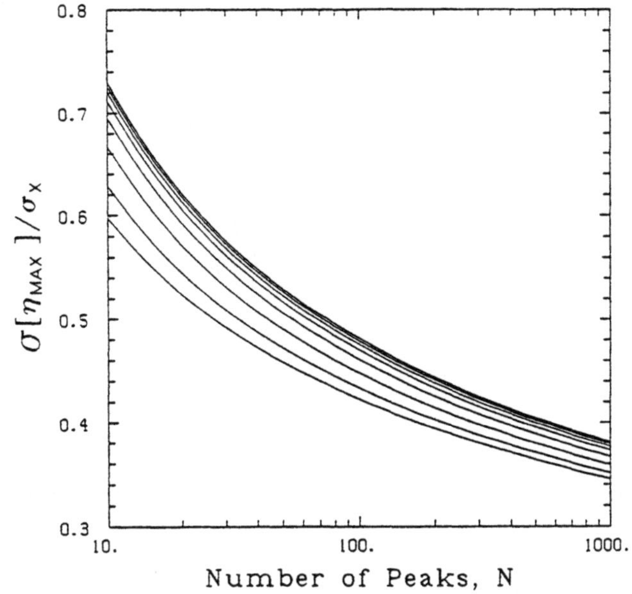

Fig. 2. Standard deviations of maximum peaks for cases $\epsilon = 0$ (the lowest curve), 0.6, 0.8, 0.9, 0.95, 0.975, 0.99 and 1.00 (the highest curve).

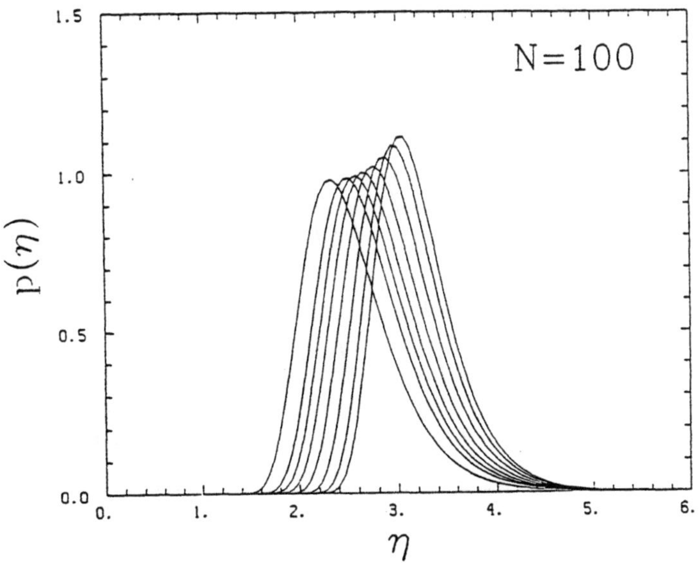

Fig. 3. Probability density functions of the maximum peak for cases $\epsilon = 0$ (the rightist curve), 0.6, 0.8, 0.9, 0.95, 0.975, 0.99 and 1.00 (the leftist curve).

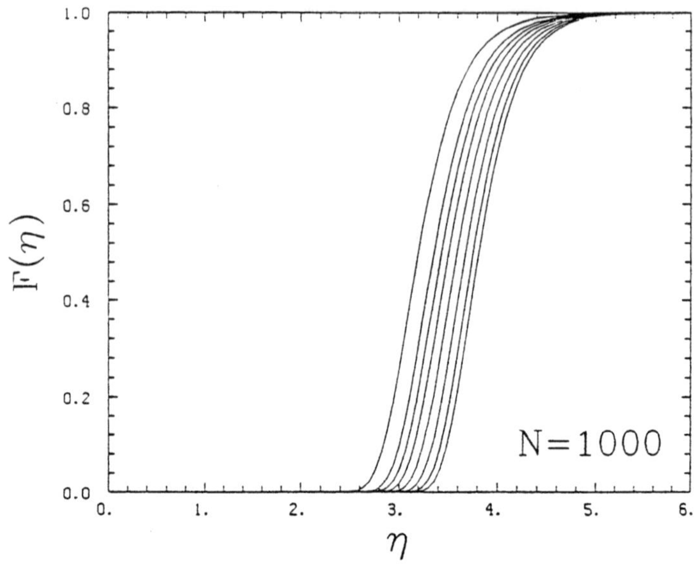

Fig. 4. Probability distribution functions of the maximum peak for cases $\epsilon = 0$ (the rightist curve), 0.6, 0.8, 0.9, 0.95, 0.975, 0.99 and 1.00 (the leftist curve).

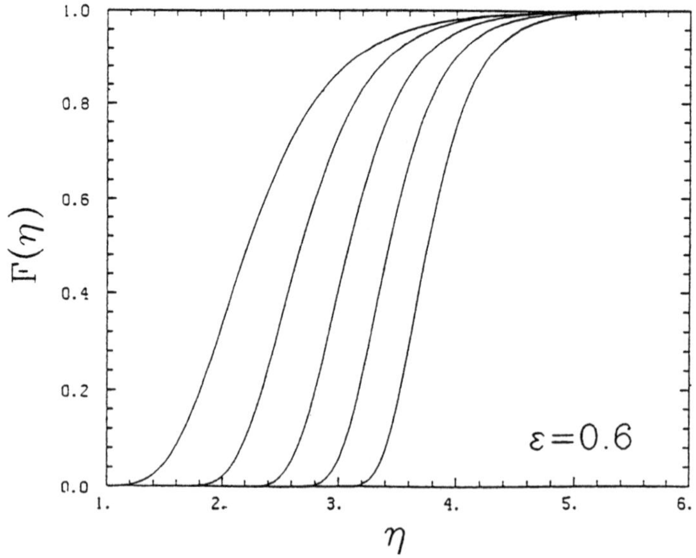

Fig. 5. Probability distribution functions of the maximum peak for cases $N = 10$ (the leftist curve), 30, 100, 300 and 1000 (the rightist curve).

$\epsilon = 0.99$		
N	Present	Eq. (19)
10	1.7641	1.5254
20	2.0804	1.8411
30	2.2524	2.0384
40	2.3633	2.1706
50	2.4562	2.2686
60	2.5110	2.3461
70	2.5688	2.4098
80	2.6098	2.4638
90	2.6524	2.5105
100	2.6876	2.5516
200	2.9209	2.8078
300	3.0496	2.9478
400	3.1329	3.0433
500	3.2028	3.1154
600	3.2524	3.1731
700	3.2865	3.2211
800	3.3228	3.2622
900	3.3736	3.2980
1000	3.4123	3.3296

Table 2. Comparison of the present analysis with Cartwright and Longuet-Higgin's formula Eq. (19)

and accurate prediction of the maximum peak among N peaks. Of course, for this kind of wide-band process having ϵ approaches one, the number of peaks is not easy to observe and it is probably not of practical importance.

Davenport [8] used an expression derived by Cartwright and Longuet-Higgins together with some of Rice's formulas to derive the following expressions for the distribution function of the maximum of a zero-mean stationary Gaussian random process within a given time interval t:

$$F_{\eta_{max}}(\eta) = \exp\left\{-\nu_0 t \left[\exp\left(-\frac{\eta^2}{2\sigma_x^2}\right)\right]\right\} \tag{21}$$

The corresponding mean value and standard deviation were also found as

$$\frac{\bar{\eta}_{max}}{\sigma_x} = \sqrt{2\ln(\nu_0 t)} + \frac{\gamma}{\sqrt{2\ln(\nu_0 t)}} \tag{22}$$

and

$$\frac{\sigma_{\eta_{max}}}{\sigma_x} = \frac{\pi}{\sqrt{6}} \frac{1}{\sqrt{2\ln(\nu_0 t)}} \tag{23}$$

where the higher-order terms are neglected and $\nu_0 = \sqrt{m_2/m_0}/\pi$ is the expected number of zero crossing rate for the random process. It is interesting to note that Eq. (21) is also a frequently used reliability approximation for the process not crossing a two-sided threshold level η within time t. This reliability is obtained based on an independent Poisson level crossing assumption. For a narrow-band process, it is not difficult to see the close relationship between the distribution function of the maximum and the reliability of no level crossing. For a wide-band process, however, there are some differences between them. This is also why the reliability analysis based on Eq. (21) may be on the unsafe side for wide-band processes.

It is well known that the expected number of zero crossing rate is very close to the expected number of peaks per unit time for a narrow-band process. Thus, Eqs. (22) and (23) can actually be deduced directly from Eqs. (14) and (15) by replacing N with $\nu_0 t$. If the band-width parameter ϵ of a process becomes larger and larger, results from Eqs. (22) and (23) become less and less accurate. It is noted that in both Cartwright and Longuet-Higgins' and Davenport's derivations, the following approximations, or their similar approximations, were used:

$$1 - F(\eta) \approx (1-\epsilon_2)^{1/2} \exp\left(-\frac{\eta^2}{2\sigma_x^2}\right) + \left[\frac{\sigma_x^3}{\eta^3} \exp\left(-\frac{\eta^2}{2\epsilon^2\sigma_x^2}\right)\right] \tag{24}$$

$$\frac{\bar{\eta}_{max}}{\sigma_x} = \int_{-\infty}^{0} [F(\eta)]^N \, d\eta + \int_0^{\infty} \left\{1 - [F(\eta)]^N\right\} d\eta \tag{25}$$

It should be pointed out that although the approximation for the distribution function in Eq. (24) usually results in a small amount of error, this error can become tremendously large when the $[F(\eta)]^N$ term in Eq. (25) is evaluated, not to mention the integration of it. The error becomes especially significant for cases that ϵ is large, N is large and/or η is small. This also indicates that Eqs. (19) and (22) which were derived using Eqs. (24) and (25) become less and less accurate as ϵ becomes larger and larger. In order to overcome this, it is suggested that the following expected number of peaks be used to calculate N and then plug it into Eqs. (14)-(17)

$$N(t) = tM = \frac{t}{\pi}\sqrt{\frac{m_4}{m_2}} \tag{26}$$

in which M is the expected number of peaks per unit time [1]. The real peak statistical quantities of the process must lie somewhere between those results from Eqs. (14) and (15) and those from Eqs. (16) and (17). Of course, m_4 may not exist in the mean-square sense such as the response of a single-degree-of-freedom oscillator to a wide-band process excitation. Also, as it was mentioned before, as the band-width measure ϵ becomes considerably large, the physical meaning behind the process becomes vague, and it is of less importance in engineering application.

Applications

Possible applications of the preceding theory can be found in various engineering fields. In the present section, two examples, one in ocean engineering field and one in wind engineering field, will be specifically described. The first example applies the foregoing theory to predict the extreme value of the wave peaks for the ship motion in irregular seas and waves generated in the towing tank. The second example applies the same theory to find the maximum peak value of the wind fluctuation as well as the maximum structural response in a random gust environment.

According to Ref. 10, the spectral density function of the pitch motion corresponding to a full-scale ship of length 520 ft obtained from model experiments carried out on the modified Mariner with a V-shape bow in irregular waves of Newmann's sea-state 7 is obtained and shown in Fig. 6. The spectral density is for light-draft condition of the ship and for a 10-knot full-scale speed. The test was conducted for 23.5 minutes and 160 maxima were observed from which the histogram of the pitch motion was constructed as shown in Fig. 7. Included also in the figure is the probability density function based on Eq. (1) and a value of $\epsilon = 0.378$ evaluated from the spectral density function. Good agreement is found between the observed and theoretical distributions.

The number of maxima during the test time of 23.5 min can be computed from Eq. (26), and the result is $N = 178$ as compared with the observed number, 160. The expected extreme value during the test time can be obtained numerically based on the theory proposed in the previous section. The predicted extreme pitch angle is 17.2 deg which can be compared with the observed value 18.9 deg. Several other comparisons of extreme values are made for the pitch motion of a ship for other test conditions, for ship vertical bow acceleration, and for irregular waves generated in a towing tank. The results are given in Table 3. As can be seen from the table, the observed extreme values fall fairly well in the vicinity of the theoretically expected extreme values.

The application of the preceding theory can also be found in the wind engineering problems. As it was described in Ref. 8 in studying the buffeting of structures by gusty wind, the largest instantaneous wind speed occurring during samples of some specific duration of a stationary wind process is of great importance. In this context, the stationary random process could consist of a turbulent velocity components, the fluctuating aerodynamic force induced on a structure thereby, or the structure's response in terms of the deflection, stress, bending moment, etc. Since climatological records commonly refer to hourly mean wind velocities, the sample duration of frequent interest is correspondingly also one hour. The notation η_{max} adopted in the previous section should therefore indicate the largest instantaneous value of the velocity, force, deflection or stress, etc. occurring during that hour. Expressions such as Eqs. (11) to (13), or Eqs. (21) to (23) then relate the probability or statistics of these largest instantaneous

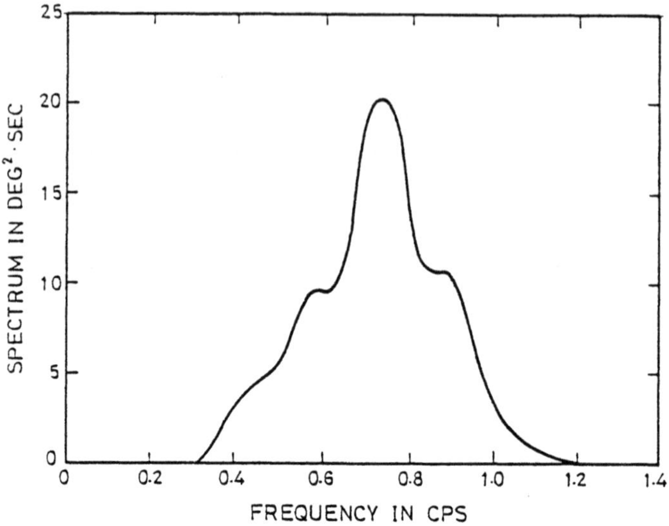

Fig. 6. Spectral density function of the pitch motion (after Ref. 10).

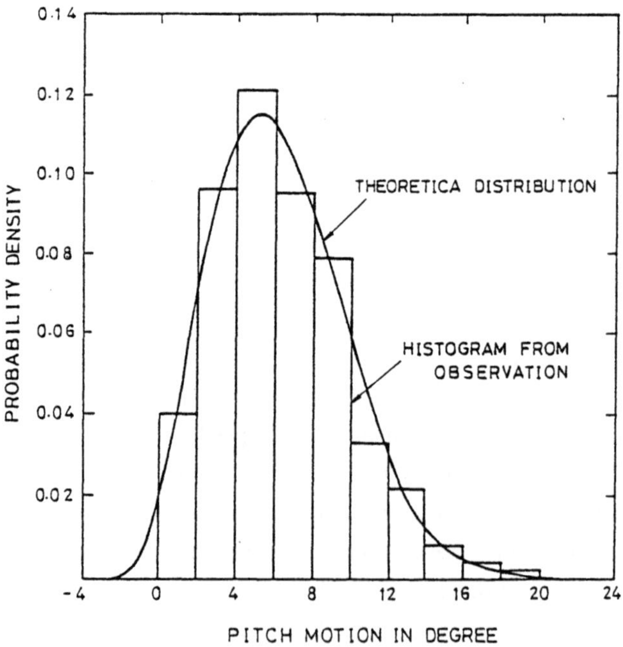

Fig. 7. Observed histogram and theoretical distribution of pitch motion.

	Test Time*	ϵ^*	Observed N^*	Theoretical N	Observed $\bar{\eta}^*_{max}$	Theoretical $\bar{\eta}_{max}$
Pitch Motion (Sea-state 7)	23.5 (min)	0.378	160	178	18.9 (deg)	17.2
Pitch Motion (Sea-state 6)	24.8	0.350	233	218	11.2 (deg)	12.8
Bow Accel. (Sea-state 7)	37.4	0.231	360	330	1.78 (g)	1.86
Generated Waves (Sea-state 5)	27.5	0.513	177	200	22.3 (ft)	21.9
Generated Waves (Sea-state 6)	24.8	0.616	282	316	32.0 (ft)	34.1
Generated Waves (Sea-state 7)	32.9	0.598	213	223	48.7 (ft)	46.8

* (after Ref. 10)

Table 3. Comparison of predicted and observed extreme peaks in ship motion

values to the mean values, power spectral densities and other statistical properties of the random processes on which they are based.

As an example, consider the gust spectrum shown in Fig. 8 as proposed by Davenport [7]. In this figure, the abscissa indicates the inverse gust wave length which is obtained by dividing frequency n by the mean wind speed μ_x. The ordinate indicates the spectrum $S(n)$ in a nondimensional form for the fluctuation portion of the gust speed. A surface drag coefficient k and the mean wind speed μ_x are used to make $S(n)$ a dimensionless quantity. In order to match his theoretical result with the 10-min real observed wind data recorded at Prestwick Airport, Scotland, Davenport has assumed in his analysis that the average mean speed was 40 ft/sec and the gust intensity (ratio of r.m.s. velocity fluctuation to mean velocity) 15%. Using the above information and applying Eq. (21) he showed that Eq. (21) would predict the gust factor reasonably well as shown by the dashed curve in Fig. 9. The gust factor is defined by meteorologists as the ratio of the maximum gust speed to the concurrent mean wind speed during a certain observed period. Other than Davenport's formula which was derived based on a narrow-band assumption, the general theory proposed in the previous section can also be used in the present example. Accordingly, after the wind spectrum is established, Eqs. (11), (12), (13) and (26) can then be applied instead of Eq. (21). After a small amount of numerical calculations, the distribution of the gust factor can be obtained.

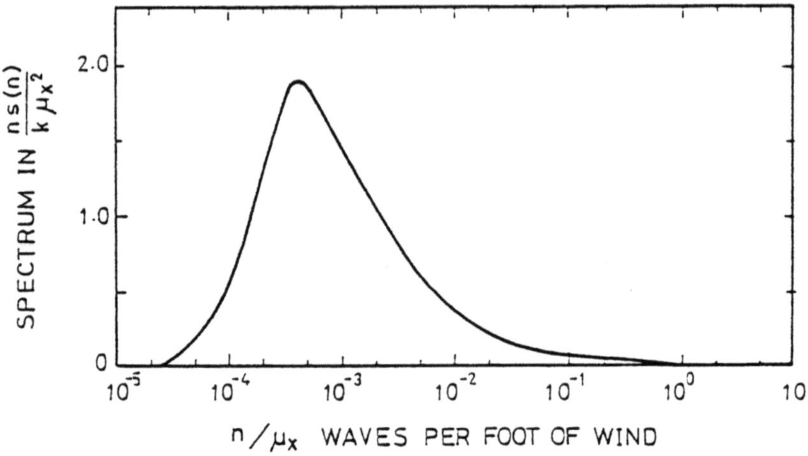

Fig. 8. Spectral density function of the fluctuation portion of the gust (after Ref. 8).

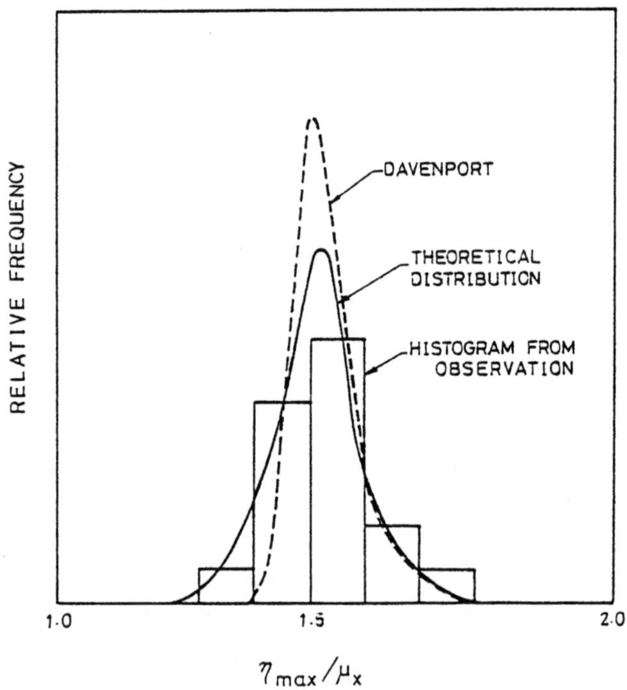

Fig. 9. Observed histogram and theoretical distribution of the gust factor.

The result is again shown in Fig. 9 in solid curve. From Fig. 9 it can be seen that the present analysis gives better prediction for the gust factor over that predicted by Davenport.

Concluding Remarks

The extreme peak distribution related to a Gaussian random process is studied. It is shown that the extreme peak distribution within a time interval can be derived directly from Rice's formulas and Gumbel's order statistics. In particular, the mean value and standard deviation of the extreme peak can be obtained analytically. Unlike some commonly used formulas which are only true for narrow-band processes, the derived result covers both narrow-band and wide-band processes.

Possible applications of the extreme peak distribution are illustrated by two examples, one in the ocean engineering field and the other in the wind engineering field. Both results agree reasonably well with results obtained by other investigators using other more complicated analyses. The results also agree quite well with experimental observations, which shows the applicability of the present analysis.

Finally, it should be mentioned that in addition to the illustrated examples, the present analysis can also be applied to the earthquake engineering field [11,12] and first-passage problems [13,14]. The latter is of special interest to many researchers in the reliability analysis of structures subjected to random excitations.

Acknowledgment

The writers are grateful for the financial support from the National Science Council of the Republic of China for the preparation of this paper.

References

1. Rice, S. O., "Mathematical Analysis of Random Noise," *Bell System Technical Journal*, 1944, **23**, 282–332, and 1945, **24**, 46–156. Reprinted in *Selected Papers on Noise and Stochastic Processes*, N. Wax, ed., Dover Publications, Inc., New York, 1954.

2. Longuet-Higgins, M. S., "On the Statistical Distribution of the Heights of Sea Waves," *Journal of Marine Research*, 1952, **11**(3), 245–266.

3. Cartwright, D. E. and M. S. Longuet-Higgins, "The Statistical Distribution of the Maxima of a Random Function," *Proceedings of the Royal Society of London*, Series A, 1956, **237**, 212–232.

4. Huston, W. B. and T. H. Skopinski, *Probability and Frequency Characteristics of some Flight Buffet Loads*, NACA TN 3733, 1956.

5. Gumbel, E. J., *Statistical Theory of Extreme Values and Some Practical Applications*, National Bureau of Standards, Applied Mathematics, Series No. 33, 1954.

6. Gumbel, E. J., *Statistics of Extremes*, Columbia University Press, New York, 1958.

7. Davenport, A. G., "The Application of Statistical Concepts to the Wind Loading of Structures," *Proceedings, Institution of Civil Engineers*, London, 1961, **19**, 449–472.

8. Davenport, A. G., "Note on the Distribution of the Largest Value of a Random Function with Application to Gust Loading," *Proceedings, Institution of Civil Engineers*, London, 1964, **28**, Paper No. 6739, 187–196.

9. Cramer, H., *Mathematical Methods of Statistics*, Princeton University Press, Princeton, New Jersey, 1946.

10. Ochi, M. K., "On Prediction of Extreme Values," *Journal of Ship Research*, 1973, **17**(1), 29–37.

11. Shinozuka, M. and J. N. Yang, "Peak Structural Response to Non-Stationary Random Excitations," *Journal of Sound and Vibration*, 1971, **16**(4), 505–517.

12. Shinozuka, M., "Methods of Safety and Reliability Analysis," in *International Conference on Structural Safety and Reliability*, ed. A. M. Freudenthal, Pergamon Press, 1972, 11–45.

13. Ditlevsen, O., *Extremes and First Passage Times with Applications in Civil Engineering*, Technical University of Denmark, Copenhagen, Denmark, 1971.

14. Ditlevsen, O., "First Outcrossing Probability Bounds," *Journal of Engineering Mechanics Division*, ASCE, February, 1984, **110**(EM2), 282–292.

Statistical Quadratization for Nonlinear Systems in Offshore Engineering

M.G. Donley
Structural Dynamics Research Corp.
2000 Eastman Drive
Milford, Ohio 45150

P. D. Spanos
L.B. Ryon Chair in Engineering
Rice University
P.O. Box 1892
Houston, Texas 77251

Summary
The statistical linearization method is often inadequate for estimating spectral properties of random responses of nonlinear systems. This is due to the fact that the spectra of responses for linear systems span only the frequency range of the excitation spectrum, whereas significant responses outside this range are possible for nonlinear systems. Therefore, as an extension of the linearization method, a statistical "quadratization" method is presented. For certain nonlinearities, the quadratization method yields notably improved system response spectral estimates. The method is developed for a nonlinear single-degree-of-freedom oscillator under stationary, Gaussian excitation. As a practical application, the proposed method is used to reliably estimate the surge response statistics of a tension leg offshore platform subject to wave and current loads.

Introduction

The statistical linearization method has proven to be a versatile analytical tool for determining the response statistics of randomly excited nonlinear systems. This method was first used to find the stationary response of single-degree-of-freedom nonlinear mechanical systems with random excitations by Caughey [1]. Later investigators [2-6] generalized the method to be applicable for treating multi-degree-of-freedom systems, nonstationary responses, and non-Gaussian responses. For a survey on linearization methods see Roberts and Spanos [7].

In certain cases the accuracy of statistical linearization is not satisfactory. Specifically, the response spectrum obtained by linearization spans only the frequency range of the excitation spectrum. It is well known, however, that for nonlinear systems, responses can occur outside the frequency range of the excitation. Consequently, some of the response statistical properties may be estimated erroneously.

The Volterra series method, see Schetzen [8], is an approximate analytical method for treating nonlinear systems which does not suffer from this drawback. This method, however, is best suited for systems with polynomial nonlinearities; whereas, in many engineering applications, the nonlinearities are not polynomials. Therefore, a statistical "quadratization" method is proposed as an improvement to the linearization method. The nonlinearity is replaced with "equivalent" polynomials up to quadratic order. The resulting nonlinear "equivalent"

I. Elishakoff · Y. K. Lin (Eds.)
Stochastic Structural Dynamics 2
New Practical Applications
2nd International Conference on Stochastic Structural Dynamics
May 9–11, 1990, Boca Raton, Florida

system is solved by the Volterra series method. The response probability distribution, which in general is non-Gaussian, is estimated by a Gram-Charlier expansion.

The method is developed for a nonlinear single-degree-of-freedom(sdof) oscillator under stationary, Gaussian excitation. Response solutions are confined to stationary responses. As a practical application, the method is used to obtain the response statistics for a tension leg offshore platform subject to wave and current forces.

Analytical Method for SDOF Oscillator

Consider the following nonlinear equation of motion of a sdof oscillator subject to a stationary, Gaussian random force f(t)

$$m\ddot{x} + c\dot{x} + kx + g(x,\dot{x}) = f(t) \tag{1}$$

In this equation, x is the displacement of the oscillator, m is the mass of the oscillator, c is the viscous damping coefficient, k is the spring stiffness, and $g(x,\dot{x})$ is a nonlinear force which may contain velocity and displacement terms. The symbol, ($\dot{}$), denotes differentiation with respect to time, t.

The force, f(t), is described mathematically by the sum of a deterministic mean part, μ_f, and a random part, which is a filtered white noise process. The latter part is expressed mathematically by a time-invariant linear transform involving the convolution of white noise with the filter impulse response function, $h_f(\tau)$. Thus, f(t) can be written as

$$f(t) = \mu_f + \int_{-\infty}^{\infty} h_f(\tau)\, w(t-\tau)\, d\tau \tag{2}$$

where w(t) is a zero-mean, Gaussian, delta-correlated (white noise) process. That is $\langle w(t)w(t+\tau)\rangle = 2\pi\delta(\tau)$, where the symbols $\langle\cdot\rangle$ and δ stands for the operator of mathematical expectation and the Dirac delta function, respectively.

In general, the nonlinear function $g(x,\dot{x})$ is non-symmetric with respect to $(x,\dot{x})$. That is

$$g(x,\dot{x}) \neq - g(-x,-\dot{x}) \tag{3}$$

Consequently, the solution of equation (1) may not have a zero mean even if μ_f is zero. Therefore, the stationary response of the oscillator is written as

$$x(t) = \mu_x + \hat{x}(t) \tag{4}$$

where μ_x is the mean of x(t), and $\hat{x}(t)$ is a stationary, zero-mean process, which in general, is not Gaussian. Substituting equation (4) into equation (1) and ensemble averaging, yields the following equation for the mean response

$$\mu_x = [\mu_f - \langle g(\mu_x+\hat{x},\dot{\hat{x}})\rangle]/k \tag{5}$$

Because of the stated shortcomings of the linearization method for predicting the response spectrum, the response, $\hat{x}(t)$, is obtained by solving the response of an "equivalent" quadratic order nonlinear system. Although this equivalent system cannot be solved exactly as an "equivalent" linearized system could, it can be solved approximately by the Volterra series method. The form of the equivalent system is constructed as follows

$$m\ddot{\hat{x}} + c\dot{\hat{x}} + k\hat{x} + \alpha_1\hat{x} + \alpha_2\dot{\hat{x}} + \alpha_3(\hat{x}^2 - <\hat{x}^2>) + \alpha_4(\dot{\hat{x}}^2 - <\dot{\hat{x}}^2>) + \alpha_5\hat{x}\dot{\hat{x}} = f - \mu_f \qquad (6)$$

where the α_i's are the quadratization coefficients. The quadratic term associated with the α_5 coefficient has been simplified to reflect the fact that for stationary responses $<\hat{x}\dot{\hat{x}}> = 0$. It is also noted that the statistical quadratization method reduces to the statistical linearization method for the special case when $\alpha_3 = \alpha_4 = \alpha_5 \equiv 0$.

The quadratization coefficients are found by minimizing the mean square error between the original system and the equivalent quadratic system. This leads to a set of linear equations involving the joint central moments of $\hat{x}$ and $\dot{\hat{x}}$. These moments are evaluated by using the Volterra series method to approximate the response of the equivalent quadratic system. Including terms up to second order in the series expansion, the following approximation for the equivalent system is obtained

$$m\ddot{\hat{x}} + c_{eq}\dot{\hat{x}} + k_{eq}\hat{x} = \hat{f}(t) = \hat{f}^{(0)} + \hat{f}^{(1)}(t) + \hat{f}^{(2)}(t) \qquad (7)$$

The coefficients α_1 and α_2 have been absorbed into the equivalent linear stiffness and damping terms respectively. The quadratic order terms are shifted to the right hand side of the equation and absorbed in the terms $\hat{f}^{(0)}$ and $\hat{f}^{(2)}(t)$.

It can be shown that the force $\hat{f}^{(1)}(t)$ is the linear filtered white noise process, and the force $\hat{f}^{(2)}(t)$ is quadratic filtered white noise. Thus, the force $\hat{f}(t)$ is a second order Volterra series with white noise as the input function. This is expressed mathematically by the equation

$$\hat{f}(t) = \hat{f}^{(0)} + \int_{-\infty}^{\infty} h_{\hat{f}}^{(1)}(\tau_1)\, w(t-\tau_1)\, d\tau_1 + \iint_{-\infty}^{\infty} h_{\hat{f}}^{(2)}(\tau_1,\tau_2)\, w(t-\tau_1)w(t-\tau_2)\, d\tau_1 d\tau_2 \qquad (8)$$

where the functions $h_{\hat{f}}^{(1)}(\tau)$ and $h_{\hat{f}}^{(2)}(\tau_1,\tau_2)$ are respectively linear and quadratic Volterra kernels for force. The kernels have the corresponding linear and quadratic transfer functions, $H_{\hat{f}}^{(1)}(\omega)$ and $H_{\hat{f}}^{(2)}(\omega_1,\omega_2)$. The forces $\hat{f}^{(1)}(t)$ and $\hat{f}^{(2)}(t)$ are referred to as the linear and quadratic forces.

Because of the linearity of the operator in equation (7), the steady-state response must also be a Volterra series and is written as

$$\hat{x}(t) = \hat{x}^{(0)} + \int_{-\infty}^{\infty} h_{\hat{x}}^{(1)}(\tau_1)\, w(t-\tau_1)\, d\tau_1 + \iint_{-\infty}^{\infty} h_{\hat{x}}^{(2)}(\tau_1,\tau_2)\, w(t-\tau_1)w(t-\tau_2)\, d\tau_1 d\tau_2 \qquad (9)$$

where $h_{\hat{x}}^{(1)}(t)$ and $h_{\hat{x}}^{(2)}(\tau_1,\tau_2)$ are respectively the linear and quadratic Volterra kernels for displacement. The associated transfer functions are $H_{\hat{x}}^{(1)}(\omega)$ and $H_{\hat{x}}^{(2)}(\omega_1,\omega_2)$.

The procedure to compute the Volterra transfer functions associated with equations (8) and (9) is given by Donley and Spanos [9]. The Volterra kernels are not computed, because all numerical computations are carried out in the frequency domain.

Response Probability Distribution

In general, the probability distribution of $\hat{x}(t)$ is not Gaussian. The exact distribution of the response is not known, but there are methods available to approximate it. Using a Gram-Charlier expansion, as described in Johnson and Kotz [10], the joint probability distribution, $p(\hat{x},\dot{\hat{x}})$, of the equivalent system can be approximated from the expansion of the joint Gaussian distribution, $\Phi(\hat{x},\dot{\hat{x}})$. Recalling that the response is stationary, the joint Gaussian distribution, $\Phi(\hat{x},\dot{\hat{x}})$, is equal to the product of two one dimensional Gaussian distributions, $\phi(\hat{x})$ and $\phi(\dot{\hat{x}})$. Truncating the Gram-Charlier expansion after third order terms gives the following equation

$$p(\hat{x},\dot{\hat{x}}) = \{ 1 - \frac{1}{6}\mu_{\hat{x}^3}\frac{\partial^3}{\partial\hat{x}^3} - \frac{1}{6}\mu_{\dot{\hat{x}}^3}\frac{\partial^3}{\partial\dot{\hat{x}}^3} - \frac{1}{2}\mu_{\hat{x}\dot{\hat{x}}^2}\frac{\partial^3}{\partial\hat{x}\partial\dot{\hat{x}}^2}\} \; \phi(\hat{x})\,\phi(\dot{\hat{x}}) \tag{10}$$

where $\mu_{\hat{x}^m\dot{\hat{x}}^n}$'s are joint central moments defined by the equation

$$\mu_{\hat{x}^m\dot{\hat{x}}^n} = < \hat{x}^m \, \dot{\hat{x}}^n > \tag{11}$$

Equation (9) is used to compute these joint central moments.

Response Spectral Density

In addition to the central moments, the response spectral density function, $S_{\hat{x}\hat{x}}(\omega)$, is often useful in gaining insight into the nature of the response. Because of the linearity of the differential operator in equation (7), the response spectral density $S_{\hat{x}\hat{x}}(\omega)$ can be written in the following well known form

$$S_{\hat{x}\hat{x}}(\omega) = | H(\omega) |^2 \, S_{\hat{f}\hat{f}}(\omega) \tag{12}$$

where $H(\omega)$ is the transfer function for the linear differential operator in equation (7). The force spectral density, $S_{\hat{f}\hat{f}}(\omega)$, can be derived from equation (8) and is given by the equation

$$S_{\hat{f}\hat{f}}(\omega) = | H_{\hat{f}}^{(1)}(\omega) |^2 + 2\int_{-\infty}^{\infty} | H_{\hat{f}}^{(2)}(\omega_1,\omega-\omega_1) |^2 \, d\omega_1 \tag{13}$$

The above force spectral density function consists of two parts. The first part is the spectral density of the linear filtered white noise force, $\hat{f}^{(1)}(t)$. The second part is the spectral density of the quadratic force, $\hat{f}^{(2)}(t)$, and involves a frequency domain convolution.

Solution Procedure

Clearly, for the implementation of the preceding solution scheme, an iterative procedure is required as the quadratization coefficients are initially unknown. The procedure is started by assuming initial values for the coefficients. Joint central moments and expectations involving the nonlinear term are computed using equations (9) and (10). New quadratization coefficients are then computed based on these statistics. This procedure is repeated until convergence.

Numerical Results for Oscillator with Quadratic Damping Nonlinearity

The discussed quadratization procedure has been used for the following system

$$\ddot{x} + \beta\dot{x} + x + \gamma|a + \dot{x}|(a + \dot{x}) = f(t) \tag{14}$$

with $\beta = .1$, and $\gamma = 1$. The parameter a is referred to as the velocity offset and its value was varied. The linear filter for the force was described by the following transfer function

$$H_f(\omega) = \begin{cases} \dfrac{1}{\sqrt{2(\omega_b - \omega_a)}} & \omega_a \le |\omega| \le \omega_b \\ 0 & \text{everywhere else} \end{cases} \tag{15}$$

The mean force, μ_f, was assumed to be equal to zero. The above linear filter yielded a block banded excitation spectral density with a standard deviation equal to one. For the example problem which follows the values, $\omega_a = 4$ and $\omega_b = 8$ were used. Note that the excitation frequencies in the interval $[\omega_a, \omega_b]$ were well distanced from the natural frequency of the system which was equal to one.

Response statistics have been computed for the nonlinear system in equation (14) using both the statistical linearization and statistical quadratization methods. Monte Carlo simulation has also been performed to assess the accuracy of these results. The simulation was performed by using the sum of harmonics method, described by Borgman [11], to produce 300 realizations.

For a velocity offset equal to .3, Figure 1 shows the one-sided form of the displacement response spectral density obtained by quadratization, linearization, and simulation. The peak at the excitation frequencies is the response due to the linear force. The low frequency peak is a resonance response due to the quadratic force. The response due to the high frequency components of the quadratic force is negligible. It can be seen that the spectra obtained by simulation and quadratization are in good agreement. The quadrat ation method, however, is computationally much more efficient, to the extent that it reduces the requisite computation time by two orders of magnitude. The spectrum from linearization is virtually identical to the spectrum from quadratization at the excitation frequencies, but cannot account for any response frequencies outside that range.

The probability distribution of the response is shown in Figure 2. The distribution obtained by simulation and quadratization are in good agreement. The vertical lines mark the computed mean responses and clearly indicate that the simulated and quadratization solutions are skewed. The linearization solution is Gaussian. It is noted that the correlation between the

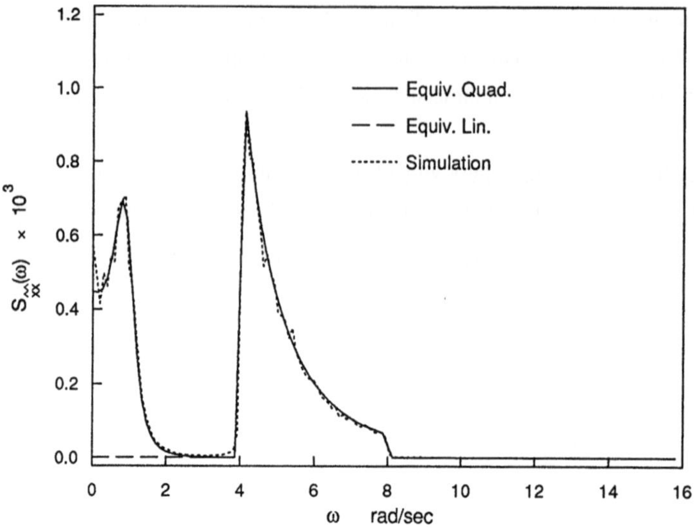

Figure 1. Displacement Response Spectral Density, a=.3

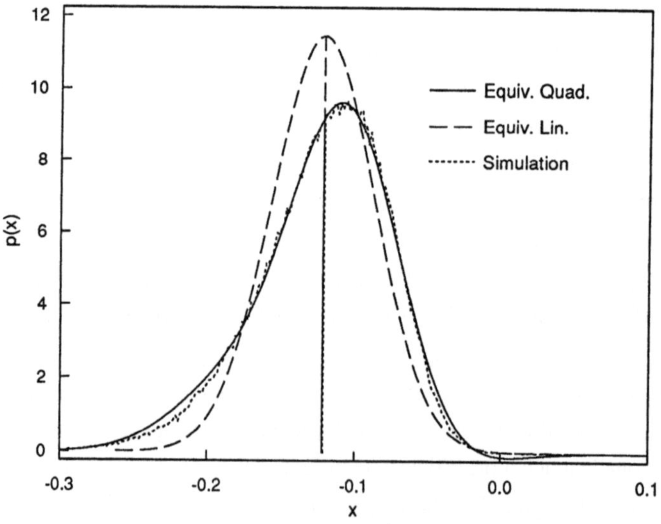

Figure 2. Displacement Response Probability Distribution, a=.3

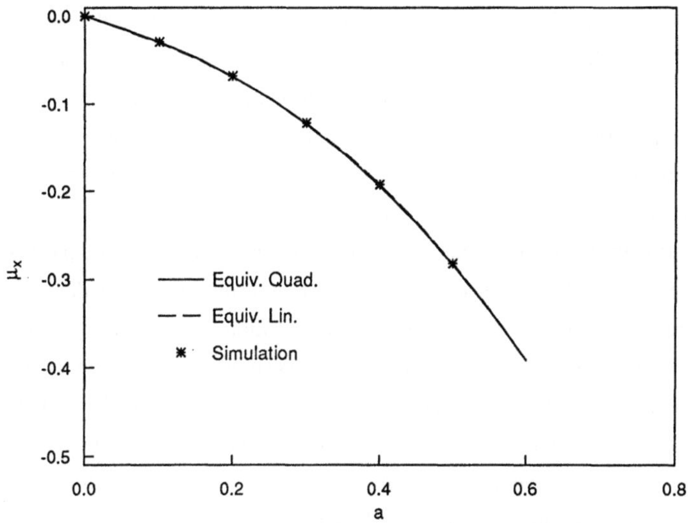

Figure 3. Mean Displacement Response

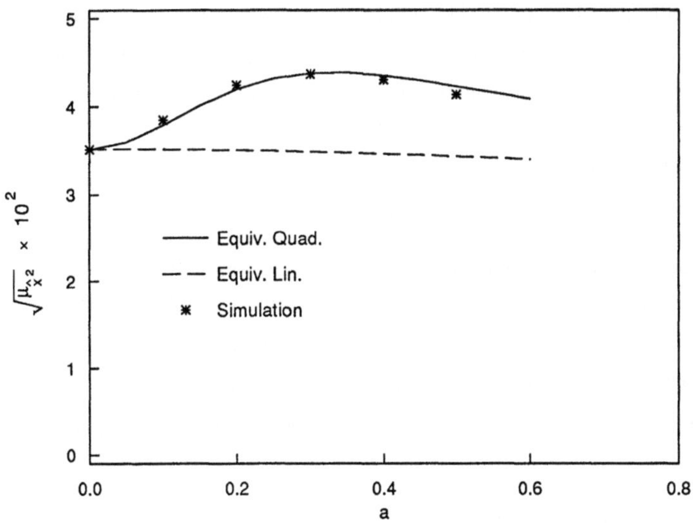

Figure 4. Displacement Response Standard Deviation

probability distribution obtained by simulation and quadratization is better near the mean than in the tails of the distribution. In fact, the tail of the distribution obtained by quadratization becomes negative over a small range. This of course is physically impossible, but can result mathematically from the Gram-Charlier expansion. Including higher order moments in the expansion may lessen the degree of this problem, but negative values for probability distributions are still possible as shown by Crandall [12].

In Figures 3 and 4, the mean, and standard deviation of the displacement are plotted versus the velocity offset. The linearization and quadratization results are nearly identical to simulation for the mean response. The response standard deviations obtained by quadratization are also quite reliable. In comparison, the linearization results significantly deviate from the simulation results, because the resonance response is not accounted for. Note that as the velocity offset tends to zero the statistics computed by linearization and quadratization converge. This is due to the fact that for zero velocity offset the nonlinearity is perfectly symmetric. Thus, the equivalent quadratic system only consists of the linear terms.

Application to Offshore Engineering

As a practical application, the statistical quadratization method has been used to obtain the response statistics for a tension leg offshore platform (TLP) subject to wave and current forces. A diagram of the idealized TLP model is shown in Figure 5. The TLP hull was assumed to be a rigid body with the vertical tethers and buoyancy forces providing linear restoring forces. The vector $\underline{x}$ represents the in plane displacements of the center of gravity relative to a fixed system with x_1, x_2, and x_3 representing surge, heave, and pitch respectively. The equation of motion for the three degree-of-freedom system is

$$\mathbf{m}\,\ddot{\underline{x}} + \mathbf{c}\,\dot{\underline{x}} + \mathbf{k}\,\underline{x} = \underline{f}_P(t) + \underline{f}_D(\dot{\underline{x}},t) \tag{16}$$

where **m**, **c**, and **k** are the mass, damping, and stiffness matrices respectively.

The linear potential force vector, $\underline{f}_P(t)$, is described by a linear transform of the wave elevation, $\eta(t)$, which is typically assumed to be a zero mean, stationary, Gaussian, random process. The nonlinearity in the system exists in the hydrodynamic drag force vector, $\underline{f}_D(\dot{\underline{x}},t)$, which is modeled by the following equation based on the generalized Morison equation

$$\underline{f}_D(\dot{\underline{x}},t) = \sum_{n=1}^{N_m} \Big[\int_{L_n} \underline{c}_{Dn}(z) \,|\, U(z) + v(x,z,t) \,|\, \big(U(z) + v(x,z,t)\big)\; dL_n \Big] \tag{17}$$

where

$$v(x,z,t) \approx u(x,z,t) - \dot{x}_1(t) \tag{18}$$

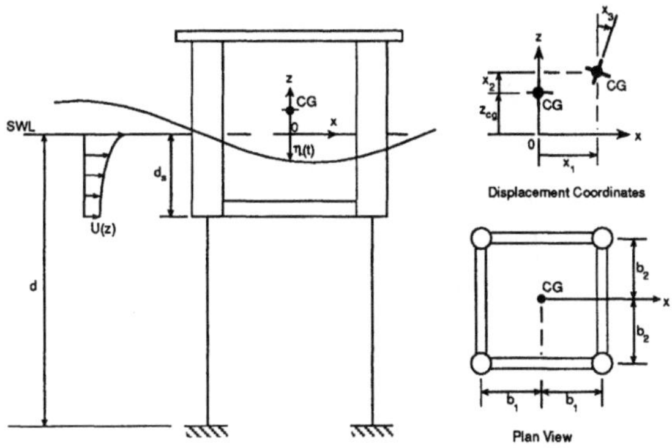

Figure 5. Diagram of Idealized Tension Leg Platform

Here N_m is the number of submerged structural members and L_n is the mean wetted length of the nth member. The vector $\underline{c}_{Dn}(z)$ contains the Morison drag parameters for the nth member. The velocity quantity U(z) + v(x,z,t) is the relative horizontal water particle velocity at the position (x-z). The current velocity, U(z), is considered to be a time invariant deterministic quantity. The random part of the relative velocity, v(x,z,t), is due to the wave field and rigid body motion. The random velocity due to the wave field, u(x,z,t), can be written as a linear transform of the wave elevation.

Numerical results have been obtained for the TLP model with a surge natural frequency of .057 rad/sec. A Pierson-Moskowitz wave spectrum with a mean period of 16.9 sec and a significant wave height of 19.2 m was used. The current velocity decreases linearly from 1.0 m/sec at the surface to .5 m/sec at the keel of the hull. Figure 6 shows the one-sided form of the power spectral density of the zero-mean part of the surge response, $\hat{x}_1$, obtained by linearization, quadratization, and simulation. One hundred realizations were used in the simulation. The high frequency peak is the wave frequency response and is due predominantly to the linear potential force. The low frequency peak is a resonance response at the surge natural frequency and is due to the drag force. It can be seen that the spectra obtained by simulation and quadratization are in good agreement for both peaks. The spectrum from linearization is nearly identical to the spectrum from quadratization in the wave frequency response region, but is zero in the low frequency region.

Conclusions

The statistical quadratization method provides a notable improvement over the statistical linearization method for computing the response statistics of an oscillator with a non-symmetric nonlinearity. The quadratization results are obtained quite efficiently and compare well with Monte Carlo simulation data. The method can be particularly application in offshore engineering. Further results can be found in Donley and Spanos [9].

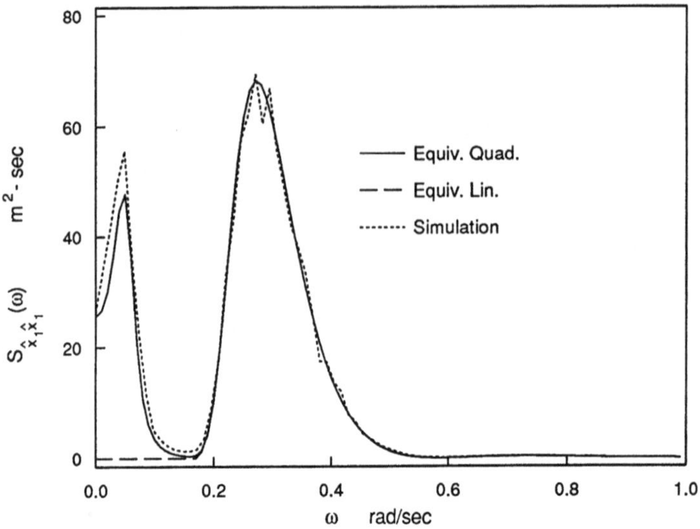

Figure 6. Surge Displacement Response Spectral Density

References

1. Caughey, T.K. Equivalent linearization techniques, *J. Acoust. Soc. of Am.* Nov 1963, **35**, 1706-1711.

2. Iwan, W.D. and Yang, I.M. Application of statistical linearization techniques to nonlinear multidegree-of-freedom systems, *J. Appl. Mech.* June 1972, **39,** 545-550.

3. Atalik, T.S. and Utku, T. Stochastic linearization of multi-degree-of-freedom non-linear systems, *Earthquake Engng. Struct. Dyns.* 1976, **4** , 411-420.

4. Spanos, P-T.D. Formulation of stochastic linearization for symmetric or asymmetric m.d.o.f. nonlinear systems, *J. Appl. Mech.* 1980, **47**, 209-211.

5. Spanos, P-T.D. Monte Carlo simulations of responses of a non-symmetric dynamic system to random excitations, *Computers and Structs.* 1981, **13,** 371-376.

6. Beaman, J.J. and Hedrick, J.K. Improved statistical linearization for analysis and control of nonlinear stochastic systems: Part I: An extended statistical linearization technique, *J. Dyn. Sys., Meas. and Con.*, March 1981, **102**, 14-21.

7. Roberts, J.B., and Spanos, P.D., *Random Vibrations and Statistical Linearization*, 1990, John Wiley, New York.

8. Schetzen, M. *The Volterra and Wiener Theories of Nonlinear Systems,* 1980, New York, John Wiley.

9. Donley, M.G., and Spanos, P.D., *Dynamic Analysis of Non-Linear Structures by the Method of Statistical Quadratization*, Lecture Notes in Engineering, New York, 1990, **57**, Springer-Verlag.

10. Johnson, N.L. and Kotz, S. *Distributions in Statistics: Continuous Multivariate Distributions*, 1972, New York, John Wiley .

11. Borgman, L.E. Ocean wave simulation for engineering design, *ASCE J. Water. & Harb. Div.* Nov 1969, 557-583.

12. Crandall, S.H. Non-gaussian closure techniques for stationary random vibration, *International Journal of Non-Linear Mechanics*, 1985, **20**, 1-8.

Probability of Liquefaction Spread

MASARU HOSHIYA

Professor of Civil Engineering
Musashi Institute of Technology
1-28-1 Tamazutsumi, Setagaya-ku, Tokyo 158, Japan

Abstract

In order to predict the potential of liquefaction spread for sandy soil surface ground, a methodology is presented which evaluates probability of fraction of liquefaction spread for a zone under study, conditional on earthquake hazard. The SPT values are considered as random variables which are spatially correlated. The method which is used in this study is a Monte Carlo simulation based on a disjoint set of sampling density functions to make easy the sampling of random variables which are , otherwise , to be sampled from a joint probability density function of general distribution.

The ground is modeled to consist of three dimensional discrete semi-zones having discrete elements in the vertical direction.

A liquefaction potential index that is widely used in the design practice in Japan to account for the distribution of the safety factor in the soil-column direction is employed as the performance function for the probability of liquefaction spread , and the numerical results are presented as a fragility curve of liquefaction spread.

Introduction

Liquefaction due to pore pressure increase in soils during seismic loading has been receiving the largest share of attention for a few decades since enormous damages have been attributed to this phenomenon in past earthquake disasters. Recently, in conjunction with the seismic risk assessment,probabilistic evaluation of liquefaction has been studied for saturated sandy soil based on the variability of soil parameters(1). However,the result offers information on probability of liquefaction at a point on a typical soil profile of subsurface ground but not on the spatial spread of liquefaction.

Since engineering soil properties may exhibit considerable

I. Elishakoff · Y. K. Lin (Eds.)
Stochastic Structural Dynamics 2
New Practical Applications
2nd International Conference on Stochastic Structural Dynamics
May 9–11, 1990, Boca Raton, Florida

variation from a point to a point even in homogeneous soil, the degree of liquefaction potential may differ at each point. In order to analyze the safety of a buried pipeline occuping a spatially wide range area of liquefaction potential, the fraction of liquefaction spread becomes an important indicator to be made clear.

As the first step, a methodology is developed for the evaluation of probability of fraction of liquefaction spread for a zone of liquefaction potential conditional on given earthquake hazard in which spatial correlations of soil parameters are taken into account. The subsurface zone under study may consist of layered soil profiles and be of several square hundred meters,being not necessarily geographically homogeneous. However the methodology in which a Monte Carlo simulation is employed as the analytical tool can apply for such a zone dependent upon the way of modelling. Although the present study intends to develope a macro-zoning map on liquefaction spread potential as the final goal, it focuses on the formulation of evaluation of probability of liquefaction spread.

It is confessed that the present study is yet crude at the preliminary stage and needs considerable improvement.

<u>Modeling of Layered Soil</u>

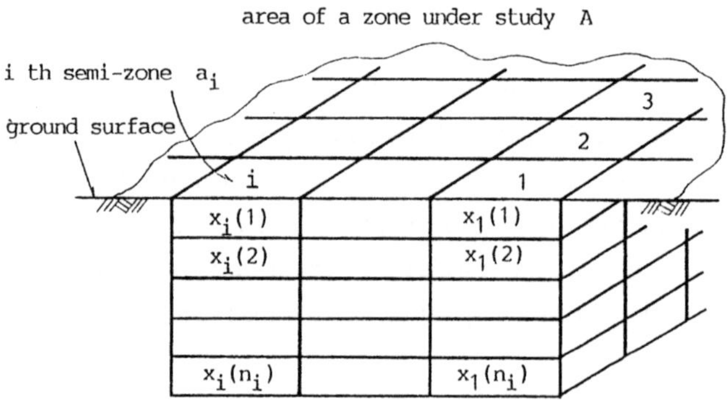

Fig.1 Model of Layered Soil

The soil profile in a zone under consideration may be modeled in terms of homogeneous layers with statistical variability of geotechnical engineering parameters, and furthermore be discretized into m semi-zones having homogeneous elements as depicted in Figure 1.

Let i th semi-zone consist of n_i elements, each of which has geotechnical engineering parameter of $x_i(j)$; $j=1,n_i$ which is treated as a random variable. In this study, only one representative parameter, that is, blowcounts (the SPT N-value) is taken as $x_i(j)$. So, random variables in a zone are expressed by the following random vectors.

$$x_i = [x_i(1), x_i(2), \ldots, x_i(n_i)] \quad (1)$$

and
$$\underline{\mathbf{x}} = [\underline{x}_1, \underline{x}_2, \ldots, \underline{x}_m] \quad (2)$$

in which the coefficient of correlation $\rho_{ij}(k,\ell)$ exists between $x_i(k)$ and $x_j(\ell)$. The area of the ground surface of i th semi-zone is a_i and the whole area of a zone is given by

$$A = \sum_{i=1}^{m} a_i \quad (3)$$

As to the size of the three dimensional extent of the zone, no comment can be made so far as the hypothetical model is concerned , but the methodology following may be hopefully applicable to appropriately modeled subsurface ground.

The idealized model depicted in Figure 1 includes random variables $x_i(j)$ whose standard deviation should reflect the effect of their local variability within each element and of their correlation between spatial averages (2),(3),(4).

Probability of Liquefaction Spread

The safety factor of the j th element of the i th semi-zone is expressed by $F_{Li}(j)$ which is usually called the liquefaction resistance factor standing for liquefaction when $F_{Li}(j)<1$ and for non-liquefaction otherwise. However, the F_L value indicates the safety of the concerned element only, and does not account for the whole soil column of the i th semi-zone.

To compensate the shortcoming, a liquefaction potential index P_{Li} was developed and widely used in the common practice of aseismic design in Japan (5),(6). The present study utilizes this index to evaluate the probability of liquefaction of semi-zones.

Figure 2 shows the soil column of the i th semi-zone in which the safety factor and depth of each element are indicated respectively by $F_{Li}(j)$ and $Z_i(j)$

The safety factor $F_{Li}(j)$ is a function of soil properties as well as earthquake hazard intensity (refer to Numerical Example), and consequently the liquefaction potential index P_{Li} is also a function of these parameters and is defined by the following expressions.

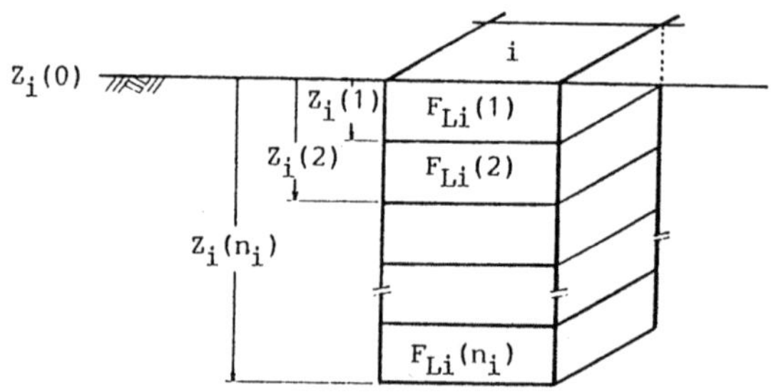

Fig.2 Soil Column

$$F_i(j) = 1-F_{Li}(j) \quad , \text{ if } F_{Li}(j) < 1$$
$$= 0 \quad , \text{ if } F_{Li}(j) \geq 1 \tag{4}$$

and
$$P_{Li}(\underline{x}_i \mid H) = \sum_{j=1}^{n_i} \int_{Z_i(j-1)}^{Z_i(j)} F_i(j)w(z)dz \tag{5}$$

where the blowcounts $\underline{x}_i$ are random variables and the hazard level H (PGA may be used) is conditionally given. The weighting function w(z) accounts for the dependency of liquefaction potential on the depth.

Since the liquefaction potential index indicates more severity of liquefaction as the value increases, the following performance function is employed with the critical index P_L^*

$$g_i\ (\underline{x}_i \mid P_L^*, H) = P_L^* - P_{Li}(\underline{x}_i \mid H) \qquad (6)$$

where $g_i \leqq 0$; Liquefaction potential exceeds the critical value.

The fraction of liquefaction spread over a zone to be investigated is defined as

$$f = 1/A \sum_{i=1}^{m} I(\underline{x}_i \mid P_L^*, H) a_i \qquad (7)$$

where a_i = the surface area of i th semi-zone.
A = the whole area of a zone given by eq(3).

$$I(\underline{x}_i \mid P_L^*, H) = 1 \ , \quad \text{if} \quad g_i(\underline{x}_i \mid P_L^*, H) \leqq 0$$
$$= 0 \ , \quad \text{otherwise}$$

Obviously $0 \leqq f \leqq 1$ and when $f=1$, the whole area is a state of liquefaction severity of at least more than or equal to P_L^* whereas when $f=0$ the whole area is under less than P_L^* . The value of f inbetween indicates the fraction of liquefaction spread of the degree of P_L^*.

Since f is a function of the random vectors of eq(1) or eq(2), f is itself a random variable. Probability of f exceeding a critical value f^* may be a quantity to indicate the extent of liquefaction spread. This probability is formally expressed by

$$P(f \geq f^* \mid P_L{}^*, H) = \int_{f \geq f^*} f(\underline{\mathbf{x}}) d\underline{\mathbf{x}} \tag{8}$$

where $f(\underline{\mathbf{x}})$ is a joint probability density function of $\underline{\mathbf{x}}$.

Integration of eq(8) is,however, rather difficult because of the complicated functional form of $f(\underline{\mathbf{x}})$ in general and of the complexity of the integration domain. Then, a Monte Carlo simulation becomes of practical use and is employed in the present study.

To avoid the difficulty of sampling $\underline{\mathbf{x}}$ from the joint probability density function, a sampling density function $h(\underline{\mathbf{x}})$ is introduced to eq(8). After arrangement, eq(8) becomes

$$P(f \geq f^* \mid P_L{}^*, H) = 1/N \sum_{\ell=1}^{N} J(\underline{\mathbf{x}}\ell \mid f^*, P_L{}^*, H)\, f(\underline{\mathbf{x}}\ell)/h(\underline{\mathbf{x}}\ell) \tag{9}$$

where $\underline{\mathbf{x}}\ell$: ℓ th sampled vector from $h(\underline{\mathbf{x}})$

$$J(\underline{\mathbf{x}}\ell \mid f^*, P_L{}^*, H) = 1 \quad , \text{if} \quad f \geq f^*$$
$$= 0 \quad , \text{otherwise}$$

Equation(9) is the basic equation utilized in the Monte Carlo simulation. It is again noted that equation(9) is the probability that the fraction f exceeds a given value f^* conditional on $P_L{}^*$ and the hazard level H, and therefore, the graph of $P(f \geq f^* \mid P_L{}^*, H)$ vs. H with parameters f^* and $P_L{}^*$ is a fragility curve of liquefaction spread. If eq(9) is incorporated with a hazard density function $p(H)$, a non-conditional probability of the fraction is obtained by

$$P(f \geq f^* \mid P_L{}^*) = \int p(H) P(f \geq f^* \mid P_L{}^*, H) dH \tag{10}$$

<u>Joint P.D.F. $f(\underline{\mathbf{x}})$ and Sampling Density Function $h(\underline{\mathbf{x}})$</u>

Because of inadequacy of statistical data in reality, available information on $f(\underline{\mathbf{x}})$ is usually limited to the means and standard deviations of elements of $\underline{\mathbf{x}}$ and their coefficients of correlation, or at most marginal density functions of $\underline{\mathbf{x}}$.

Under this situation, the usefulness of the Nataf model with prescribed marginal density functions and covariances is demonstrated in engineering field(7) which is herein utilized for $f(\underline{\mathbf{x}})$.

The joint density function of this type is given by

$$f(\underline{\mathbf{x}}) = \prod_{j=1}^{n_1} f_{x1(j)}\{x1(j)\} \prod_{j=1}^{n_2} f_{x2(j)}\{x2(j)\} \dots \prod_{j=1}^{n_m} f_{xm(j)}\{xm(j)\}$$

$$\times \frac{\psi\,(\underline{\mathbf{Z}},\mathbf{R'})}{\prod_{j=1}^{n_1} \psi\,\{Z_1(j)\} \prod_{j=1}^{n_2} \psi\,\{Z_2(j)\} \dots \prod_{j=1}^{n_m} \psi\,\{Z_m(j)\}} \tag{11}$$

where $Z_i(j) = \Phi^{-1}[F_{xi(j)}\{xi(j)\}$; standard normal variables, $F_{xi(j)}\{xi(j)\}$ = probability distribution function of $x_i(j)$, $\psi\,\{z_i(j)\}$ = standard normal probability density function and $\psi\,(\underline{\mathbf{Z}},\mathbf{R'})$ = joint normal probability density function of zero means, unit standard deviations and correlation matrix $\mathbf{R'}$. The dimension of

$\psi\,(\underline{\mathbf{Z}},\mathbf{R'})$ is $\sum_{i=1}^{m} n_i$

$$\text{and} \quad \underline{\mathbf{Z}} = [Z_1(1),Z_1(2),\dots,Z_1(n_1)/Z_2(1) \dots/,\dots,/Z_m(1),Z_m(2)\dots Z_m(n_m)]^T.$$

The correlation matrix $\mathbf{R'}$ consists of apparent coefficients of correlation $\rho'_{ij}(k,\ell)$ given by $\rho\,{}_{ij}(k,\ell) = \rho_{ij}(k,\ell) * F$ in which F, satisfying $F \geqq 1$ is a function of $\rho_{ij}(k,\ell)$ and the marginal distributions of $x_i(k)$ and $x_j(k)$. The values or the formula of F are prepared in tables(7).

It is quite difficult to directly sample $\underline{\mathbf{x}}$ from eq(11), and therefore a sampling density function $h(\underline{\mathbf{x}})$ is employed as a special type of independency.

$$h(\underline{\mathbf{x}}) = \prod_{j=1}^{n_1} h_{x1(j)}\{x1(j)\} \prod_{j=1}^{n_2} h_{x2(j)}\{x2(j)\} \ldots \prod_{j=1}^{n_m} h_{xm(j)}\{xm(j)\} \qquad (12)$$

There may be efficient sampling schemes available(8) by suitably choosing $h_{xi(j)}(\ )$ so that events satisfying the condition of f greater than f^* are more often sampled and consequently the reduction of the variance of estimated $P(f \geq f^* \mid P_L^*, H)$ is attained.

However, putting aside this discussion,the present study postulates that $h_{xi(j)} = f_{xi(j)}$. In this case, the part of the right hand side of eq(9) is simplified using eqs (11) and (12) as follows.

$$\frac{f(\underline{\mathbf{x}})}{h(\underline{\mathbf{x}})} = \frac{\psi\ (\underline{\mathbf{Z}},\mathbf{R'})}{\prod_{j=1}^{n_1} \psi\ \{Z_1(j)\} \prod_{j=1}^{n_2} \psi\ \{Z_2(j)\} \ldots \prod_{j=1}^{n_m} \psi\ \{Z_m(j)\}} \qquad (13)$$

Numerical Example

For a demonstration purpose , a hypothetical surface ground is numerically analyzed whose surface area is 130(= 40 m + 50 m + 40 m) * 130 square meters. Thus , the model consists of nine semi-zones having five elements and the depth is uniformly 5 meters.

The coefficients of correlation between SPT values of each element are assumed to be an exponentially decreasing function of separation distance Δ ;

$$\gamma(\ \Delta\ ) = \exp\ (-\ \Delta\ /\ \ell\ )$$

where ℓ = 50 m in the horizontal direction and ℓ = 1.4 m in the vertical direction. Also , the SPT values are assumed lognormal random variables in the study. Because of the hypothetical model , detailed description may not be much meaningful.

Figure 3 shows fragility curves of liquefaction spread.

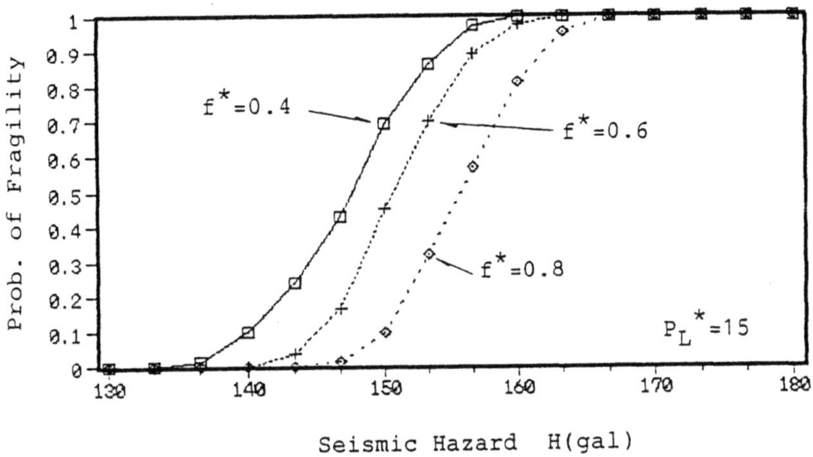

Fig.3 Fragility Curves of Liquefaction Spread

Concluding Remarks

A methodology is presented which evaluates probability of fraction of liquefaction spread , conditional on earthquake hazard , in which a Monte Carlo simulation method is used. SPT values are treated as random variables which are spatially distributed.

An application of the method may be found in the preparation of a zoning map of liquefaction spread , which is the author's research topic in the future.

References

1. M. Shinozuka, S. Kishimoto and K. Ohtomo, Fragility Curves for Liquefaction, Proc. of 1st Japan-US Workshop on Liquefaction,Large Ground Deformation and their Effects on Lifeline Facilities, Japan-US Cooperative Research and Collaboration, Nov., 1988,PP.230 ∽ pp.238.

2. E.H. Vanmarcke, Probabilistic Modeling of soil Profiles, Jour. of Geotech. Eng., ASCE, Vol.103, No.GT11, Nov., 1977, pp.1227∽ pp.1246.

3. A. Asaoka and D.A-Grivas, Spatial Variability of the Undrained Strength of Clays, Jour. of Geotech.Eng.,ASCE,Vol. 108, Nov.GT5, May,1982, pp.743∽pp.755.

4. W.H. Tang, Probabilistic Evaluation of Penetration Registances, Jour. of Geotech. Eng., ASCE, Vol.105, No.GT10,Oct.1979, pp.1173 ∽ pp.1191.

5. T. Iwasaki,et. al., Microzonation for Soil Liquefaction Potential Using Simplified Method, Proc. of the 3rd Inter, Earthq. Microzonation Conf., Seattle, 1982.

6. Highway Bridge Design Specification, Japan Highway Association, 1980.

7. Pei-Ling Liu and A.Der Kiureghian, Multivariate Distribution Models with Prescribed Marginals and Covariances, Jour. of Probabilistic Eng. Mech.,1986, Vol.1, No.2, pp.105 ∽ pp.112

8. G.I Schuëller, C.G. Bucher, U. Bourgund and W. Quypornprasert, On Efficient Computational Schemes to Calculate Structural Failure Probabilities, Jour.of Probabilistic Eng. Mech., 1989, Vol.4, No.1,pp.10 ∽ pp.18.

Nonstationary Random Vibrations: Approximate Spectral Formulation Versus Exact Theory

Hans Irschik, Franz Ziegler

Technical University Vienna,
Department of Civil Engineering,
A-1040, Vienna, Austria.

Summary

A simple approximation is presented for the nonstationary probabilistic analysis of slightly damped linear oscillators excited by earthquakes. The method is based on the spectral moment approach. It applies for the case of wide-banded input power spectral density functions with slowly varying, arbitrarily distributed deterministic time modulations. The formulation is presented in dimensionless form, and is compared to various exact and approximate results from the literature, concerning barrier-upcrossing rates and failure probabilities.

Introduction.

In earthquake engineering,the input ground motion most realistically is modelled as a non-white,non-stationary random process,see e.g. Clough,Penzien[1].Such a process commonly is characterized following the theory of evolutionary power spectra by Priestley[2]. Hence, an evolutionary (one-sided) power spectrum $G(t,\omega)$ of the strong motion phase is introduced,Vanmarcke[3],where G is constructed from a zero-mean stationary random process with spectral power density $G_0(\omega)$ and a slowly varying deterministic envelope $A(t,\omega)$, such that $G(t,\omega)= A(t,\omega)^2 G_0(\omega)$. Usually, the frequency dependence of A is not considered to be important,i.e., a uniformly modulated random process is connected to the seismological modelling. Thereby, the time evolution of A is assumed to be of a box-car type, or to be exponentially varying, $A(t)=C(\exp(-\alpha t)- \exp(-\beta t))$, $\beta>\alpha$, Shinozuka,Sato[4]. Furthermore, $G(t,\omega)$ is assumed to be wide-banded, where a band-limited white noise or the well-known Kanai-Tajimi[5] representation frequently is used for G_0 in the engineering literature,Vanmarcke[3]. (For an incorporation of more recent seismological formulations in the evolutionary power concept, see e.g.Safak[6], Quek,Teo,Balendra[7] and the literature cited there.)

I. Elishakoff · Y. K. Lin (Eds.)
Stochastic Structural Dynamics 2
New Practical Applications
2nd International Conference on Stochastic Structural Dynamics
May 9-11, 1990, Boca Raton, Florida

Important theoretical benchmarks for the probabilistic analysis of structures due to a non-stationary excitation are presented by Howell,Lin[8], like the exact formulations for the evolutionary power spectral density $G_y(t,\omega)$ of the response $y(t)$ of a linear oscillator with a given impulse-response function, for its time-dependent variance $\sigma_y^2(t)$, for the average rate of level-up-crossings $\nu_a(t)$ and for other response measures. They are based on the pioneering work of Rice[9], see also Shinozuka,Yang[10],Roberts[11] and Howell[12]. Only the assumptions of a Gaussian input and of the existence of an evolutionary input power enter this general formulas (gathered in the Appendices of Ref.8), which require the evaluation of the variances and covariances for y, dy/dt and d^2y/d^2t.

No exact relations have yet been developed for the crucial problem of the first-excursion failure probability $F(a,t)$, i.e., the probability that the structural response passes out of prescribed barrier levels $\pm a$ (the safety bounds) for the first time within the time interval $0\leq\tau\leq t$. Various methods with different degrees of approximation are offered in the literature. Some of these proceed by connecting ν_a, the expected number of upcrossings per unit time of level a at time t,regardless if there are any up-crossings prior to t,in an approximate manner to the decay rate, or hazard function,$\alpha(a,t)$. α then enters the probability that the structural response remains within the barrier levels $\pm a$ for $0\leq\tau\leq t$ in the form:

$$L_a(t)= \exp(-\int_0^t \alpha(a,\tau)\, d\tau), \tag{1}$$

see e.g. Yang[13], Corotis,Vanmarcke,Cornell[14], Lutes,Chen,Tzuang[15] and Safak[6]. This procedure is motivated by an extension from stationary results, see Vanmarcke[16] for a review of the stationary counterpart to Equ.(1). Quiet initial conditions have been assumed in Equ.(1), and $F(a,t)= 1-L_a(t)$. α can be considered as the average number of barrier upcrossings of level a during a time increment dt at time t, given no upcrossings prior to t, see Ref.9.

Among the approximate formulations for the probabilistic analysis of nonstationary structural response, special attention has been given to results derived by Vanmarcke and co-workers, Refs.3,14,and Corotis,Vanmarcke[17], Corotis,Marshall[18], Vanmarcke[19]. This approximation makes use of the spectral moments λ of the evolutionary spectrum $G_y(t,w)$ of the structural response by inserting them into the formulas for the stationary case. Thus, with

$$\lambda_j(t)= \int_0^\infty \omega^j \; G_y(t,\omega) \; d\omega, \qquad j=0,1,2,\dots. \tag{2}$$

where

$$\sigma^2_y(t)= \lambda_0(t), \tag{3}$$

a central frequency

$$\Omega_y(t)=[\lambda_2(t)/\lambda_0(t)]^{1/2} \tag{4}$$

and a bandwidth-parameter of the response

$$\delta_y(t)= [1-\lambda_1(t)^2/\lambda_0(t)\lambda_2(t)]^{1/2} \tag{5}$$

are evaluated, which enter into

$$\nu_a(t)= \nu_0(t) \; \exp(-a^2/ \; 2\sigma_y^2(t)), \; \nu_0= \Omega_y/2\pi, \tag{6}$$

$$n_a(t)= (2\pi)^{1/2} \; \delta_y(t) \; \nu_a(t) \; a/\sigma_y(t), \tag{7}$$

and finally into

$$\alpha(a,t)= 2\nu_a(t)[1-\exp(-n_a(t)/2\nu_a(t))]/[1-\nu_a(t)/\nu_0(t)]. \tag{8}$$

In Equs.(2)-(5), the index y has been supressed for the spectral moments λ_j. Note,that Equ.(3) for σ_y^2 is exact,see Ref.8. Furthermore, in the *stationary* case there is λ_2 the variance of dy/dt and δ_y can be interpreted as the ratio of the standard deviations of the slope of the envelope and of dy/dt, Vanmarcke[19], where the envelope definition of Cramer,Leadbetter[20] has been used. The latter corresponds to the one given in Ref.9. n_a in Equ.(7) is the mean rate of envelope crossings, and Equ.(8) has been derived assuming the level-crossings to be a two-stage Markov-process, Ref.19.

The case of a lightly damped modal linear oscillator suddenly exposed to a wide-banded stationary excitation, $A(t)=H(t)$, H denoting the Heaviside function,has been treated in Ref.14 by means of this spectral moment approximation. Possible divergence of the higher spectral moments has been noted, if the excitation contains significant motion for frequencies more than 10 times the natural frequency ω_n of the oscillator, and the convergent parts of the integrals corresponding to Equ.(2) have been taken only in Ref.14. For the practically important situation of small damping levels ζ and intermediate or low natural frequencies , the following satisfactory approximations finally have been listed in Ref.3:

$$\sigma_y^*(t) = G_0(\omega_n)\pi/4\omega_n^3\zeta_t, \tag{9}$$

and

$$\Omega_y^*(t) = 2\pi\nu_0^*(t) = \omega_n, \tag{10a}$$
$$\delta_y^*(t) = (4\zeta_t/\pi)^{1/2}, \tag{10b}$$

where the fictitious time dependent damping is

$$\zeta_t = \zeta \;/\; [1-\exp(-2\omega_n\zeta t)]. \tag{11}$$

The case of an exponentially varying time envelope A has been treated in Ref.18 for the SDOF oscillator, where a numerical integration of the time-dependent spectral moments,Equ.(2), has been performed, using an upper cut-off frequency of four times the oscillator's natural frequency. The well- known broad-band input and small-band output approximation for slightly damped linear oscillators with intermediate natural frequencies enters the analysis of Refs.14,18, taking a constant $G_0(\omega_n)$ instead of using the original distribution $G(\omega)$.

A practically appealing advantage of the basic form of the spectral approximation, Equs.(2)-(8), furthermore, is that it can be applied easily to more complex problems of earthquake engineering, where the frequency response function of the structural response component of interest is given only numerically. This advantage is due to the fact that no time-derivatives of the evolutionary output power spectrum G_y have to be performed for a calculation of ν_a according to Equ.(8), contrary to the exact formulation, Ref.8, and that the application of Equ.(9) is straightforward. A suitable manipulation of the underlying formulas of Priestley[2] leads to a computationally efficient evaluation of G_y itself, where the numerical integration is in the frequency domain only, and the influence of the distribution of $A(t)$ is taken into account analytically, Hoellinger[21]. For a review with respect to advanced structural applications, see Ziegler[22]. The spectral approximation has also been used for the evaluation of the response statistics of the linear solution part of randomly excited elastoplastic oscillators, see Irschik, Ziegler[23] for a review.

From a theoretical point of view, the possible inappropriateness of the above spectral moment formulation has been pointed out, e.g., by Yang[13], Krenk, Madsen, Madsen[24]. Only recently, numerical results have been presented by Bucher, Lin[25], indicating (by comparison to exact results for ν_a according to Ref.8) that the spectral moment approach may lead to erroneous solutions in some nonstationary cases.

The purpose of the present contribution is two-fold: At first, an approximation is presented for the nonstationary response statistics of a SDOF (modal) oscillator due to a wide-banded excitation process. This approximation is based on the spectral moment approximation, and leads to extremely simple formulas applying to Heaviside as well as to exponentially varying modulations $A(t)$. Besides the theoretical complexity, the numerical effort for the calculation of the failure probability in common formulations is generally quite high, even in the simplest case of a SDOF oscillator, including, e.g., infinite series of Gamma-functions, Yang[26] and Ref.7, or of confluent hypergeometric functions, Lennox, Fraser[27], evaluations of Bessel functions, Krenk[28], Solomos, Spanos[29], Yang[13], of error functions, Refs.6,8,13, and Langley[30], the solution of integral equations, Ref.29, and, last but not least, contour integration or numerical integration in the frequency domain for the cited exact or approximate formulations starting from evolutionary power spectral functions. From a practical point of view, therefore, there is a need for simple approximations, which are versatile and lead to a quick first estimation of the results. The present approximation makes use of Equs.(10) for the central frequency and the bandwidth-factor of the response, also in case of non-Heaviside modulation functions $A(t)$. Recently, a similar type of approximations has been derived by Igusa[32] for white-noise excitations, using a more com-

plex formula for the bandwidth parameter. The present paper gives a similarity analysis, where the response variance is calculated according to a pertinent formula derived by Spanos[31], which takes into account the special form of $A(t)$, and is appropriate for wide-banded, non-white input power spectral densities $G_0(\omega)$. The approximated central frequency, bandwidth-factor and response variance finally enter the expression for the survival probability, Equ.(1), in a dimensionless form.
In a second step, this approximation is compared to results from the literature, where the papers of Yang[26] and Solomos,Spanos[29] are used, which contain analytical as well as results derived from simulations.

Despite of its low numerical effort, a standard programmable pocket computer suffices, the present approximation turns out to give satisfactory solutions in all of the examples considered below.

As a by-product, it is shown, that the original spectral moment approximation, when used in the form suggested by their founders, see especially Refs.3,14, leads to reliable results even in the case when it has been claimed to fail numerically by Bucher,Lin[25].

<u>A Simple Approximation Based on the Spectral Moment Formulation</u>.
As a measure for the bandwidth of the response process of a linear oscillator when subjected to a suddenly imposed stationary excitation, $A(t)= H(t)$, a band-width factor $\bar{\delta}(t)$ has been introduced by Krenk,Madsen, Madsen[24], which corresponds to the correlation coefficient of dy/dt and the Hilbert transform of y. It is seen from the extended set of formulas presented in the paper by Langley[30], which refers to the methods used in Ref.24, that $(1-\bar{\delta}(t)^2)^{1/2}$ indeed plays the role of δ_y in Equ.(5),see Equ.(52) of Ref.30. Remarkably, the results presented in Fig.5 of Ref.24 show that the stationary limit of $\bar{\delta}$ is reached after a few cycles of motion, and that it is only slightly influenced by the spectral distribution of $G_0(\omega)$. A solution for $\bar{\delta}$ very similar to the one for $A(t)= H(t)$ in Ref.24 has been obtained by Langley[30],when studying oscillations due to a time modulation $A(t)= C\ t\ \exp(-\beta t)$, see Fig.8 of Ref.30.

Heuristically, it is thus concluded from an inspection of Fig.8 of Ref.30 and Fig.5 of Ref.24 that the influence of the frequency contents of $G_0(\omega)$ as well as of the time variation of $A(t)$ approximately may be neglected for the evaluation of $\delta_y(t)$.

The same conclusion can be drawn for $\nu_0(t)$, which gives the average rate of zero-crossings, see Equ.(6) for the spectral moment approximation: From the exact results presented in Fig.2 of Bucher,Lin[25], where a Kanai-Tajimi representation of $G_0(\omega)$ and an exponentially varying modulation $A(t)= C\,[\exp(-\alpha t)-\exp(-\beta t)]$ is considered, it is seen that the constant value $\omega_n/2\pi$ of Equ.(10a), which has been developed for a wide-banded excitation with $A(t)= H(t)$, is approached already after a few cycles of the motion. The same conclusion recently has been drawn by Igusa[32]. In Ref.32,however, a more complex formulation for the bandwidth-parameter has been given. In case of exponentially varying modulations $A(t)$, the difference between this latter, more accurate approximation and δ^*_y of Equ.(10.b) turns out to be comparatively small.

For a simple approximation concerning slightly damped oscillators with low and intermediate natural frequencies due to wide-banded excitation, it is recommended,therefore, to take the expressions of the spectral moment method for $A(t)= H(t)$ given in Equ.(10), Ω^*_y and δ^*_y, regardless of the special actual form of the (slowly varying) time modulation $A(t)$. Clearly then, the influence of the frequency contents and the non-Heaviside time modulation A upon the response measures ν_a, n_a and α, taken from Equs.(6)-(8), will be expressed only through the zero-th spectral moment $\lambda_0= \sigma_y^{\,2}$. The quality of this approximation is checked through the numerical examples presented below.

In order to use a further simplification, which turns out to be extremly useful, $\sigma_y(t)$ is calculated in the approximate form suggested by Spanos[31]

$$\sigma^2_y(t)= \sigma^2_s\ \Phi(\omega_n t), \tag{12}$$

where $\sigma_s= \sigma^*(t=\infty)$,see Equ.(9), and

$$\Phi(t^*)= 2\zeta\ \exp(-2\zeta t^*)\int_0^{t^*} A^2(\tau^*)\ \exp(\ 2\zeta\tau^*)\ d\tau^*, \tag{13}$$

is derived by combining Equs.(29),(40) of Ref.31. $t^*=\omega_n t$ stands for the oscillator's eigentime. Equ.(13) can be readily evaluated for common functions $A(t)$ in closed form, see Ref.31 for various illustrations. The transformation from $A(t)$ to $A(t^*)$ in Equ.(13)

is introduced for the sake of a similarity analysis. Equ.(12) has been derived by Spanos[31] using the assumptions of slowly varying time modulations and the wide-band input, small-band output approximation as mentioned above.
Defining the dimensionless barrier

$$a^* = a / \sigma_s = a / (\pi G_0(\omega_n) / 4\zeta\omega_n^3), \tag{14}$$

a scaling frequently used in the literature, the following simple non-dimensional expressions are finally obtained:

$$\nu_a(t^*)/f_n = \exp(-a^{*2} / 2\Phi(t^*)), \tag{15a}$$

$$n_a(t^*)/f_n = (2\pi)^{1/2} \, \delta_y(t^*) \, (a^* / \Phi(t^*)^{1/2}) \exp(-a^{*2} / 2\Phi(t^*)), \tag{15b}$$

where $f_n = \omega_n/2\pi$ is the natural frequency measured in Hz. (Introducing a semi-empirical factor b, suggested in Ref.3, which lies between 1 and 1.2, δ_y^* may occasionally be replaced by $(\delta_y^*)^b$.) Equ.(15) is inserted into Equ.(8) to provide the decay-rate in the non-dimensional form $\alpha(a^*, t^*)/f_n$.

This approximate similarity analysis, which is tested in the next section, demonstrates that, for many practical situations, the failure probability mainly is influenced by the time variation of A and,of course, by a^*, the eigentime and the viscous damping, while -given a wide-banded excitation- the influence of $G_0(\omega)$ is expected to be basically of minor influence. The natural frequency ω_n explicitly enters only through the transformation from $A(t)$ to $A(t^*)$. Multiplying A by a proportionality factor C is equivalent to a re-scaling of the barrier a^*.

Results.

The approximation,discussed above,is compared to results derived by Yang[26] for SDOF linear oscillators with a natural frequency of 1Hz and viscous damping ratios of 2 and 5 percent, respectively. The excitation is white noise with intensity G_0,and time modulation is specified by $A(t)=3.06(\exp(-0.251t)-\exp(-0.628t))$. Fig.1 shows the time evolution of the variance σ^2_y/G_0, compared to the results illustrated in Fig.2 of Ref.26. The results turned out to be indistinguishable. Fig.2,3 show the time evolution of the failure probability $F(a^*, NT_0/2)$ for various values of a^* and $\zeta=0.05$ and $\zeta=0.01$, respectively. It is $T_0=1/f_n$, and N gives the numbers of half-cycles. Comparison is made to nine approximations presented in Figs.5,6 of Ref.26, which have been derived on the basis of the point process approach by Yang,Shinozuka[33]: (1) simple upper bound, (2) upper bound II, (3)Poisson approximation, (4) Markov approximation, (5) non-approaching excursion, (6)maximum entropy approximation, (7) and (8) lower bounds, (9) modified non-approaching excursions. The present results are given in Figs.2,3, and in the following Figures, as *bolt* lines. For the evaluation of Equ.(1), numerical integration has been performed according to the trapezoidal rule. Fig.4 gives the asymptotic values ,$N=\infty$, which are those of the highest practical interest, when compared to the results of Figs.7,8 of Ref.26,where also simulation results have been shown. It is seen that the asymptotic values are matched quite well by the present approximation, while F is underestimated during the first few cycles of motion.

In addition, comparison is made to analytical results and simulations presented by Solomos,Spanos[29] for SDOF linear oscillators with a natural frequency of 1Hz and a viscous damping of $\zeta=0.01$ and $\zeta= 0.02$, respectively. A Kanai-Tajimi spectrum with characteristic frequency $\omega_g=8\pi$ and ground damping $\zeta_g=0.8$ has been used. Analytical solutions of Ref.29 have been derived according to a discrete extreme-point process approach and to an integral equation solution, both methods being based on approximating the response amplitude by a Markov process. Figs.5-7 show the survival probability ,Equ.(1), for the nondimensional barriers $a^*=2$ and $a^*=3$ as a function of the cycles of motion, when compared to the results of Figs.3,4 and 6-10 of Ref.29. $b=1.1$ has been used for $\zeta=0.01$, and $b=1$ otherwise.

In Fig.5, there is $A(t)=H(t)$, while in Fig.6 the fast-rise time modulation $A(t)=10.13(\exp(-0.25t)-\exp(-0.5t))$ has been used . Fig.7 gives the results for the slow-rise modulation $A(t)=5.934(\exp(-0.124t)-\exp(-0.25t))$. In case of the Heaviside-type switching, results fit well. For the exponential variations, the survival probability is overestimated during the first few cycles of motion, and underestimated later on, the results beeing generally quite satisfactory. The slow-rise modulation shows a better fit than the fast-rise one.

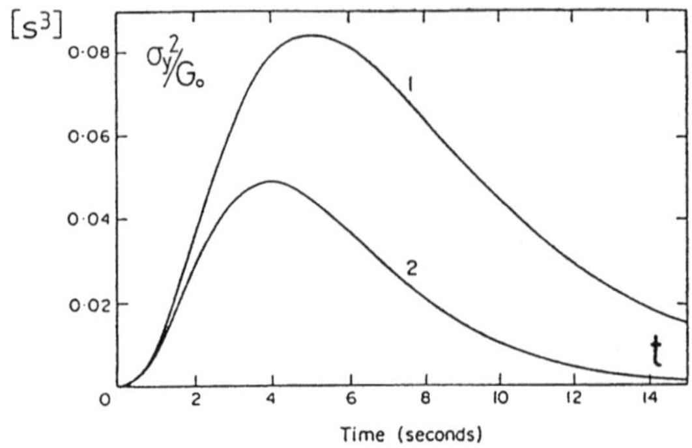

Figure 1: Time evolution of σ^2_y/G_0, when compared to Fig.2 of Ref.26.: (1): ζ=0.02, (2): ζ=0.05.

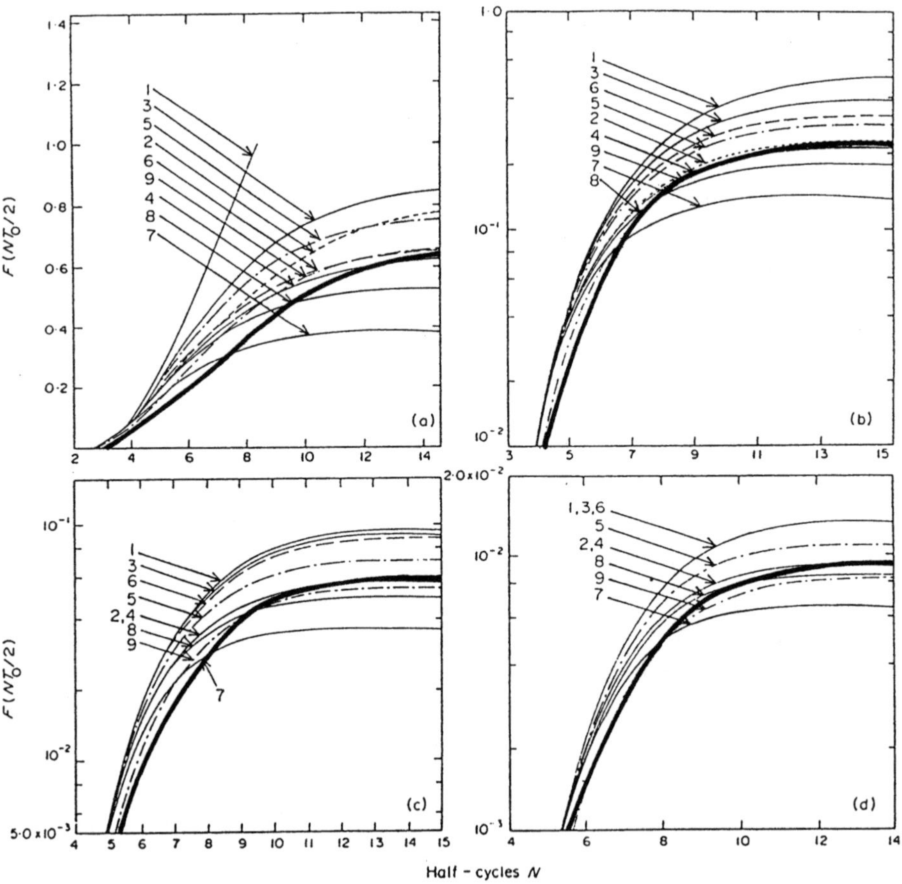

Figure 2: Failure probability F as a function of the number of half-cycles N_* for ζ= 0.05, when compared to Fig.5 of Ref.26: (a): a^* = 1.5, (b): a^* = 2.0, (c): a^* = 2.5, (d): a^* = 3.0. Bolt line: Present solution.

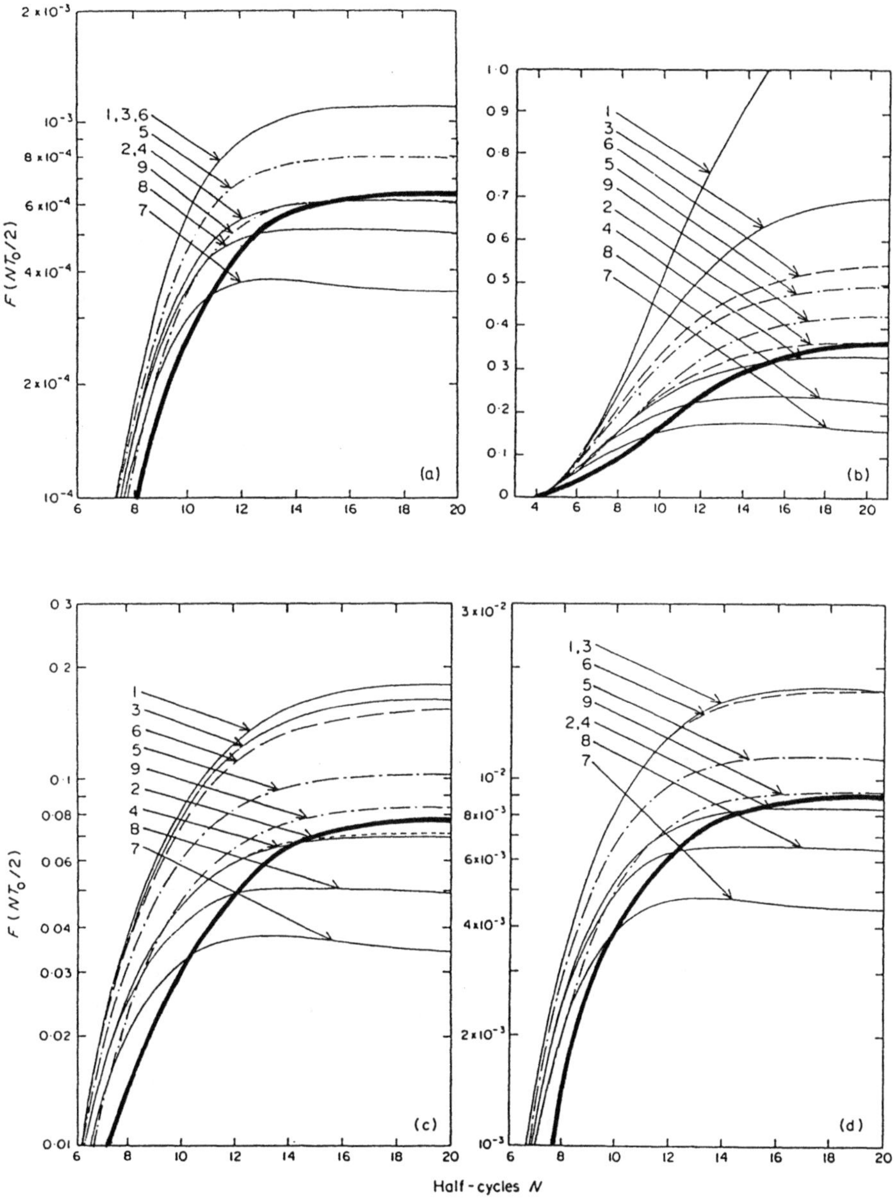

Figure 3: Failure probability F as a function of the number of half-cycles N for $\zeta = 0.02$, when compared to Fig.6 of Ref.26; (a): $a^* = 3.0$, (b) $a^* = 1.5$, (c): $a^* = 2.0$, (d): $a^* = 2.5$. Bolt line: Present solution.

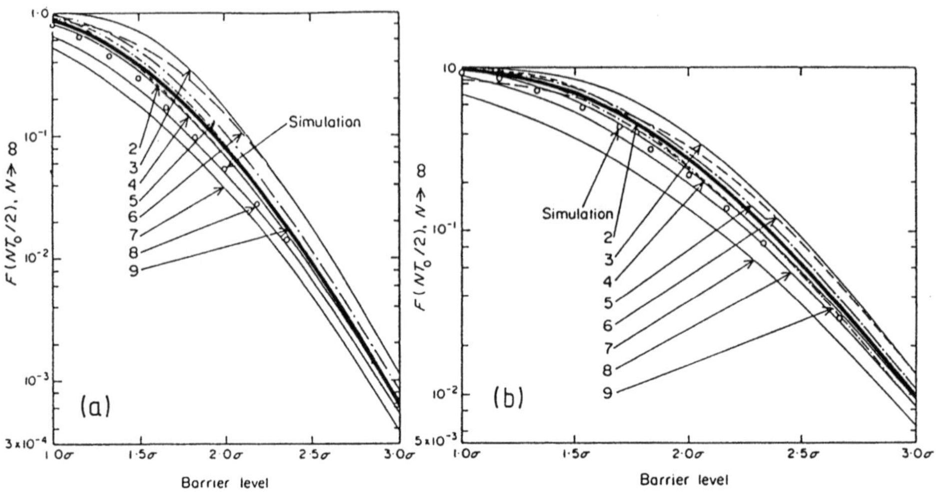

Figure 4: Asymptotic values for the failure probability, $F(a^*, N=\infty)$, when compared to Figs.7,8 of Ref.26: (a): $\zeta=0.05$, (b): $\zeta=0.02$. Bolt line: Present solution.

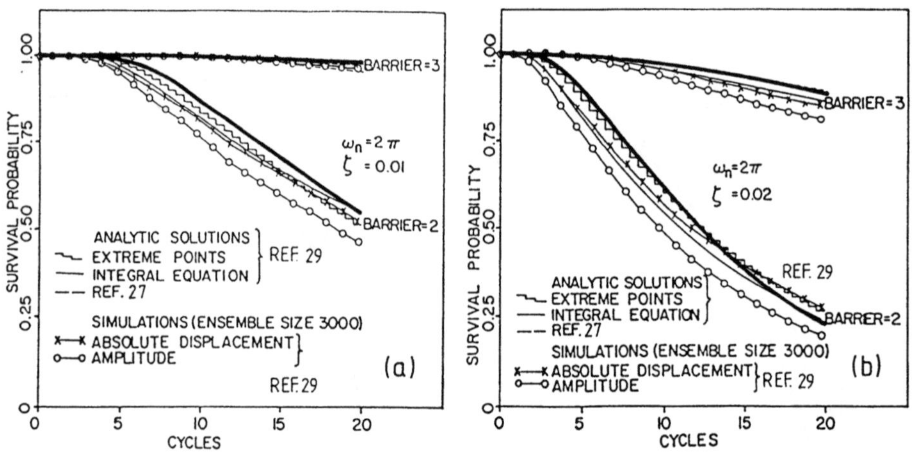

Figure 5: Survival probability L_a as a function of cycles of motion, when compared to Figs.3,4 of Ref.29: (a): $\zeta=0.01$, (b): $\zeta=0.02$. Bolt line: Present solution. Heaviside-type time modulation.

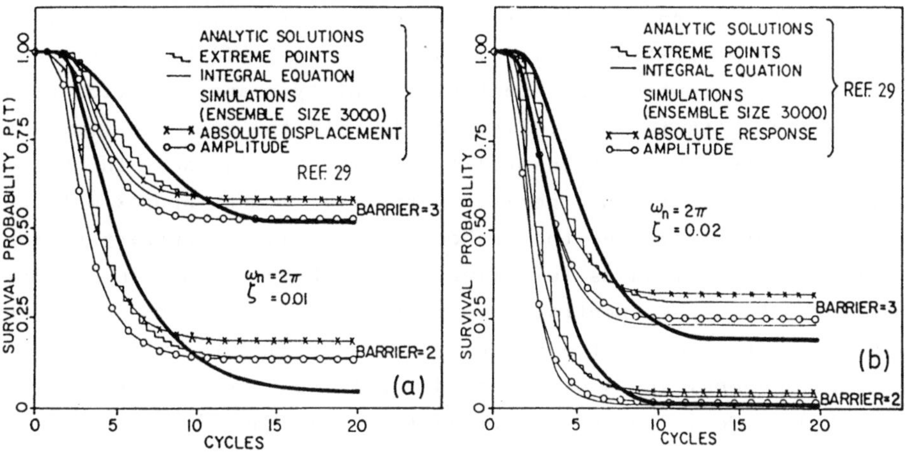

Figure 6: Survival probability L_a as a function of cycles of motion, when compared to Figs.7,8 of Ref.29: (a): $\zeta=0.01$, (b): $\zeta=0.02$. Bolt line: Present solution, Fast-rise time evolution.

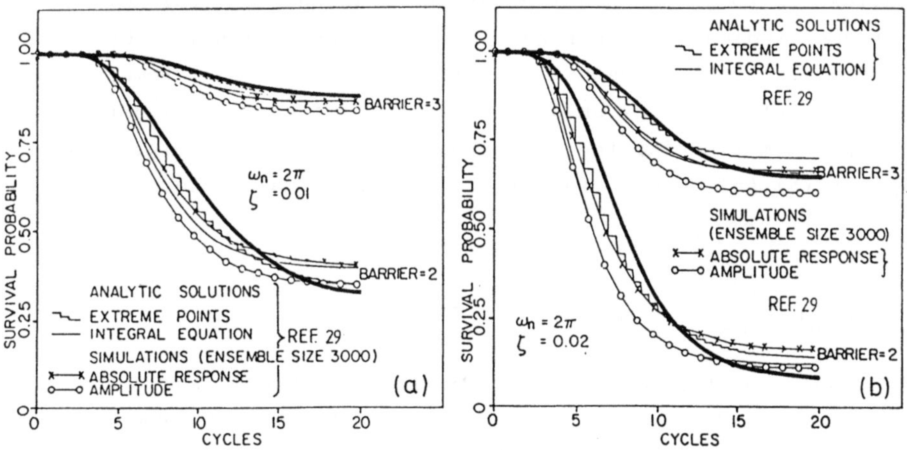

Figure 7: Survival probability L_a as a function of cycles of motion, when compared to Figs.9,10 of Ref.29: (a): $\zeta=0.01$, (b): $\zeta=0.02$. Bolt line: Present solution. Slow-rise time evolution.

Finally, results for the upcrossing rate ν_a, Equs.(6),(15a), are compared to the exact results of Bucher, Lin[25], where a SDOF oscillator with a natural frequency of $1/2\pi$ Hz and a damping $\zeta=0.05$ subject to a Kanai-Tajimi spectrum with $\omega_g=15$ rad/sec and $\zeta_g=0.6$ has been treated. A barrier in the form $a = 1\sigma_{y,\max}$ is used, where $\sigma_{y,\max}$ is the maximal value of $\sigma_y(t)$. Fig.8 shows the comparison of the present method to the results of Fig.4 of Ref.25, where a time modulation $A(t)=\exp(-0.15t)-\exp(-0.2t)$ has been used. Ecxellent agreement of the spectral moment type approximation with the exact results is found. Fig.9 deals with the same problem, but a Heaviside-type modulation $A(t)=1-\exp(-10000t)$ is used. In this latter case, Bucher,Lin[25] derived a spectral moment approximation of oscillatory type, see the dashed line in Fig.9, contrary to the exact solution, given as full line. It has been claimed in Ref.25, therefore, that the spectral moment approach may lead to erroneous results. In this case of a slightly damped oscillator with low natural frequency and a wide-banded input, however, it has been suggested in Ref.3 to use Equ.(10a) for ν_a in Equ.(6). This spectral moment approach, which is included in the present approximation as a special case, and has not been considered by Bucher, Lin[25], gives the bolt line in Fig.9. It is seen that the coincidence to the exact result is again quite satisfactory. What remains, is to put the failure of the original spectral moment approach, indicated in Ref.25, into the right dimension: Obviously, Fig.9 deals with the rather academic example of an extremely weak structure of $1/(2\pi)$ Hz on a relatively stiff soil. Considering the approximation for the zero-crossing rate in the original form of Equs.(5),(6), where, following Corotis,Vanmarcke,Cornell[14]

$$\lambda_2(t) \doteq \frac{\pi G(\omega_n)}{4\omega_n\zeta}\,(1-\exp(-2\omega_n\zeta t)) + G(\omega_n)\,\frac{\sin^2(\omega_1 t)}{\omega_1}\,\exp(-2\omega_n\zeta t)\,\left(\frac{\omega_u}{\omega_n}\right), \tag{16}$$

it is seen that the influence of oscillatory terms is controlled by the ratio ω_u/ω_n . ω_1 is the damped natural frequency, and ω_u is the maximum frequency for which there is significant power in the broad band excitation. Re-interpreting the time-scale in Fig. 9 as a non-dimensional time $\omega_n t$, and the frequency scale as a non-dimensional upcrossing rate $\nu_{\sigma_{y,max}}/\omega_n$, the results become independent of the natural frequency and depend only on the ratio ω_u/ω_n. Using $\omega_u/\omega_n = 10$ in Equ.(16), which is a reasonable value for commonly encountered engineering structures, the "exact" solution in Fig. 9 now is matched with high accuracy. However, returning to the example structure of a more academic nature in Ref.25, where $\omega_n = 1$ rad/s and the characteristic ground frequency $\omega_g = 15$ rad/s, a much larger value of ω_u/ω_n should be used in Equ.(16), which then fails.

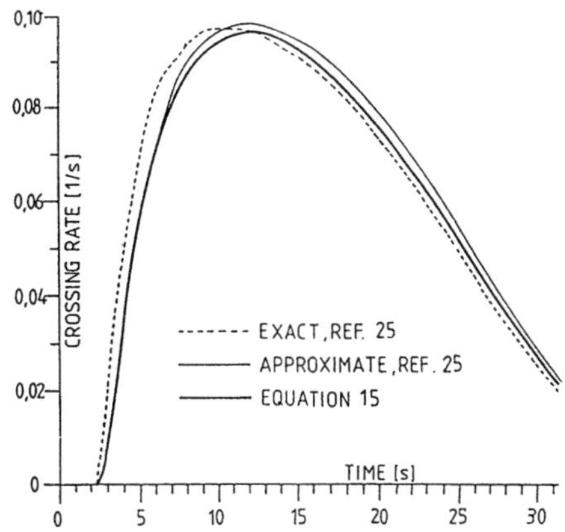

Figure 8: Average rate of $1\sigma_{y,max}$ upcrossings, when compared to Fig.4 of Ref.25. Bolt line: Present solution. Exponentially varying time modulation.

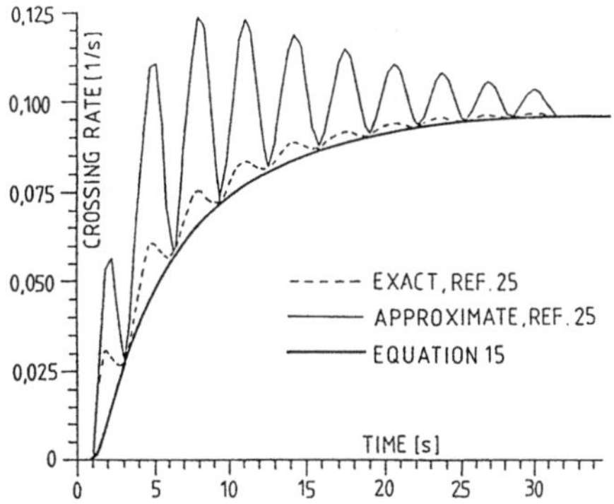

Figure 9: Average rate of $1\sigma_{y,max}$ upcrossings, when compared to Fig.6 of Ref.25. Bolt line: Present solution. Dashed line: Spectral moment solution of Ref.25. Heaviside-type time modulation.

Conclusion.

A simple approximation, based on the well-known spectral moment formulation, has been presented for pertinent nonstationary stochastic response measures of slightly damped linear oscillators. The method is characterized by an extremely low numerical effort, and proved to work quite satisfactory in various examples studied previously in the literature. The formulation has been given in nondimensional form, illustrating the importance of time modulation, barrier scaleing, viscous damping and eigentime with respect to a unified representation of nonstationary up-crossing rates and failure probability

Acknowledgement.
This work is sponsored by the Austrian 'Fonds zur Foerderung der wissenschaftlichen Forschung', grant: P6998-TEC,1989-1990. The authors are grateful to Mr. Ch. Adam for his help in preparing the figures of this paper.

References

/1/ Clough, R.W., Penzien, J., Dynamics of Structures, McGraw-Hill, New York 1975.

/2/ Priestley, M.B., Evolutionary Spectra and Nonstationary Processes, J. Roy. Statistical Society, 1965, 27, 204-228.

/3/ Vanmarcke, E.H., Structural Response to Earthquakes in: Seismic Risk and Engineering Decision (Eds. C. Lomnitz and E. Rosenblueth), Elsevier Publishing Co., Amsterdam, London, New York 1976.

/4/ Shinozuka, M., Sato, Y., Simulation of Nonstationary Random Processes, J. Eng. Mech. Div. Proc. ASCE 1967, 93, 11-40.

/5/ Kanai, K., An Empirical Formula for the Spectrum of Strong Earthquake Motions, Bull. Earthquake Res. Inst., Univ. of Tokyo 1961, 39, 85-95.

/6/ Safak, E. Analytical Approach to Calculation of Response Spectra from Seismological Models of Ground Motion, Earthquake Engineering and Structural Dynamics 1988, 16, 121-134.

/7/ Queck, S.T., Teo, Y.-P., Balendra, T., Nonstationary Structural Response with Evolutionary Spectra using Seismological Input Model, Earthquake Engineering and Structural Dynamics 1990, 19, 275-288.

/8/ Howell, L.J. and Lin, Y.K., Response of Flight Vehicles to Nonstationary Atmospheric Turbulence, AIAA-Journal, 1971, 9, 2201-2207.

/9/ Rice, S.O., Mathematical Analysis of Random Noise, Bell Syst. Tech. J., 1944, 23, 282-332, 1945, 24, 46-156.

/10/ Shinozuka, M., Yang, J.N., Peak Structural Response to Nonstationary Random Excitations, J. Sound Vibr. 1971, 16, 505-517.

/11/ Roberts, J.B., Structural Fatigue Under Non-Stationary Random Loading, J. Mechanical Engineering Science, 1966, 8, 392-405.

/12/ Howell, L.J., Response of Flight Vehicles to Nonstationary Atmospheric Turbulence, Ph. D. Dissertation, University of Illinois, Urbana, Ill., 1971.

/13/ Yang, J.N., Nonstationary Envelope Process and First Excursion Probability, J. Struct. Mech. 1972, 1, 231-248.

/14/ Corotis, R.B., Vanmarcke, E.H., Cornell, C.A., First Passage of Nonstationary Random Process, J. Eng. Mech. Div., Proc. ASCE 1972, 98, 401-414.

/15/ Lutes, L.D., Chen, Y.-T., Tzuang, S.-H., First-Passage Approximations for Single Oscillators, J. Eng. Mech. Div., Proc. ASCE 1980, 106, 1111-1124.

/16/ Vanmarcke, E.H., Some Recent Developments in Random Vibrations, Applied Mechanics Reviews, 1979, 32, 1197-1203.

/17/ Corotis, R.B., Vanmarcke, E.H. , Time-Dependent Spectral Contents of System Response, J. Eng. Mech. Div., Proc. ASCE 1975, 101, 623-637.

/18/ Corotis, R.B., Marshall, T.A., Oscillator Response to Modulated Random Excitation, J. Eng. Mech. Div., Proc. ASCE 1977, 103, 501-513.

/19/ Vanmarcke, E.H., On the Distribution of the First-Passage Time for Normal Stationary Random Processes, J. Appl. Mech., Trans. ASME 1975, 42, 215-220.

/20/ Cramer, H., Leadbetter, M.R., Stationary and Related Stochastic Processes, Wiley, New York 1967.

/21/ Höllinger, F., Time Harmonic and Nonstationary Stochastic Vibrations of Arch Dam-Reservoir-Systems, Acta Mechanics 1983, 49, 153-167.

/22/ Ziegler, F., Random Vibrations: A Spectral Method for Linear and Nonlinear Structures, Probalistic Engineering Mechanics 1987, 2, 92-99.

/23/ Irschik, H., Ziegler, F., Evolutionary Damage Prediction of Discrete and Continuous Elastoplastic Structures with Random Earthquake Excitation in: Computational Mechanics of Reliability Analysis (Belytschko, T., Liu, W.R., eds.) Lausanne, Washington: Elmepress Int. 1989, 432-449.

/24/ Krenk, S., Madsen, H.O., Madsen, P.H., Stationary and Transient Response Envelopes, J. Eng. Mech. Div., Proc. ASCE 1983 , 109, 263-278.

/25/ Bucher, C.G., Lin, Y.K., A Note on Spectral moments in Nonstationary Random Vibration, Probabilistic Engineering Mechanics, 1988, 3, 53-55.

/26/ Yang, J.-N., First Excursion Probability in Nonstationary Random Vibration, J. Sound Vibr., 1973, 27, 165-182.

/27/ Lennox, W.C., Fraser, D.A., On the First-Passage Distribution for the Envelope of a Nonstationary Narrow-Band Stochastic Process, J. Appl. Mech., Trans. ASME 1974, 41, 793-797.

/28/ Krenk, S., Nonstationary Narrow-Band Response and First-Passage Probability, Journal of Applied Mechanics, Trans. ASME 1979, 46, 919-924.

/29/ Solomos, G.P., Spanos, P.-T.D., Structural Reliability under Evolutionary Seismic Excitation, Int. J. Dynamics and Earthquake Engineering 1983, 2, 110-116.

/30/ Langley, R.S., Structural Response to Non-stationary Non-white Stochastic Ground Motions, Earthquake Engineering and Structural Dynamics 1986, 14, 909-924.

/31/ Spanos, P.-T.D., Probabilistic Earthquake Energy Spectra-Equations, J. Eng. Mech. Div., Proc. ASCE 1980, 106, 147-159.

/32/ Igusa, T., Characteristics of Nonstationary White Noise: Applications, J. Eng. Mech. 1989, 115, 1919-1934.

/33/ Yang, J.N., Shinozuka, M., On the First Excursion Probability in Stationary Narrow-Band Random Vibration, J. Appl. Mech., Trans. ASME, Part I: 1971, 38, 1017-1022, Part II: 1972, 39, 733-738.

FFT-Based Simulation of Multi-Variate Nonstationary Random Processes

Ahsan Kareem
Department of Civil Engineering
University of Notre Dame
Notre Dame, IN 46556-0767
Yousun Li
Shell Development Company
Houston, Tx 77001

Summary

A technique based on the fast Fourier transform (FFT) is developed to simulate a multi-variate nonstationary Gaussian random process with a prescribed evolutionary spectral description. The utilization of the FFT algorithm has been made possible by a stochastic decomposition technique. The decomposed spectral matrix is expanded into a weighted summation of basic functions and time-dependent weights which are simulated by the FFT algorithm. A general procedure is presented to express the spectral characteristics of the multi-variate uni-dimensional process in terms of the desired expansion. The effectiveness of the proposed technique is demonstrated by means of three examples with different evolutionary spectral characteristics derived from past earthquake events. The closeness between the target and the corresponding estimated correlation structure suggests that the simulated time series reflect the prescribed probabilistic characteristics extremely well. The simulation approach is computationally efficient, particularly for simulating large numbers of multiple-correlated nonstationary random processes.

Introduction

The simulation of stationary random processes has been done primarily by the Monte Carlo approach for applications in various aspects of computational mechanics. The simulation can be realized by decomposing the prescribed spectral matrix by means of the Cholesky or eigensystem decomposition and subsequent utilization of a summation of trigonometric series with statistically independent phase angles (Shinozuka, 1971). It has been noted that the summation of the trigonometric series may be carried out by utilizing the FFT algorithm which dramatically reduces the computational effort. Although the summation of trigonometric function approach requires a significant computational effort in terms of computer time, the method is applicable to both stationary and nonstationary processes (Shinozuka and Jan, 1972). As noted earlier, the second approach utilizing the FFT algorithm is computationally very

I. Elishakoff · Y. K. Lin (Eds.)
Stochastic Structural Dynamics 2
New Practical Applications
2nd International Conference on Stochastic Structural Dynamics
May 9–11, 1990, Boca Raton, Florida

efficient and it has been used in a wide range of applications with particular reference to simulation of multi-variate wind velocity fluctuations (e.g., Shinozuka, Vaicaitis, and Asada, 1976; Kareem and Dalton, 1982; and Shinozuka et al., 1989). In contrast with the direct summation of the trigonometric series, the FFT approach has been limited primarily to the simulation of stationary processes.

The parametric times series modelling of random processes has gained recent popularity due to the computational efficiency and ability to simulate long duration of time series without requiring large computer storage (Gersch and Yonemoto, 1977; Samaras et al., 1985; Naganuma et al., 1987; Spanos and Mignolet, 1987; and Li and Kareem, 1989). While the foregoing studies have concentrated on the simulation of stationary random processes, a limited effort has been devoted to the simulation of nonstationary processes (e.g., Kozin and Nakajima, 1980; Gersch and Kitagawa, 1985; Cakmak et al., 1985; Hoshiya et al., 1984; and Deodatis and Shinozuka, 1988). The simulation of ground motion records has also been accomplished by filtered white noise and filtered Poisson models (Shinozuka & Deodatis, 1988).

This paper demonstrates a FFT-based procedure for the simulation of multi-variate random processes with prescribed evolutionary probabilistic characteristics. A stochastic decomposition technique, summarized in this paper, facilitates utilization of the FFT algorithm. The choice of a simulation procedure depends on the accuracy in terms of the closeness of the prescribed target and the corresponding estimated probabilistic structure of the random process and computational efficiency in terms of computer run time. The computation of the correlation functions of nonstationary processes entails ensemble averaging of the sample time histories. Therefore, the demand on computational efficiency in the simulation of nonstationary processes is more significant in comparison with the stationary processes. Compared to other simulation techniques for nonstationary multi-variate random processes with large size vectors, e.g., the direct summation of trigonometric functions, and AR models, the proposed method offers, in many cases, substantial computational efficiency.

The theoretical background of random processes with nonstationary probabilistic characteristics is omitted here for the sake of brevity. For additional background, a sample of related references is provided: Priestly (1967), Spanos and Solomos (1983), Mark (1986), Lin and Yong (1985), Li and Kareem (1988), Sun and Kareem (1989), Madsen and Krenk (1982), and Borino et al. (1988).

Theoretical Background

The present simulation scheme is based on a stochastic decomposition technique. A brief introduction is provided here for the sake of completeness. Central to this technique is the decomposition of a set of random processes into component random subprocesses, the relationship between any two of which is either fully coherent or noncoherent. Utilizing this concept, a vector of correlated random processes, $\mathbf{y}(t)$ (N x 1) can be decomposed into random subprocesses

$$y_i(t) = \sum_{\mu=1}^{\Lambda} y_{i/\mu}(t) , \qquad (1a)$$

or in the vectorial form,

$$\mathbf{y}(t) = \sum_{\mu=1}^{\Lambda} \mathbf{y}_{/\mu}(t) , \qquad (1b)$$

in which the sub-process vector $\mathbf{y}_{/\mu}(t)$ is defined as $\mathbf{y}_{/\mu}(t) = [y_{1/\mu}(t), y_{2/\mu}(t), \ldots, y_{N/\mu}(t)]^T$, and Λ is the decomposition order. The relationship between any two sub-processes, $y_{i/\mu}(t)$ and $y_{i/\lambda}(t)$, is either fully-coherent (if $\lambda = \mu$) or noncoherent (if $\lambda \neq \mu$). In the frequency domain, the sub-processes are expressed in terms of decomposed spectra, $D_{i/\mu}(f)$, which is related to the sub-process in such a manner that the cross-power spectral density function between $y_{i/\mu}(t)$ and $y_{j/\mu}(t)$ is given by

$$G_{i/\mu\, j/\mu}(f) = D_{i/\mu}(f)\, D^*_{j/\mu}(f) . \qquad (2)$$

Typical elements of the spectral matrix of the parent process $\mathbf{y}(t)$ are given by

$$G_{y_i y_i} = \sum_{\mu=1}^{\Lambda} D_{y_{i/\mu}}(f)\, D^*_{y_{i/\mu}}(f) \text{ , and} \qquad (3)$$

$$G_{y_i\, y_j}(f) = \sum_{\mu=1}^{\Lambda} D_{y_{i/\mu}}(f)\, D^*_{y_{j/\mu}}(f) . \qquad (4)$$

The corresponding spectral matrix is described by

$$\mathbf{G}(f) = \mathbf{D}(f)\, \mathbf{D}^*(f) , \qquad (5)$$

where $\mathbf{G}(f)$ represents (N,N) cross-spectral density matrix, $\mathbf{D}(f)$ is (N,Λ) decomposed spectral matrix, and * represents conjugation.

Let us express each subprocess in terms of its decomposed spectrum

$$y_{i/\mu}(t) \overset{\text{Re}}{=} \int_0^{+\infty} \sqrt{2}\, D_{i/\mu}(f)\ \exp(j\, 2\pi f t)\ dZ_\mu(f)\,, \tag{6}$$

in which $dZ_\mu(f)$ represents an orthogonal increment which, satisfies the following orthogonality requirement

$$E[dZ_\mu(f_m)\, dZ_\lambda(f_n)] = \delta_{nm}\, \delta_{\lambda\mu}\ df\,,$$

and

$$E[dZ_\mu(f)] = 0\,. \tag{7}$$

Following Eq. (1) and introducing the FFT in the preceding equation, a stationary random process can be simulated by utilizing the expression below

$$y_i(n\,\Delta t) \overset{\text{Re}}{=} \sum_{m=0}^{N_t/2} \Big[\sum_{\mu=1}^{\Lambda} \sqrt{2\,\Delta f}\ D_{i/\mu}(m\,\Delta f)\, \varepsilon_\mu(m) \Big] \exp\!\Big(j\, 2\pi \frac{nm}{N_t}\Big)\,, \tag{8}$$

in which N_t is the total number of time intervals, and Δt and Δf are time and frequency resolutions, respectively,

$$\Delta f = \frac{1}{N\,\Delta t}\,, \tag{9}$$

and $\varepsilon_\mu(m)$ is a zero mean white noise process such that

$$E\,[\varepsilon_\mu(m)\, \varepsilon_\lambda(n)] = \delta_{nm}\, \delta_{\lambda\mu}\,. \tag{10}$$

Evolutionary Spectral Description

Many random environmental load effects have nonstationary characteristics, i.e., their frequency contents and/or amplitudes are time-dependent. Typical examples of such loadings are atmospheric turbulence, seismic excitation and evolutionary sea states. A representative analytical expression for the evolutionary power spectral density is given by

$$G(f,t) = |A(f,t)|^2\, K(f)\,. \tag{11}$$

The evolutionary spectral matrix $\mathbf{G}(f,t)$ may be decomposed into time-dependent decomposed spectra

$$\mathbf{G}(f,t) = \mathbf{D}(f,t)\, \mathbf{D}^*(f,t)\,. \tag{12}$$

A sub-process can be expressed in terms of the decomposed spectrum

$$y_{i/\mu}(t) \overset{Re}{=} \int_0^\infty \sqrt{2}\; D_{i/\mu}(f,t) \exp(j\, 2\pi f t)\; dZ_\mu(f)\,. \tag{13}$$

Recall that each parent process is a summation of decomposed sub-processes, therefore, the time sample function for each parent process is given by

$$y_i(t) \overset{Re}{=} \sqrt{2} \sum_{\mu=1}^{\Lambda} \int_0^\infty D_{i/\mu}(f,t)\; \exp(j\, 2\pi f t)\; dZ_\mu(f)\,. \tag{14}$$

A nonstationary process may be simulated by evaluating Eq. (14). In the following section, the FFT technique is implemented in the simulation scheme to enhance the computational efficiency.

Simulation By FFT Technique

The decomposed spectrum may be expressed as a product of frequency- and time-dependent functions

$$D_{i/\mu}(f,t) \approx \sum_{r=1}^{N_r} A^{(r)}(t)\; \Phi_{i/\mu}^{(r)}(f)\,, \tag{15}$$

or in a matrix form,

$$\mathbf{D}(f,t) \approx \sum_{r=1}^{N_r} A^{(r)}(t)\; \mathbf{\Phi}^{(r)}(f)\,. \tag{16}$$

In the preceding equations, $D_{i/\mu}(f,t)$ can be viewed as weighted summation of $\Phi_{i/\mu}^{(r)}(f)$. The definition of the sum of decomposed spectra suggests that subprocesses exist that correspond to the decomposed spectrum $\Phi_{i/\mu}^{(r)}(f)$

$$\phi^{(r)}_{i/\mu}(t) = \int_0^\infty \sqrt{2}\ \Phi^{(r)}_{i/\mu}(f)\ \exp(j\ 2\pi f t)\ dZ_\mu . \tag{17}$$

These subprocesses $\Phi^{(r)}_{i/\mu}(f)$ and $\Phi^{(p)}_{j/\lambda}(f)$ are fully coherent if $\mu = \lambda$. The sample time function of $y_{i/\mu}(t)$ is expressed by

$$y_{i/\mu}(t) = \sum_{r=1}^{N_r} A^{(r)}(t)\ \phi^{(r)}_{i/\mu}(t) , \tag{18}$$

and consequently the time history of the target process is given by

$$y_i(t) = \sum_{r=1}^{N_r} \Big(A^{(r)}(t) \sum_{\mu=1}^{\Lambda} \phi^{(r)}_{i/\mu}(t) \Big). \tag{19}$$

The previous equation permits the simulation of a nonstationary vector process. Due to the nonstationary nature of the process, which is often characterized by a short duration time history, the frequency resolution may not be high (i.e., $\Delta f = \frac{1}{N_t\, \Delta t}$; N_t = total number of points to be simulated and Δt = time increment). Therefore, the frequency resolution may be improved by deriving $\mathbf{D}(f,t)$ based on averaging the spectral matrix over a frequency interval Δf

$$\mathbf{D}(m\,\Delta f, n\,\Delta t)\ \mathbf{D}^*(m\,\Delta f, n\,\Delta t) = \frac{1}{\Delta f} \int_{(m-\frac{1}{2})\Delta f}^{(m+\frac{1}{2})\Delta f} \mathbf{G}(f, n\,\Delta t)\ df . \tag{20}$$

The procedure for simulating a multi-variate random process is summarized here. First, transpose the spectral matrix into a decomposed spectral matrix. Second, determine $A^{(r)}(n\,\Delta t)$ and $\Phi^{(r)}_{i/\mu}(m\,\Delta f)$ utilizing the decomposed spectral matrix. Third, simulate a complex white noise vector $\varepsilon_\mu(m)$ with $\mu = 1, 2, \ldots, \Lambda$, and $m = 0, 1, 2, \ldots, N_{t/2}$, such that the real and imaginary parts of $\varepsilon_\mu(m)$ are independent, zero mean Gaussian white noise processes. Then by invoking the FFT algorithm

$$\phi^{(r)}_i(n\,\Delta t) \overset{Re}{=} \sum_{m=0}^{N_t/2} \sqrt{2\,\Delta f}\ \Big[\sum_{\mu=1}^{\Lambda} \Phi^{(r)}_{i/\mu}(f)\ \varepsilon_\mu(m) \Big] \exp\Big(j\ 2\pi \frac{nm}{N_t}\Big) . \tag{21}$$

In this manner, the FFT algorithm is utilized ($N \times N_r$) times to simulate a vector process ($N \times 1$). Finally, the ith element of the desired vector process is given by

$$y_i(n\,\Delta t) = \sum_{r=1}^{N_r} A^{(r)}(n\,\Delta t)\,\phi_i^{(r)}(n\,\Delta t)\,. \qquad (22)$$

General Matching Procedure

In the following, a general matching procedure is presented to describe the decomposed spectra (Eq. (15)) in terms of a polynomial. This procedure is suitable for a given nonstationary spectral description of a multi-variate random process in either an analytical or a numerical form.

An error minimization procedure is utilized. Let

$$D_{i/\mu}(f,t) = \sum_{r=1}^{\infty} A^{[r]}(t)\;\Phi_{i/\mu}^{[r]}(f) \qquad \mu = 1, 2, \ldots, \Lambda \qquad (23)$$

in which $A^{[r]}(t)$ is an orthogonal function

$$\int_0^{t_{max}} A^{[r]}(t)\,A^{[s]}(t)\;dt = \delta_{rs}\,, \qquad (24)$$

and t_{max} is the maximum time length to be simulated.

The number of summation terms in Eq. (23) are generally truncated to a finite value N_r. Let $y_i'(t)$ represent the error caused by such a truncation

$$y_i'(t) = y_i(t) - \hat{y}_i(t) \qquad (25)$$

where $y_i(t)$ and $\hat{y}_i(t)$ represent simulation based on infinite and finite terms in Eq. (23), respectively. Utilizing the stochastic decomposition approach, it can be shown that the power spectral density function of $y_i'(t)$ is given by

$$G_i'(f,t) = \sum_{\mu=1}^{\Lambda} \Big[D_{i/\mu}(f,t) - \sum_{r=1}^{N_r} A^{[r]}(t)\;\Phi_{i/\mu}^{[r]}(f)\Big]\Big[D_{i/\mu}(f,t) - \sum_{r=1}^{N_r} A^{[r]}(t)\;\Phi_{i/\mu}^{[r]}(f)\Big]\,. \qquad (26)$$

The spectrum of the error caused by truncation of summation in Eq. (23) to a finite value is minimized. This is accomplished by considering the integral of Eq. (26) over time, t_{max}

$$\varepsilon(f) = \int_0^{t_{max}} G_i'(f,t)\ dt\ . \tag{27}$$

Minimizing the preceding function and subsequently invoking orthogonality (Eq. (24)), leads to the following description of $\Phi_{i/\mu}^{[r]}(f)$

$$\Phi_{i/\mu}^{[r]}(f) = \int_0^{t_{max}} D_{i/\mu}(f,t)\ A^{[r]}(t)\ dt\ . \tag{28}$$

It is noted that the preceding minimization provides results that are exactly similar to those obtained in the expansion of a function in terms of a series of orthogonal polynomials. The trigonometric and Legendre polynomial functions have been utilized to describe $A^{[r]}(t)$ in this study.

<u>Examples</u>

In this section, examples concerning ground motion are utilized to demonstrate the effectiveness of the simulation technique presented here. The simplest example is presented first which involves the N-S component of the December 30, 1934 El Centro Earthquake (Liu, 1970). The single-point evolutionary spectral density of the N-S component was analytically described by Deodatis and Shinozuka (1988)

$$G_{ii}(f) = \left[\frac{\exp(-at) - \exp(-bt)}{\max[\exp(-at) - \exp(-bt)]} \right]^2 \hat{K}(f)\ , \tag{29}$$

where a = 0.25 and b = 0.5, and $\hat{K}(f)$ is the Kanai-Tajimi spectrum, expressed as

$$\hat{K}(f) = 4\pi\ S_o \frac{f_g^4 + 4\zeta_g^2\ f_g^2\ f^2}{(f^2 - f_g^2)^2 + 4\zeta_g^2\ f_g^2\ f^2}\ , \tag{30}$$

in which the parameters are $S_o = 0.1\ cm^2\ sec^{-3}$, $f_g = 15/2\pi$ Hz, and $\zeta_g = 0.25$. To illustrate the multi-variate feature we are using an extension based on the empirical relationship for the cross-spectral density function given by Harichandran and Vanmarcke (1986),

$$G_{ij}(f) = G_{ii}(f)\ \rho(v_{ij},f)\ \exp(-\ j\ 2\pi f\ v_{ij}/V_{ij})\ , \tag{31}$$

where v_{ij} denotes the distance between the ith and j^{th} locations, V_{ij} denotes the apparent wave propagation velocity along the direction between the ith and j^{th} locations, and $\rho(v_{ij},f)$ is the coherence function, given by

$$\rho(v_{ij},f) = a \exp\Big[-\frac{2v_{ij}}{\alpha\, \theta(f)} (1 - a + \alpha a)\Big] + (1 - a) \exp\Big[-\frac{2v_{ij}}{\theta(f)} (1 - a + \alpha a) \Big] \tag{32}$$

and

$$\theta(f) = k \left[1 + (\frac{f}{f_0})^b\right]^{-1/2}, \tag{33}$$

in which the parameters are: $a = 0.736$, $\alpha = 0.147$, $k = 5210$, and $f_0 = 1.09$.

It can be shown that the decomposed spectral matrix of **G**(f,t) can be conveniently formulated analytically. However, here the results are reported for the general computational procedure. Details following the analytical approach for some specific spectral descriptions can be found in Li and Kareem (1989). The simulation involves two locations 150 meters apart and V is taken to be equal to 2000 m/sec. Equation (15) was matched numerically and a set of 1000 sample time series was generated for each location. The estimated and target correlation functions for the N-S component of the December, 1934 El Centro Earthquake are compared in Figs. (1 and 2). The target values are shown in solid continuous lines. It is noted that the comparisons are very good for time lags as large as 0.1 second.

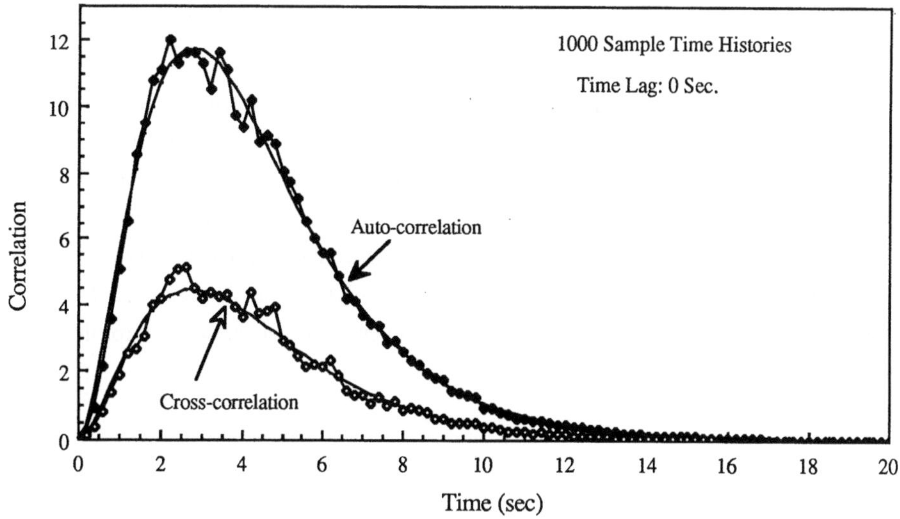

Fig. 1 Correlation Function of N-S Component 1934 El Centro

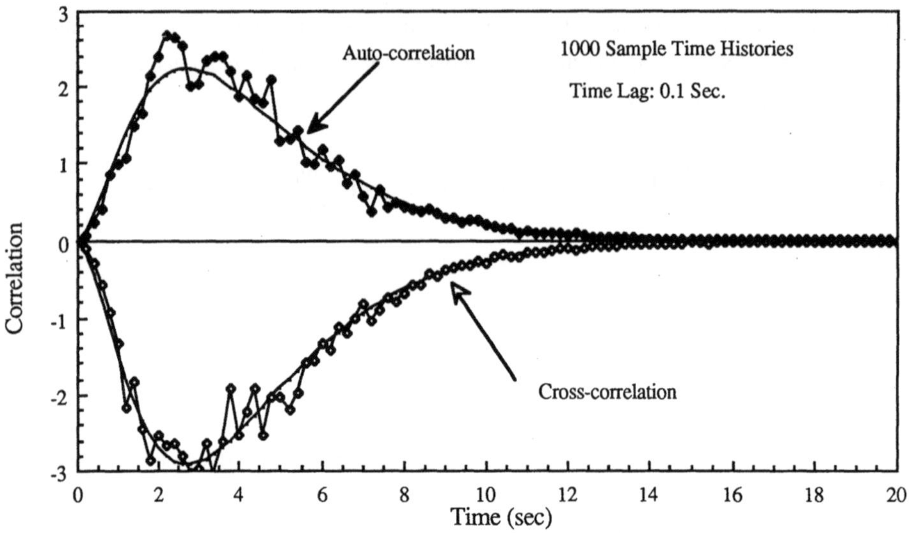

Fig. 2 Correlation Function of N-S Component 1934 El Centro

A relatively more complicated example relates to the E-W component of the December 30, 1934 El Centro Earthquake (Liu, 1970). The single-point-evolutionary spectrum was analytically described by Deodatis and Shinozuka (1988)

$$G_{ii}(f,t) = \left\{ \frac{\exp(-at) - \exp(-bt)}{\max[\exp(-at) - \exp(-bt)]} \sqrt{\hat{K}_1(f)} + \exp\left[-\frac{(t-m)^2}{2\sigma^2}\right] \sqrt{\hat{K}_2(f)} \right\}^2, \quad (34)$$

where $\hat{K}_1(f)$ and $\hat{K}_2(f)$ satisfy the Kanai-Tajimi spectrum (Eq. (30)) with the parameters being the same as stated previously with the exception that $f_g = 30/2\pi$ Hz in $\hat{K}_2(f)$, and a and b have the same values as the preceding example, and m = 5.0 sec. and σ = 1.0 sec. The cross-spectral density function follows Eq. (31). Only the results based on the general matching procedure are presented here. The correlations obtained from the simulated E-W component of the December, 1934 El Centro Earthquake are given in Figs. (3 and 4). Again, the comparison between the estimated and target correlations is excellent. N_r is equal to seven for this and the previous example

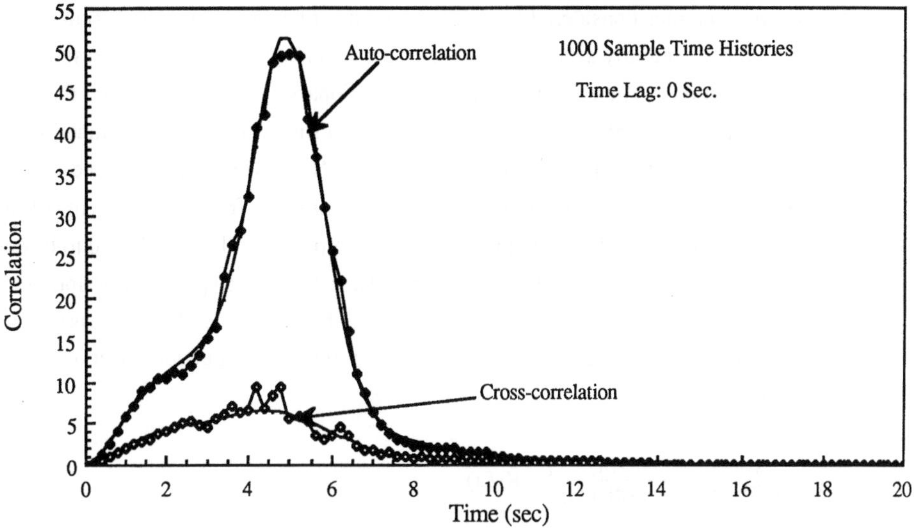

Fig. 3 Correlation Function of N-S Component 1934 El Centro

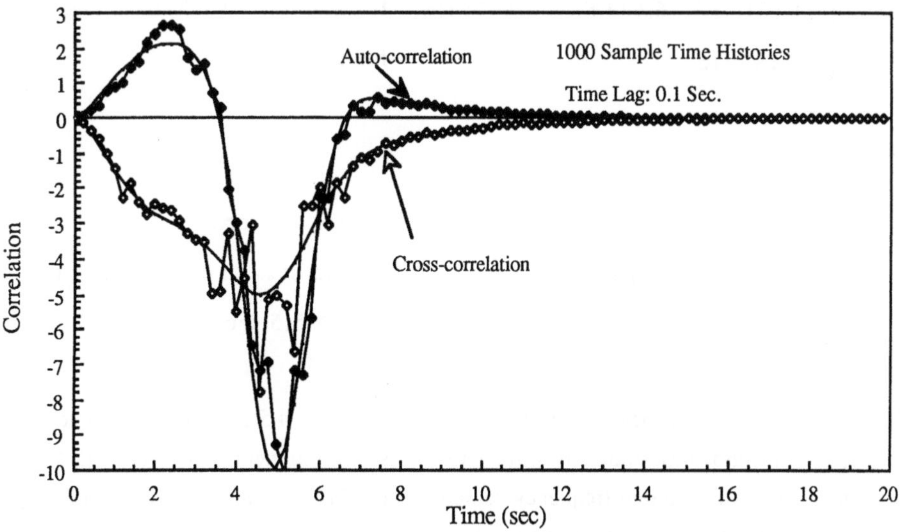

Fig. 4 Correlation Function of E-W Component 1934 El Centro

In the two examples considered here, the time- and frequency-dependent components could be separated conveniently as a consequence of the functional form of the corresponding target spectral descriptions. There is a large class of evolutionary spectral descriptions that may not be amenable to a convenient separation into a time- and a frequency-dependent function. A typical example is the evolutionary spectrum of the 1964 Niigata Earthquake. The ground motion time history has a peculiar feature that, after 7 seconds, the frequency contents of the signal changes and it exhibits a dominant low frequency component. This is believed to be due to the liquefaction of the ground [Deodatis and Shinozuka, 1988]. The spectral description for the evolutionary spectrum is essentially given by the same form as the earlier two cases, with the exception that the parameters f_g and ζ_g in $\hat{K}(f,t)$ are time-dependent (Deodatis and Shinozuka, 1988)

$$G_{ii}(f) = \left[\frac{\exp(-at) - \exp(-bt)}{\max[\exp(-at) - \exp(-bt)]} \right]^2 \hat{K}(f,t) \,, \tag{35}$$

$$f_g(t) = \begin{cases} 2.476 \text{ Hz} & \text{for } t \le 4.5 \text{ sec.,} \\ (4.1316\, t'^3 - 6.474\, t'^2 + 2.467) \text{ Hz} & \text{for } 4.5 \text{ sec.} < t < 5.5 \text{ sec.,} \\ 0.3183 \text{ Hz} & \text{for } t \ge 5.5 \text{ sec.} \end{cases} \tag{36}$$

and

$$\zeta_g(t) = \begin{cases} 0.64 & \text{for } t \le 4.5 \text{ sec.,} \\ (1.250\, t'^3 - 1.875\, t'^2 + 0.64) & \text{for } 4.5 \text{ sec.} < t < 5.5 \text{ sec.,} \\ 0.015 & \text{for } t \ge 5.5 \text{ sec.} \end{cases} \tag{37}$$

with $t' = (t - 4.5)$. This means that during time periods $t \le 4.5$ sec. and $t \ge 5.5$ sec., the evolutionary spectral description is similar to the N-S component of the 1934 El Centro Earthquake, but with different frequency contents. The abrupt change is completed during a one-second time period. However, at $t = 4.5$ sec. and 5.5 sec., the spectrum is continuous up to the first-order time derivative.

The results of the numerical simulation of the 1964 Niigata Earthquake for two locations 75 meters apart (V = 1500 m/sec.) are shown in Fig. (5). In this case, N_r is equal to fifteen. The correlation and the cross-correlation functions estimated from samples of 500 time series with various time lags up to 0.06 seconds are compared with the target ones in Fig. (5). The results demonstrate a good agreement with the target functions and a good comparison is particularly noted for the rapid transition from the high to low frequency contents. A time history of the simulated earthquake is shown in Fig. (6). A similar time history record is reported by Deodatis and Shinozuka (1988) using an AR model. In the low frequency portion of the signal, both the present simulation and the one reported by Deodatis and Shinozuka (1988), did not exhibit secondary oscillation present in the actual recording. This is attributed to the form of spectral description used in these studies. An adjustment in the spectral description would introduce introduce this secondary oscillation.

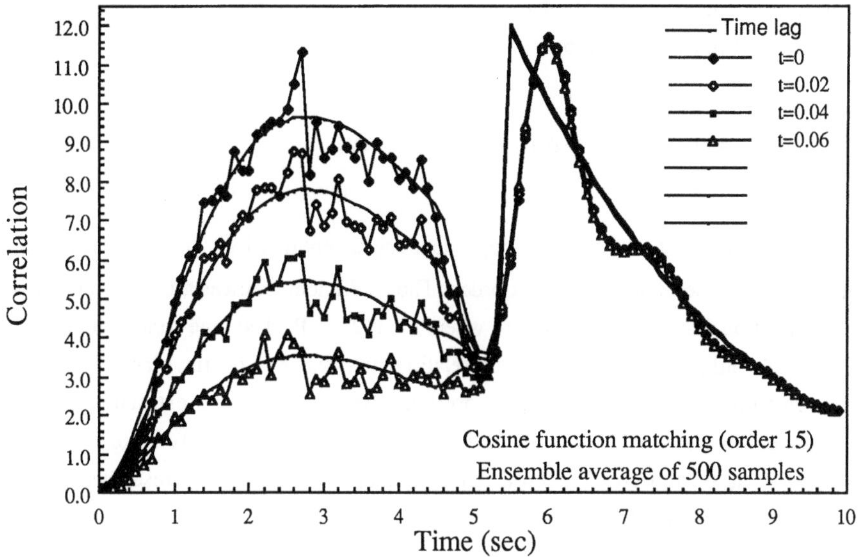

Fig. 5 Correlation Function of Ground Motion (1964, Niigata)

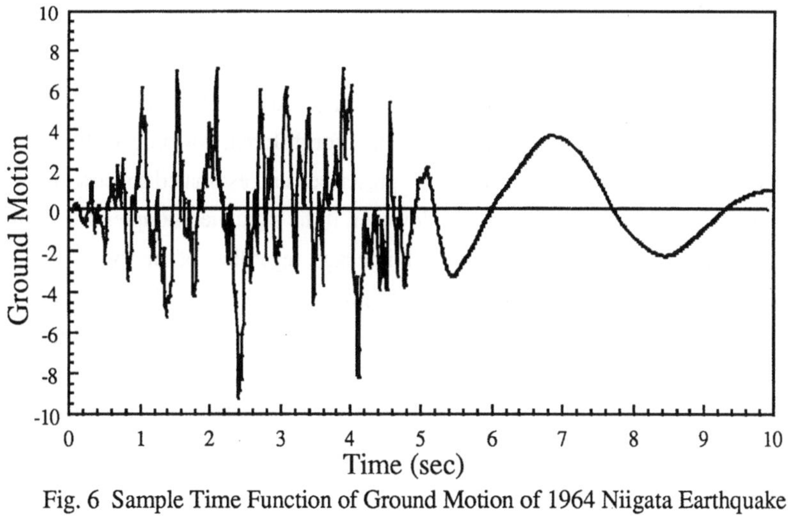

Fig. 6 Sample Time Function of Ground Motion of 1964 Niigata Earthquake

These examples have demonstrated the effectiveness of the general numerical matching procedure. In conjunction with the analytical matching procedure, the FFT-based simulation technique presented here offers a computationally efficient and effective means of generating multi-variate random processes.

Comparison With Auto Regressive Models

A procedure to simulate uni-variate nonstationary processes utilizing auto-regressive (AR) models has been reported by Deodatis and Shinozuka (1988). This concept can be expanded to include simulation of multi-variate evolutionary processes. For a multi-variate process, the AR model is given by

$$y(n\,\Delta t) + \sum_{r=1}^{P} \mathbf{A}_r(n\,\Delta t)\, y[(n-r)\,\Delta t] = \mathbf{B}_0(n\,\Delta t)\, \boldsymbol{\varepsilon}(n\,\Delta t), \tag{38}$$

where $\mathbf{A}_r(n\,\Delta t)$ and $\mathbf{B}_0(n\,\Delta t)$ are time-dependent coefficient matrices, and $\boldsymbol{\varepsilon}(n\,\Delta t)$ is a vector of white noise processes, and P is the AR order. The coefficient matrices are determined from the cross-correlation matrices $\mathbf{R}_y[(r-1)\,\Delta t]$, where $r = 1, 2,\ldots, P$. The time series simulated by the AR model should have the exact cross-correlation matrices up to the time lag of $(P-1)\,\Delta t$, and at the remaining time lags, the cross-correlation matrices may be estimated through the maximum entropy approach. Hence, only an AR model with a suitably large P will have an evolutionary auto- and cross-spectral density function close to the prescribed ones. The evolutionary spectral density matrix may be approximated by

$$\mathbf{G}(f,t) \approx \Big[\, \mathbf{A}_0(t) + \sum_{r=1}^{P} \mathbf{A}_r(t) \exp(j\, 2\pi\, r\, f\, \Delta t) \Big]\, \Delta t\,. \tag{39}$$

In Fig. (7), the target and estimated spectral density functions of the E-W component of the 1934 El Centro (at t = 5.5 sec.) are presented for different AR orders. It is noted, judging from the closeness between the analytically prescribed and estimated spectra, that for the E-W component, P = 39 provides a good fit. However, if the simulated samples are required only to have a correct correlation function for small time lags, regardless of the spectral description one may utilize a low AR model. Deodatis and Shinozuka (1988) used AR order equal to 3 for simulating the referenced ground motion at $\Delta t = 0.01$ sec. and demonstrated exact correlation at time lags of 0.01 and 0.02 seconds. Based on the results in Fig. (7), these models may not provide a good match to the target spectral density functions. The straightforward application of the AR models for the evolutionary processes may have limitations. For example, in Fig. (8), plots of the target and estimated spectra are provided for the 1964 Niigata Earthquake at time t = 5.5 seconds. Only the spectra derived from AR equal to 3 and 4 are provided, since a straightforward derivation of the higher order models introduces numerical singularity. A distinct departure of the estimated spectra from the target is noted which is characterized by a sharp spiked shape. The difficulty in matching higher-order models also results from the fact that the Nyquist frequency $\left(\frac{1}{2\,\Delta t} = 50 \text{ Hz}\right)$ is much higher than the peak spectral frequency.

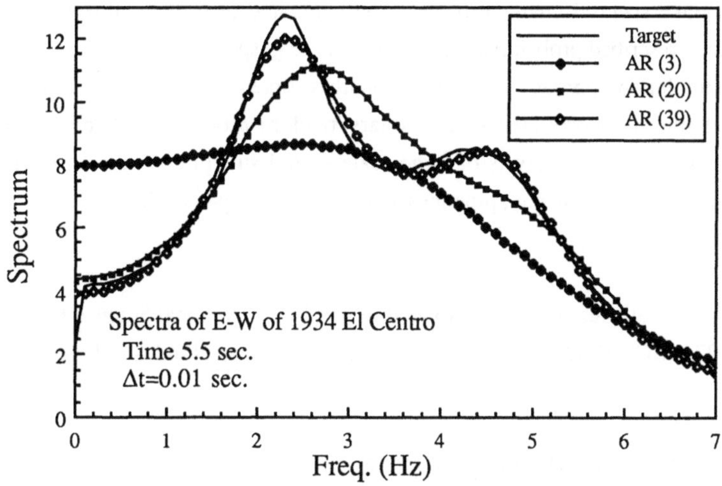

Fig. 7 Spectra Estimated from AR Model

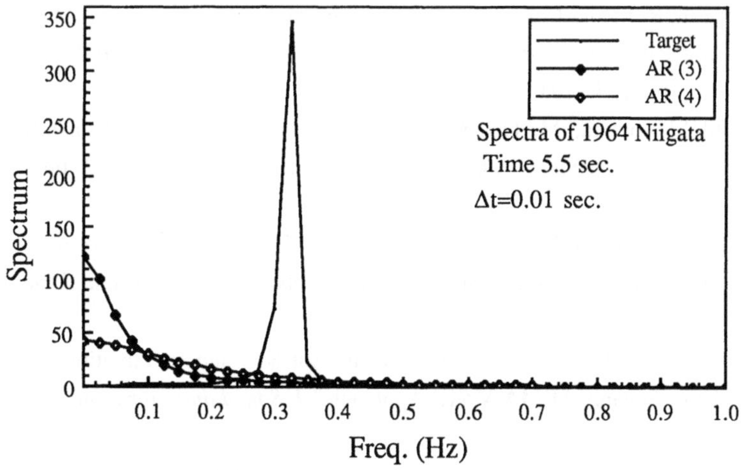

Fig. 8 Spectra Estimated From AR Models

The preceding comparisons further demonstrate the effectiveness of the proposed FFT-based simulation procedures for multi-variate nonstationary random processes.

Conclusions

A computationally efficient FFT-based procedure is developed to simulate multi-variate evolutionary random processes. A stochastic decomposition technique is implemented which facilitates application of the FFT algorithm for simulation. The effectiveness of the proposed technique is demonstrated by means of three examples utilizing analytical description of ground motion encompassing characteristics of actual earthquakes. The proposed technique offers matching procedures for both analytical forms and numerical values at discrete frequencies describing earthquake spectral characteristics. The simulated records exhibit an excellent agreement with the prescribed probabilistic characteristics, e.g, spectral density and correlation functions. The simulation approach is computationally efficient, particularly for simulating large numbers of multiple-correlated nonstationary random processes. Applications are immediate in the time domain analysis of large-scale spatial structures spanning over large spatial areas to seismic excitation, e.g., pipelines, dams, and long-span bridges.

Acknowledgements

The support for this research was provided in part by the National Science Foundation Grant BCS-9096274 (BCS-8352223) and matching funds from several industrial sponsors.

References

Borino, G., Paola, M. DI, and Muscilina, G. (1988), "Non-Stationary Spectral Moments of Base Excited MDOF Systems," *Earthquake Engineering and Structural Dynamics*, Vol. 16, pp. 745-756.

Cakmak, A. S., Sherit, R. I., and Ellis, G. (1985), "Modelling Earthquake Ground Motions in California Using Parametric Time Series Methods," *Soil Dynamics and Earthquake Engineering*, 4(3), pp. 124 - 131.

Deodatis, G., and Shinozuka, M. (1988), "Auto-Regressive Model for Nonstationary Stochastic Processes," *Journal of Engineering Mechanics*, Vol. 114, No. 12, pp. 1995 - 2012.

Gersch, W., and Kitagawa, G. (1985), "A Time Varying AR Coefficient Model for Modelling and Simulating Earthquake Ground Motion," *Earthquake Engineering and Structural Dynamics*, 13(2), pp. 124 - 131.

Gersch, W., and Yonemoto, J. (1972), "Synthesis of Multivariate Random Vibration Systems: A Two-Stage Least Squares ARMA Model Approach," *J. Sound and Vibration*, 52(4), pp. 553 - 565.

Harichandran, R. S., and Wang, W. (1988), "Response of Simple Beam to Spatially Varying Earthquake Excitation," *Journal of Engineering Mechanics*, Vol. 114, No. 9, pp. 1526 - 1541.

Hoshiya, M., Ishii, K., and Nagata, S. (1984), "Recursive Covariance of Structural Response," *J. Engineering Mechanics*, ASCE, Vol. 110, No. 12, pp. 1743 - 1755.

Kareem, A., and Dalton, C. (1982), "Dynamic Effects of Wind on Tension Leg Platforms," OTC 4229, *Proceedings of Offshore Technology Conference*, Houston, Texas.

Kozin, R., and Nakajima, F. (1980), "The Order Determination Problem for Linear Time-Varying AR Models," *IEEE Trans. Auto Control*, AC-25(2), pp. 250 - 251.

Li, Y. (1988), "Stochastic Response of a Tension Leg Platform to Wind and Wave Fields," *Ph.D. Dissertation*, Department of Mechanical Engineering, University of Houston.

Li, Y., and Kareem, A. K. (1989), "On Stochastic Decomposition and its Application in Probabilistic Dynamics," *Proceedings of the 5th International Conference on Structural Safety and Reliability* (ICOSSAR '89), San Francisco.

Li, Yousun, and Kareem, A. (1988), "Recursive Modelling of Dynamic Systems," to appear in *Journal of Engineering Mechanics*, ASCE.

Li, Yousun, and Kareem, A. (1989), "ARMA Systems in Wind Engineering," *Proceedings of the Sixth U.S. National Conference on Wind Engineering*, March 8-10, Houston, Texas, (Editor: A. Kareem).

Li, Yousun, and Kareem, A. (1989), "Simulation of Multi-Variate Nonstationary Random Processes by FFT," Department of Civil Engineering *Tech. Report* No. UHCE89-9.

Lin, Y. K., and Yong, Y. (1985), "On Modelling Earthquakes as Nonstationary Random Processes," *Structural Safety and Reliability, Vol. II*, ICOSSAR '85 (Editors: Konishi, Ang, and Shinozuka).

Liu, S. C. (1970), "Evolutionary power spectral density of strong motion earthquakes," *Bull. Seism. Soc. Am.* 60(3), pp. 891 - 900.

Madsen, P. and Krenk, S. (1982), "Stationary and Transient Response Statistics," *Journal of the Engineering Mechanics Division*, ASCE, Vol. 108, No. 8.

Mark, W. D. (1986), "Power Spectrum Representation for Nonstationary Random Vibration," *Random Vibration -- Status and Recent Developments*, Elsevier, pp. 211 - 240.

Naganuma, T., Deodatis, G., and Shinozuka, M. (1987), "An ARMA Model for Two-Dimensional Processes," *Journal of Engineering Mechanics*, ASCE, 113(2), pp. 234 - 251.

Priestly, M. B. (1967), "Power Spectral Analysis of Non-stationary Random Processes," *J. Sound Vib.*, 6:86 - 97.

Samaras, E., Shinozuka, M., and Tsurui, A. (1985), "ARMA Representation of Random Processes," *Journal of Engineering Mechanics*, ASCE, Vol. 111, pp. 449 - 461.

Shinozuka, M. (1971), "Simulation of Multivariate and Multidimensional Random Processes," *The Journal of the Acoustical Society of America*, Vol. 49, pp. 357 -367.

Shinozuka, M. (1972), "Monte Carlo Solution of Structural Dynamics," *Computer and Structures*, Vol. 2, pp. 855 - 874.

Shinozuka, M., and Jan, C.-M. (1972), "Digital Simulation of Random Processes and its Application," *Journal of Sound and Vibration*, 25(1), pp. 111 - 128.

Shinozuka, M., Vaicaitis, R., and Asada, H. (1976), "Digital Generation of Random Forces for Large-Scale Experiments," *Journal of Aircraft*, Vol. 13, No. 6, June, pp. 425 - 431.

Shinozuka, M., Deodatis, G. (1988), "Stochastic Process Models for Earthquake Ground Motion," *Probabilistic Engineering Mechanics*, Vol. 3, No. 3, pp. 114 - 123.

Shinozuka, M., Yun, C-B., and Sey, H. (1989), "Stochastic Methods in Wind Engineering," Proceedings of the *Sixth U.S. National Conference on Wind Engineering*, Houston, TX, March 8-10 (A. Kareem: Editor).

Spanos, P. D., and Mignolet, M. P. (1987), "Recursive Simulation of Stationary Multivariate Random Processes -- Part I and Part II," *Transactions of the ASME*, Vol. 54, pp. 674 - 680 and 681 - 687.

Spanos, P. D., and Solomos, G. P. (1983), "Markov Approximation to Nonstationary Random Vibration," *J. Engineering Mechanics*, ASCE, Vol. 109, No. 4, pp. 1134 - 1150.

Sun, Wei-Joe, and Kareem, A. (1989), "Response of MDOF Systems to Nonstationary Random Excitation," *Engineering Structures*, Vol. 11, April.

Wittig, L. E., and Sinha, A. K. (1975), "Simulation of Multicorrelated Random Processes Using FFT Algorithm," *Journal of Acoustic Society of America*, Vol. 58, pp. 630 - 633.

Predicting the Fatigue Life of Offshore Structures by the Single-Moment Spectral Method

Curtis E. Larsen
formerly Senior Structures Engineer, Loads and Dynamics Section, Johnson Space Center, NASA, Houston, TX 77058

currently Senior Structural Engineer, Engineering Development, Hudson Engineering Corporation, P.O. Box 218218, Houston, TX, 77218

Loren D. Lutes
Professor of Civil Engineering,
Texas A & M University, College Station, TX 77843-3136

Summary

The dependence of fatigue damage accumulation on power spectral density (psd) is investigated for Gaussian random processes representing stresses in offshore structures. This involves extensive computer simulation of representative stress time histories and related analyses to predict the time of fatigue failure. The recently introduced single-moment (SM) spectral method is shown to be very attractive for predicting fatigue failure. The SM method predicts the damage accumulation rate based only on a simple calculation from the psd curve (a moment integral). The accuracy of the SM method and of other spectral methods is studied by comparing results with ones calculated from the much more expensive alternative of using simulated stress time histories, rainflow anlysis, and Palmgren-Miner calculations.

The SM method is shown to be even simpler than most spectral methods, and all spectral methods are much simpler than any method involving simulation of stress time histories. The accuracy of the SM method is shown to be quite good for a wide variety of situations. Furthermore, there are situations in which the SM method is significantly more accurate than other simple spectral methods, including the very commonly used Rayleigh approximation.

It appears that the SM method may provide a valuable tool for predicting the fatigue life of offshore structures subject to complicated time histories of loading, as is required both in the design of new structures and in the reassessment of aging structures. In order to allow a proper balance between fatigue life and initial cost, there is a need for a fatigue prediction method which is both accurate and easy to apply, and the SM method appears to be adequate on both bases.

Key Words

Damage accumulation, offshore structures, rainflow analysis, random processes, spectral methods, stochastic fatigue

I. Elishakoff · Y. K. Lin (Eds.)
Stochastic Structural Dynamics 2
New Practical Applications
2nd International Conference on Stochastic Structural Dynamics
May 9–11, 1990, Boca Raton, Florida

Introduction

A number of procedures exist for predicting fatigue life of structures subject to variable amplitude loading. The simplest approaches relate the concept of damage accumulation to either a time history of the load or stress process, or to the power spectral density, psd, of the process. In the time domain, cycles or half-cycles are identified, usually by the rainflow technique [1,2], and the damage per cycle is accumulated using a Palmgren-Miner summation based on an appropriate constant amplitude S-N curve. This technique requires that a time history of stress be available, either as recorded from a similar structure in a similar loading environment or as generated by simulation from an appropriate psd. In the frequency domain, several procedures exist for estimating the damage from a few parameters evaluated directly from the psd. These so-called spectral methods are attractive because loadings and structural dynamic behavior are often characterized by psd's and harmonic transfer functions, producing a psd for the response stress process. The need to measure or simulate a stress time history is thus avoided. The Rayleigh or narrowband approximation is the most commonly used of the spectral methods. This method is discussed in the following section and compared to several other spectral approaches, including the recently introduced single-moment method [3].

Spectral Methods

The Rayleigh approximation derives its name from the assumption that the stress ranges are given by twice the Rayleigh distributed amplitude of a Gaussian process. The rate of occurrence of stress cycles is taken as the rate of occurrence of zero-crossings, ν. Assuming a constant amplitude S-N curve of the form:

$$N = KS^{-m} \tag{1}$$

and the Palmgren-Miner hypothesis, the expected rate of damage accumulation is [4]:

$$\dot{D}_{RAY} = \frac{\nu}{K}(2\sqrt{2}\sigma_x)^m \Gamma(1 + m/2) \tag{2}$$

This expression can be re-written in terms of spectral moments as:

$$\dot{D}_{RAY} = \frac{2^{\frac{3m}{2}}}{2\pi K}\Gamma(1+m/2)\lambda_0^{\frac{m-1}{2}}\lambda_2^{1/2} \tag{3}$$

in which:

$$\lambda_0 = \sigma_x^2 = \int_0^\infty G(\omega)d\omega$$

$$\lambda_2 = \sigma_{\dot{x}}^2 = \int_0^\infty \omega^2 G(\omega)d\omega$$

$G(\omega)$ = one-sided psd of stationary zero-mean stress process $\{X(t)\}$

and:

$$\nu = \frac{1}{2\pi}\sqrt{\frac{\lambda_2}{\lambda_0}}$$

The assumptions of Rayleigh distributed stress ranges and one stress range per zero-crossing strictly apply only to very narrowband Gaussian processes, but this equation is often applied in many other situations, for example when the stress process is not narrowband.

A correction to the Rayleigh approximation to account for bandwidth effects has been suggested by Wirsching and Light [4,5]. Comparing the results of rainflow analysis of simulated time histories to Rayleigh predictions, they determined empirically that the ratio between the two could be written as a function of the parameter m of the S-N curve and the irregularity factor, ν/η_p:

$$\text{W \& L correction factor} = \frac{\text{Rainflow Damage Rate}}{\text{Rayleigh Damage Rate}} \tag{4}$$

$$= a(m) + [1 - a(m)](1-\epsilon)^{b(m)}$$

in which:

$$a(m) = 0.926 - 0.33m$$

$$b(m) = 1.587m - 2.323$$

$$\epsilon = (1-\alpha_2^2)^{1/2}$$

and:

$$\alpha_2 = \frac{\nu}{\eta_p} = \frac{\lambda_2}{\sqrt{\lambda_0 \lambda_4}}$$

This method, thus, gives a damage estimate depending on the three spectral moments λ_0, λ_2, and λ_4 (with λ_0 and λ_2 appearing in both the Rayleigh estimate and the correction to the Rayleigh estimate). Subsequent simulation studies have indicated, however, that this approximation may be inadequate in some situations [6]. This conclusion is further substantiated by the results of the current study as discussed in a subsequent section.

At least two other corrections to the Rayleigh approximation have since been suggested, each using spectral moments of the general form:

$$\lambda_b = \int_0^\infty \omega^b G(\omega) d\omega \tag{5}$$

Lutes et al. [6] proposed that a correction might be written as a function of a bandwidth parameter of the form:

$$\alpha_b = \frac{\lambda_b}{\sqrt{\lambda_0 \lambda_{2b}}} \tag{6}$$

with b empirically related to m. Such a method would use a total of four spectral moments λ_0, λ_2, λ_b, λ_{2b}.

Ortiz and Chen [7] have also suggested a Rayleigh correction, using the five spectral moments λ_0, λ_2, λ_4, λ_b, λ_{b+2}. :

$$\dot{D}_{OC} = \frac{\beta_b^m}{\alpha_2} \dot{D}_{RAY} \tag{7}$$

in which:

$$\beta_b = \left(\frac{\lambda_2 \lambda_b}{\lambda_0 \lambda_{2+b}} \right)^{1/2}$$

The suggested relation of b to the S-N parameter m is:

$$b = 2/m \tag{8}$$

Equation 7 can also be written as:

$$\dot{D}_{OC} = \frac{2^{3m/2}}{2\pi K}\Gamma(1+m/2)\lambda_2^{\frac{m-1}{2}}\lambda_4^{1/2}\left(\frac{\lambda_{2/m}}{\lambda_{2+2/m}}\right)^{m/2} \tag{9}$$

Note that the λ_0 term in the correction cancels out with the λ_0 term in the Rayleigh approximation so that the damage rate in this method actually depends on the four spectral moments λ_2,λ_4, $\lambda_{2/m}$, and $\lambda_{2+2/m}$.

Improvements in the accuracy of the fatigue prediction are generally realized by these methods, as is to be expected from the use of four parameters to characterize the stress process, rather than only the two used in the Rayleigh method. However, these methods are not as simple to apply as the Rayleigh method and have less physical motivation.

A new approximation recently introduced by Lutes and Larsen [3] abandons the Rayleigh approximation and estimates the damage accumulation based on a single moment (SM) of the psd. The form of the approximation is:

$$\dot{D}_{SM} = \frac{2^{3m/2}}{2\pi K}\Gamma(1+m/2)(\lambda_{2/m})^{m/2} \tag{10}$$

The SM technique was formulated after examination of the trends observed in data from extensive simulation and rainflow analysis, particularly for bimodal psd's [8]. In all the cases studied, the SM technique was found to be at least as accurate as the Rayleigh approximation, and a significant improvement over the Rayleigh method was noted in cases where the Rayleigh results differ widely from simulation results [3]. The discovery of a spectral method involving only one moment and also significantly improving on the Rayleigh approximation is somewhat surprising. It is contrary to the trend until this time which has been to correct the Rayleigh results by including increasing numbers of spectral moments [4,6,7,9]

Note that equation 9 for the Ortiz and Chen approximation contains a term of $\lambda_{2/m}$ and can be re-written as :

$$\dot{D}_{OC} = \frac{2^{3m/2}}{2\pi K}\Gamma(1+m/2)(\lambda_{2/m})^{m/2}\frac{\lambda_2^{\frac{m-1}{2}}\lambda_4^{1/2}}{(\lambda_{2+2/m})^{m/2}} \tag{11}$$

Thus, the difference between the Ortiz and Chen method and the single-moment method is the term:

$$\frac{\lambda_2^{\frac{m-1}{2}}\lambda_4^{1/2}}{(\lambda_{2+2/m})^{m/2}} \tag{12}$$

Note that this term, because it involves the $\lambda_{2+2/m}$ and λ_4 moments of the psd, is more sensitive to the high frequency content of the psd than the Rayleigh approximation.

It is also interesting to note that for a very narrowband psd, the Rayleigh approximation is considered to be quite good and it gives a damage estimate proportional to $\omega\sigma_x{}^m$. Any single moment of the narrowband psd gives a term proportional to the variance of the process as:

$$\lambda_b = \int_0^\infty \omega^b G(\omega)d\omega \propto \omega^b\sigma_x^2 \tag{13}$$

This term must be raised to the m/2 power if a single moment term is to be obtained which is proportional to $\sigma_x{}^m$, in order to agree with the Rayleigh result for narrowband cases. The only single moment which will give ω to the first power when raised to the m/2 power is $\lambda_{2/m}$ (b=2/m in equation 13). Thus, the SM method does agree with the Rayleigh approximation for very narrowband processes, and no single moment other than $\lambda_{2/m}$ could give the correct dependence on both ω and σ_x.

In the following sections the damage predicted by four spectral methods (the Rayleigh approximation, the Wirsching and Light correction factor, the Ortiz and Chen approximation, and the SM approximation) is compared to the results of rainflow analysis of Gaussian simulations for a number of unimodal and bimodal psd's. The formulae of the four spectral techniques compared are summarized in Table 1. These four techniques were chosen for differing reasons. The Rayleigh technique is the most widely used method and the Wirsching and Light correction

Chen, and SM methods represent the latest known attempts at improvements on the Rayleigh method.

Simulation Studies

Gaussian time histories were simulated for the five psd types shown in Figures 1 and 2. These psd's have been chosen to represent a variety of situations which may occur in practice by including both unimodal and bimodal psd's, and by including slowly decaying high frequency "tails" as well as truncated approximations. Although these general forms were chosen as possible representations of the stress in an offshore structure, they also seem to be applicable to a broad range of problems.

The psd of Fig. 1(a) is a band-limited white noise type of unimodal psd, herein referred to as a one-block psd. It is characterized by a constant level, G, over the frequency range ω_1 to ω_2. For each time history simulated, G was chosen to be $1/(\omega_2-\omega_1)$ so that $\sigma_x{}^2 = 1$. A more realistic unimodal psd shape, previously studied by Wirsching and Light [4], is shown in Fig. 1(b). It has a form given by:

$$G_x(\omega) = (sf)\left(\frac{3\beta}{\omega^4}\right)\exp\left(\frac{-\beta}{\omega^3}\right) \text{ for } \omega \le \omega_c \tag{14}$$

When this psd is not truncated ($\omega_c = \infty$), the fourth moment, λ_4, is not finite and the irregularity factor is zero ($\alpha_2 = 0$).

Another more realistic unimodal psd is shown in Fig. 1(c), with a convenient form given by:

$$G_x(\omega) = \frac{(sf)}{c\sqrt{2\pi}}\exp\left[-\frac{(\omega-u)^2}{2c^2}\right] \text{ for } \omega \le \omega_c \tag{15}$$

in which:

$$u = \text{center frequency}$$

$$c = \text{a shape factor}$$

This form is analogous to a Gaussian (normal) probability density function, and it has finite values for all spectral moments. A scale factor, sf, is included in each of these forms so that $\sigma_x{}^2 = 1$ for any value of ω_c. In both cases sf=1 for $\omega_c = \infty$. Table 2 gives the parameter values for the various simulations from these unimodal cases.

Simple bimodal psd cases were formed by the superposition of two one-block psd's, as in Fig. 2(a). These two-block psd's are characterized by the levels G_1 and G_2 over the frequency ranges ω_1 - ω_2 and ω_3 - ω_4. These parameters are related by the frequency ratio:

$$r = \frac{\omega}{\omega} = \frac{\omega_4}{\omega_2} \tag{16}$$

and the area ratio:

$$b = \frac{A_1}{A_2} = \frac{G_1(\omega_2 - \omega_1)}{G_2(\omega_4 - \omega_3)} \tag{17}$$

in which A_1 is the area of block 1, A_2 is the area of block 2, and $A_1 + A_2 = 1$ so that $\sigma_x{}^2 = 1$.

A more realistic bimodal psd, also studied by Wirsching [4,11] is shown in Fig. 2(b). It has a form given by:

$$G_x(\omega) = (sf)\frac{7793}{T_D^4\omega^5}\exp\left(\frac{-1948}{T_D^4\omega^4}\right)\frac{1}{[1-(\frac{\omega}{\omega_N})^2]^2 + 4\zeta^2(\frac{\omega}{\omega_N})^2} \tag{18}$$

in which:

$$T_D = \text{ dominant wave period}$$

$$\omega_N = \frac{2\pi}{T_N} = \text{dominant natural frequency}$$

$$T_N = \text{ dominant structural period}$$

The parameter values for the various simulations from the bimodal psd's are given in Table 3.

Analysis of Results

The results of Gaussian simulations for each of the psd types studied are compared in Figures 3 through 8 with each of the four selected spectral methods for values of m=3 and m=7. In each figure, the abscissa of the data points is the expected damage rate as computed by rainflow analysis of the simulation data, and the ordinate is the damage rate as estimated by a spectral method. Thus, perfect correspondence between a spectral method and simulation is indicated by data lying on a line with a positive 45 degree slope. Any deviations of the data from this line indicate inaccuracies in thespectral methods. The data are shown on log-log plots so that a deviation from the 45 degree line is directly related to a percentage error.

Figures 3 (m=3) and 4 (m=7) show the comparisons for the unimodal psd's of Figure 1 and Table 1. Fig.3(a) shows that the error for the Rayleigh approximation is quite small for most of the unimodal cases, but that it does approach a 17 percent over-estimation of the damage rate for the most wide-band one-block psd's (the right-hand portion of curve 1). The Wirsching and Light approximation, Fig. 3(b), exhibits a different trend, usually under-estimating the damage rate by about 10 percent, but it does more closely approach the simulation data for the most wide-band one-block psd's.

The Ortiz and Chen approximation, as shown in Fig.3(c), shows very good agreement with simulation for psd types 1(b) and 1(c) (curves 2 and 3), which are quite narrow-band even when truncated at a high frequency. However, this approximation tends to greatly over-estimate the damage rate for the wide-band one-block psd's, with errors approaching 92 percent (the right-hand portion of curve 1). This behavior is somewhat predictable because the λ_4 spectral moment term used in the method emphasizes the contribution to the damage rate of the high frequency content of the psd.

In contrast, the SM method, Fig. 3(b), provides a quite accurate estimate of the damage rate for all the unimodal psd's studied, with a maximum error of about negative 6 percent.

In general, the effects described for m=3 are more pronounced for m=7, with the Rayleigh approximation being in error by as much as 37 percent, and Wirsching and Light's method often estimating damage 25 percent too low. Again, the Ortiz and Chen technique is very accurate except for the one-block psd's, where the over-estimation approaches 160 percent for the most wide-band case (the right-hand portion of curve 1). The SM method is still quite good, the worst errors being about 6 percent for the one-block psd's.

The sensitivity of the Rayleigh, and the Ortiz and Chen methods to high-frequency in the psd is to be expected, although the magnitude of the error for the Ortiz and Chen method is somewhat startling. For the relatively narrowband unimodal psd's the Ortiz and Chen approximation is very good, but for these cases the Rayleigh approximation is also quite accurate and it is simpler to evaluate. For all the unimodal psd cases studied, the SM method is consistently more accurate than the Rayleigh approximation, as well as being even simpler to evaluate.

The next set of results considered are the two-block bimodal cases of Figure 2(a). The data are presented in the same form as for the unimodal cases, with the results for m=3 shown in Figure 5 and for m=7 in Figure 6. The families of curves shown are for constant values of the frequency ratio, r, with varying values of the area ratio, b, as given in Table 3. The trends noted above for the various spectral techniques are reinforced by this extensive set of simulation data.

The Rayleigh approximation is quite good, as expected, in rather narrowband cases (very small or very large values of b, or small values of r), but it exhibits errors of as much as 154 percent for m=7 in cases with large r values and intermediate b values (curve C). The Wirsching and Light technique again under-estimates the damage for some narrowband cases, but also over-estimates the damage in most other cases, although not to as large an extent as the Rayleigh method.

The Ortiz and Chen approximation also shows behavior consistent with that observed from the one- block cases. The improvement over the Rayleigh results is very good for a number of cases, but the contribution of a small high-frequency component of the psd drastically affects the damage estimate due to the high order spectral moments in the formula. The error for the worst cases (r=15, b=0.001, appearing in the left-hand portion of curve C) approaches 160 percent for m=3 and almost 300 percent for m=7.

For the SM method, errors approaching negative 12 percent occur for m=3 (r=8, b=0.1), while for m=7 the damage may be under-estimated by as much as about 23 percent (r=10, b=0.1). It may be noted that in the worst case for the SM method (r=10, b=0.1, and m=7), the Rayleigh, Wirsching and Light, and Ortiz and Chen methods have errors of +59 percent, +11 percent, and -25 percent, respectively. The error in the SM method for the case of r=15, b=0.1, and m=7 is also nearly -23 percent. For this case the Rayleigh, Wirsching and Light, and Ortiz and Chen methods have errors of +148 percent, +52 percent, and -29 percent, respectively.

The last psd type considered is the smooth bimodal psd of Fig. 2(b). Figure 7 compares the spectral methods to simulation for m=3, and Figure 8 shows the comparison for m=7. In each of the figures, curve 1 indicates the data for the constant damping ratio cases of Table 3, while curves 2 and 3 show the data for the two sets of constant period ratio cases of Table 3.

The Rayleigh approximation over-estimates the damage rate by as much as 67 percent for m=3 (τ=0.07, ζ=0.08, curve 3), as shown in Fig. 7(a), while for the same case for m=7, Fig. 8(a), the error reaches 53 percent. The Wirsching and Light correction, Figures 7(b) and 8(b), again tends to under-estimate the damage, by as much 24 percent for m=7 (τ=0.6, ζ=0.02, curve 1), but for other cases the damage is over-estimated by as much as 38 percent for m=3 (τ=0.07, ζ=0.08, curve 3).

The Ortiz and Chen method, Figures 7(c) and 8(c), again shows good agreement with the simulation results, except for cases for which the psd has a small high frequency component. In these cases (τ=0.05, ζ=0.02, curve 1) the error can be quite significant, reaching 67 percent for m=3 and 112 percent for m=7.

The SM method, because it does not include the higher order spectral moments of the other methods, is not prone to over-estimating the damage rate in the small, high frequency psd cases. The trend is rather to under-estimate the damage rate (τ=0.08, ζ=0.02, curve 1, and τ=0.07, ζ=0.08, curve 3), by about 15 percent for m=3, and by as much as 26 percent for m=7.

Another set of assumptions that can be considered a spectral method is the peak approximation, in which the stress ranges are assumed to be proportional to the positive peaks of a Gaussian process [12,5]. The damage estimate by this method depends on the three spectral moments $\lambda_0, \lambda_2, \lambda_4$. This method is even more sensitive to the high frequency content of the psd than the Ortiz and Chen approximation [12,13], so it was not considered in this study.

Summary and Conclusions

Simulation data for several unimodal and bimodal psd's demonstrate the considerable errors which may result from the use of the common Rayleigh approximation method, or the correction to this method proposed by Wirsching and Light. A third spectral method proposed by Ortiz and Chen has been shown to result in improved accuracy for a number of cases, but to give very large errors for psd's which are wide-band or have a small, high-frequency component.

The SM method has been shown to be in quite good agreement with simulation data for all the psd types studied. In some particular cases, other methods give more accurate results than the SM method, but overall the SM method seems to be the most reliable of the techniques studied. The SM method does not give perfect fatigue damage predictions, but the worst-case errors noted from the SM method were always considerably less than the worst case errors for any of the other spectral methods considered. Applying the SM method is significantly simpler than any of the other spectral methods as it only involves one integral of the psd, compared to other methods which require three or four spectral moments.

References

1. Matsuishi, M., and Endo, T. Fatigue of metals subject to varying stress, Presented to the Japan Society of Mechanical Engineers, Fukuoka, Japan. 1968.

2. Downing, S. D., and Socie, D. F. Simple rainflow counting algorithms, Int. J. Fatigue, 1982, 4(1), 31-40.

3. Lutes, L. D., and Larsen, C. E. Improved spectral method for variable amplitude fatigue prediction, J. Struct. Div., ASCE, 1990, 116(4), 1149-1164.

4. Wirsching, P. H., and Light, M. C. Fatigue under wide band random stress, J. Struct. Div., ASCE, 1980, 106(7), 1593-1607.

5.Wirsching, P. H., and Light, M. C. Probability based fatigue design criteria for offshore structures, Amer. Petroleum Inst., 1979.

6. Lutes, L. D., et al. Stochastic fatigue damage accumulation, J. Struct. Div., ASCE, 1984, 110(11), 2585-2601.

7. Ortiz, K., and Chen, N. K. Fatigue damage prediction for stationary wideband processes, Proc. Fifth Int. Conf. on Application of Statistics and Probability in Soil and Struct. Engrg, 1987.

8. Larsen, C. E. Random process simulation for stochastic fatigue analysis, thesis presented to Rice University at Houston, Tex., in partial fulfillment of the requirements for the degree of Doctor of Philosophy, 1987.

9. Perng, H.-L. Damage accumulation in random loads, thesis presented to University of Arizona at Tucson, Arizona, in partial fulfillment of the requirements for the degree of Doctor of Philosophy, 1989.

10. Vandiver, J. K. The sensitivity of fatigue life estimates to variations in structural natural periods, modal damping ratios, and directional spreading of the seas, Proc. Third Int. Conf. on Behavior of Offsshore Structures (BOSS), M.I.T., Cambridge, Mass., 1982, Vol. 2, 279-292

11. Wirsching, P. H., and Prasthofer, P. H. Preliminary dynamic assessment of deepwater platforms, J. Struct. Div., ASCE, 1976, 102(7), 1447-1462.

12. Nelson, D. V. Cumulative fatigue damage in metals, thesis presented to Stanford University, at Stanford, Calif., in partial fulfillment of the requirements for the degree of Doctor of Philosophy, 1978.

13. Lutes, L. D. Use of the Rice peak distribution to predict variable amplitude fatigue damage, Technical Report PR-5631, Exxon Production Res. Co., Houston, Tex. 1981.

Table 1. Summary of Spectral Techniques Studied

Rayleigh	$K\dot{D}_{RAY} = \frac{2^{\frac{3m}{2}}}{2\pi}\Gamma(1+m/2)\lambda_0^{\frac{m-1}{2}}\lambda_2^{1/2}$
Wirsching & Light	$K\dot{D}_{WL} = K\dot{D}_{RAY}\left[a(m)+[1-a(m)](1-\epsilon)^{b(m)}\right]$ $a(m) = 0.926 - 0.33m$ $b(m) = 1.587m - 2.323$ $\epsilon = (1-\alpha_2^2)^{1/2}$
Ortiz & Chen	$K\dot{D}_{OC} = \frac{2^{\frac{3m}{2}}}{2\pi}\Gamma(1+m/2)(\lambda_{2/m})^{m/2}\frac{\lambda_2^{\frac{m-1}{2}}\lambda_4^{1/2}}{(\lambda_{2+2/m})^{m/2}}$
Single-Moment	$K\dot{D}_{SM} = \frac{2^{\frac{3m}{2}}}{2\pi}\Gamma(1+m/2)(\lambda_{2/m})^{m/2}$

Table 2. Unimodal psd's for which Time Histories were Simulated

One-block Fig. 1(a)	Eq. 14 Fig. 1(b) $\beta = 0.01$	Eq. 15 Fig. 1(c) $u = 5.0$, $c = 1.0$
Curve 1 Figs. 3-4	Curve 2 Figs. 3-4	Curve 3 Figs. 3-4
ω_2/ω_1	ω_c	ω_c
6/5	0.392	6.0
11/9	0.587	7.0
7/5	0.783	7.5
3/2	0.979	8.0
14/9	1.958	9.0
16/9	3.916	
2/1	9.787	
7/3		
8/3		
10/3		
4/1		
7/1		

Table 3. Bimodal psd's for which Time Histories were Simulated

Two-block, Fig. 2(a)													Eq. 18, Fig. 2(b)		
Values of b													$T_D = 5.0$		
r=1.5	r=2.0	r=3.0	r=4.0	r=5.0	r=6.0	r=7.0	r=7.5	r=8.0	r=9.0	r=10.0	r=11.0	r=15.0	$\zeta = 0.02$	$\tau = 0.07$	$\tau = 0.7$
Figures 5 & 6													Figures 7 & 8		
Curve 1	Curve 2	Curve 3	Curve 4	Curve 5	Curve 6	Curve 7	Curve 8	Curve 9	Curve 0	Curve A	Curve B	Curve C	Curve 1	Curve 2	Curve 3
													$\tau = \frac{T_N}{T_D}$	ζ	ζ
0.02	0.006	0.0016	0.00075	0.0004	0.00026	0.000185	0.018	0.00018	0.00009	0.01	0.000075	0.00004	0.05	0.005	0.005
0.15	0.21	0.01	0.21	0.01	0.21	0.001	0.133	0.001	0.001	0.05	0.001	0.0001	0.06	0.01	0.01
0.3	0.39	0.11	0.39	0.04	0.39	0.01	0.18	0.01	0.01	0.1	0.01	0.001	0.07	0.02	0.02
0.44	0.59	0.25	0.575	0.1	0.575	0.1	0.32	0.1	0.1	0.17	0.1	0.01	0.08	0.04	0.04
0.67	1.0	0.33	1.0	0.2	1.0	0.21	0.46	0.21	0.21	0.32	0.21	0.10	0.09	0.08	0.08
1.0	2.0	0.47	1.87	0.39	1.85	0.39	0.62	0.39	0.39	0.46	0.39	0.21	0.1		
1.5	3.0	0.63	2.8	0.58	2.8	0.575	1.0	0.577	0.576	0.62	0.575	0.39	0.125		
2.0	6.2	1.0	5.5	1.0	5.3	1.0	1.7	1.0	1.0	1.0	1.0	0.574	0.15		
4.0	41	1.5	88	1.85	108	1.84	2.3	1.83	1.83	1.7	1.83	1.0	0.175		
21		3.0		2.5		2.8	3.0	2.7	2.7	2.3	2.7	1.83	0.2		
		7.0		5.0		5.2	7.5	5.2	5.2	3.0	5.2	2.7	0.233		
		70		101		115	116	120	122	10	127	5.2	0.267		
										125		132	0.3		
													0.333		
													0.367		
													0.4		
													0.45		
													0.5		
													0.6		
													0.7		
													0.8		
													0.9		
													1.0		

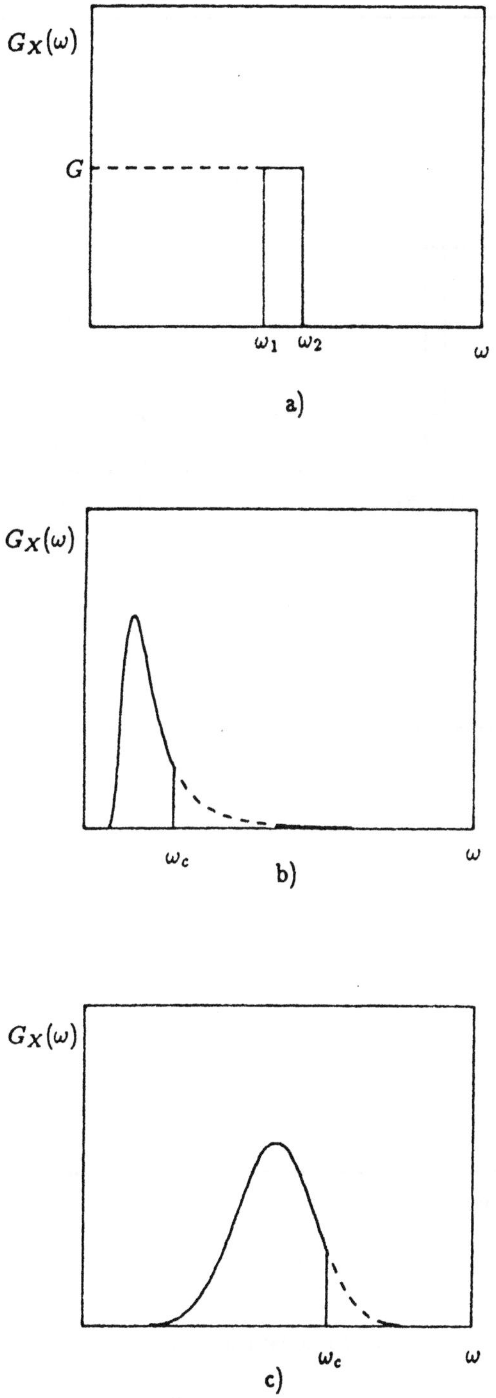

Figure 1. Unimodal psd's Studied

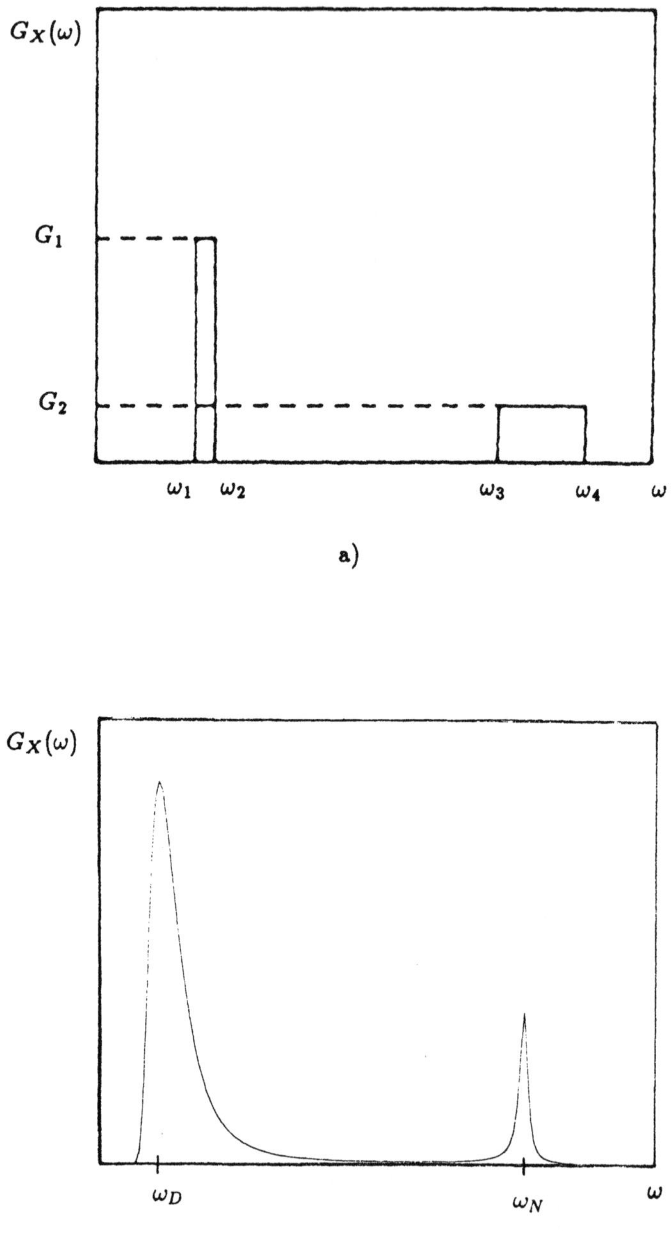

Figure 2. Bimodal psd's Studied

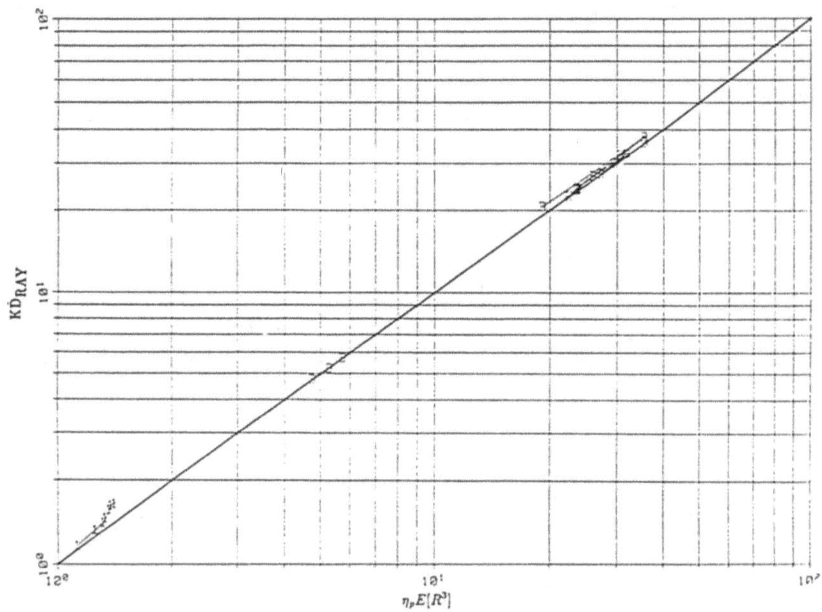

Figure 3(a) Rayleigh Damage Rate vs. Simulation Result Damage Rate, Unimodal psd's, m=3

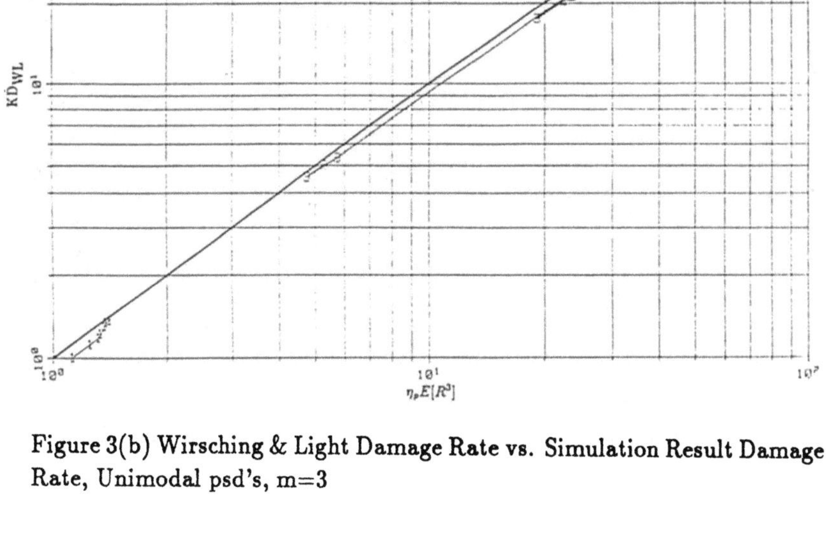

Figure 3(b) Wirsching & Light Damage Rate vs. Simulation Result Damage Rate, Unimodal psd's, m=3

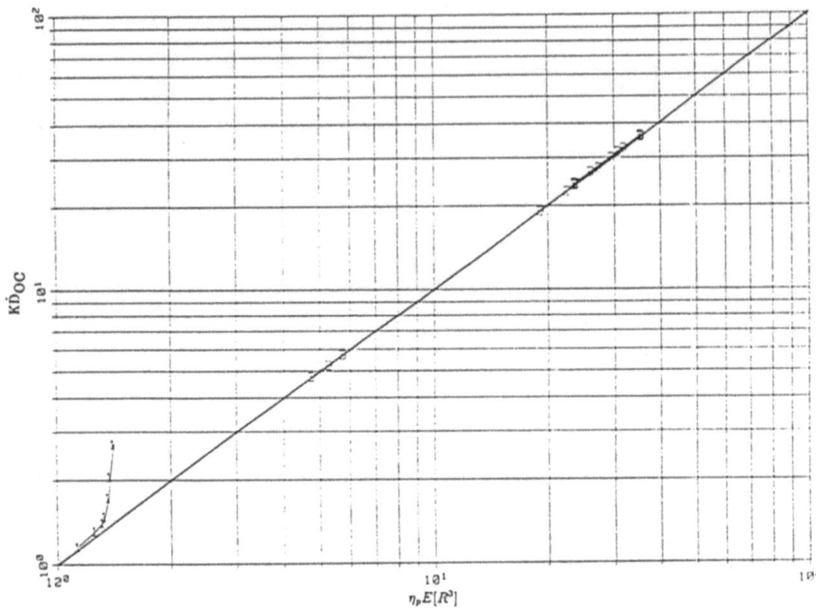

Figure 3(c) Ortiz & Chen Damage Rate vs. Simulation Result Damage Rate, Unimodal psd's, m=3

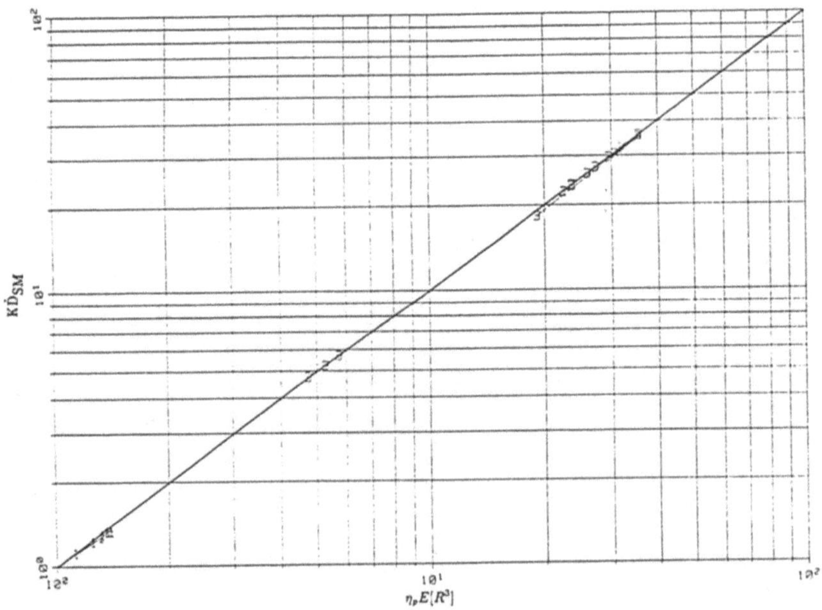

Figure 3(d) SM Damage Rate vs. Simulation Result Damage Rate, Unimodal psd's, m=3

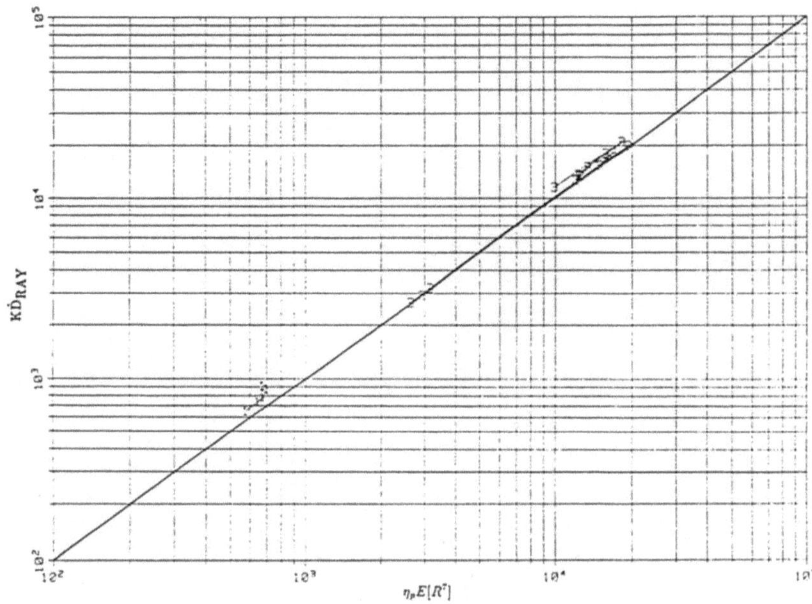

Figure 4(a) Rayleigh Damage Rate vs. Simulation Result Damage Rate, Unimodal psd's, m=7

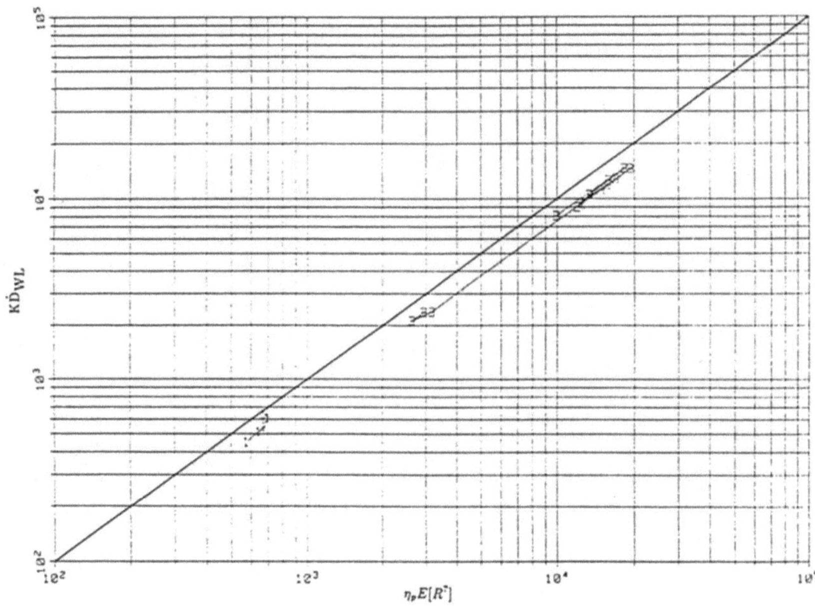

Figure 4(b) Wirsching & Light Damage Rate vs. Simulation Result Damage Rate, Unimodal psd's, m=7

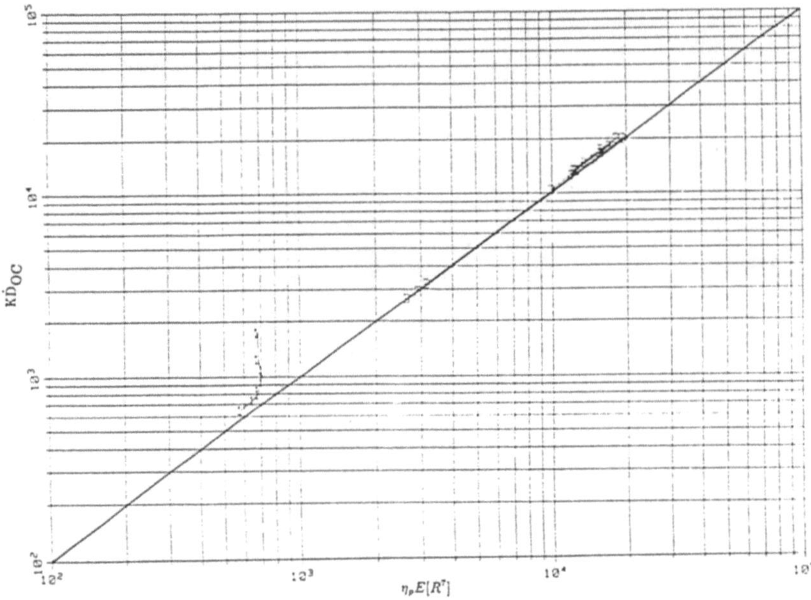

Figure 4(c) Ortiz & Chen Damage Rate vs. Simulation Result Damage Rate, Unimodal psd's, m=7

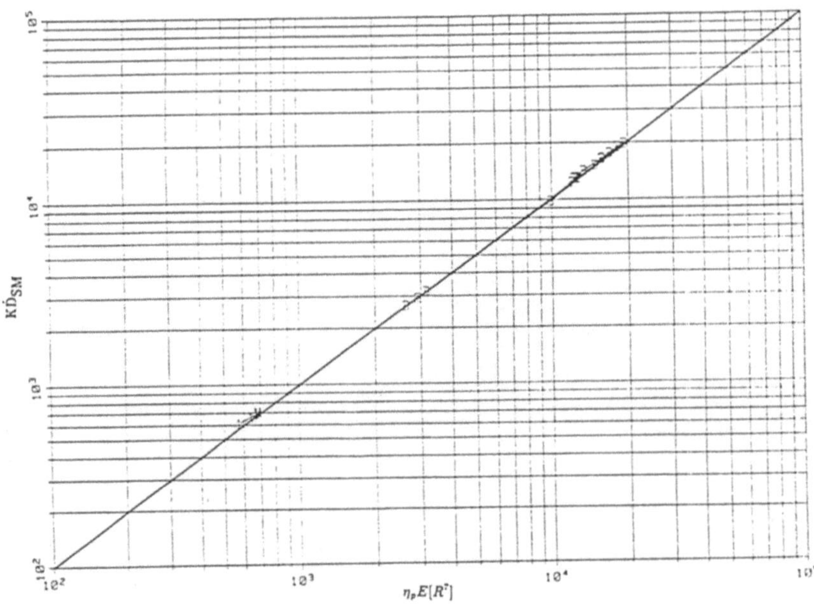

Figure 4(d) SM Damage Rate vs. Simulation Result Damage Rate, Unimodal psd's, m=7

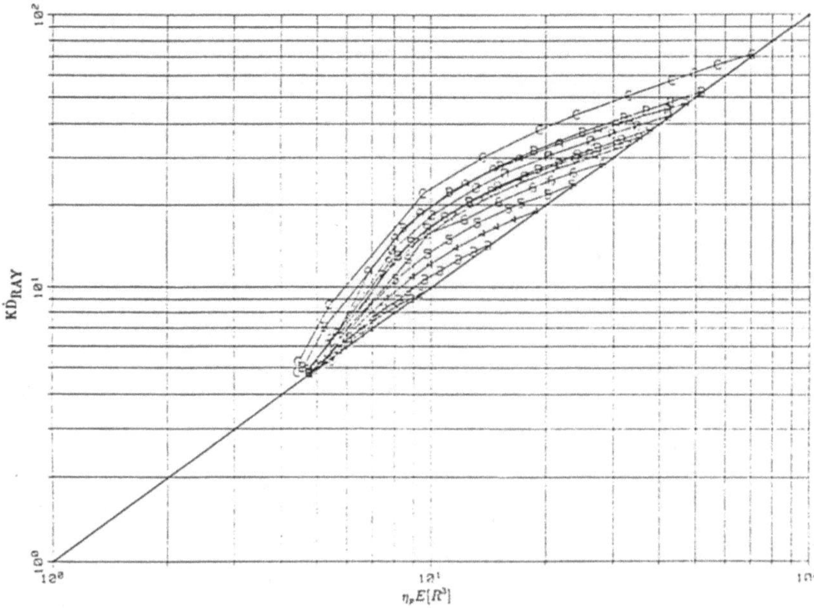

Figure 5(a) Rayleigh Damage Rate vs. Simulation Result Damage Rate, Two-block psd's, m=3

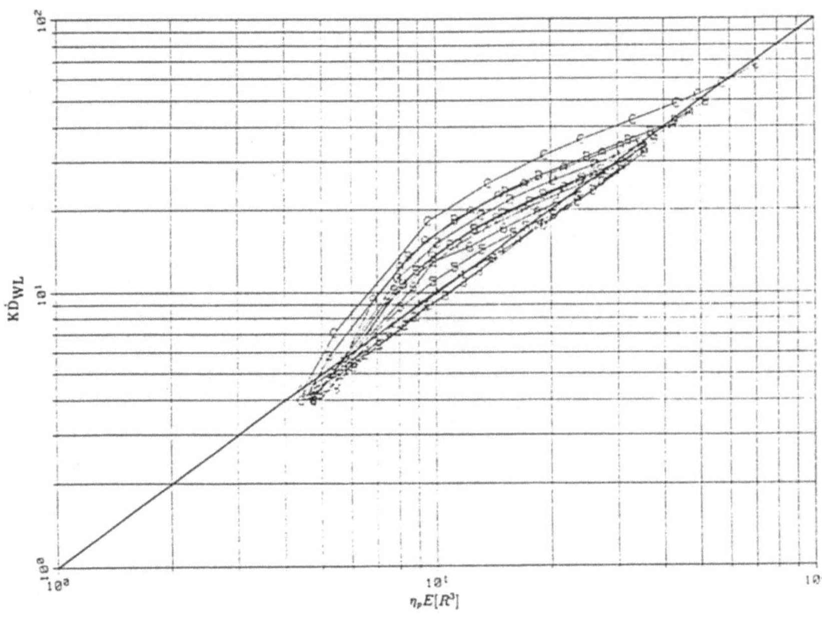

Figure 5(b) Wirsching & Light Damage Rate vs. Simulation Result Damage Rate, Two-block psd's, m=3

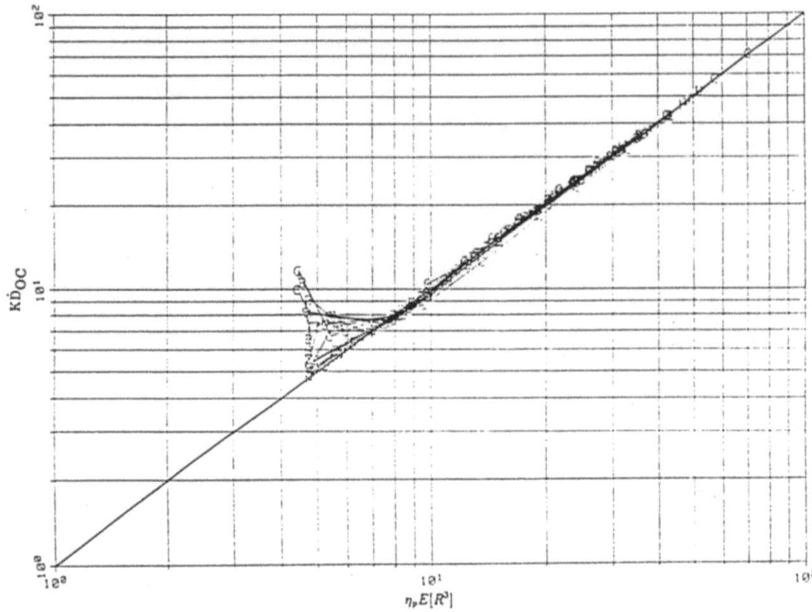

Figure 5(c) Ortiz & Chen Damage Rate vs. Simulation Result Damage Rate, Two-block psd's, m=3

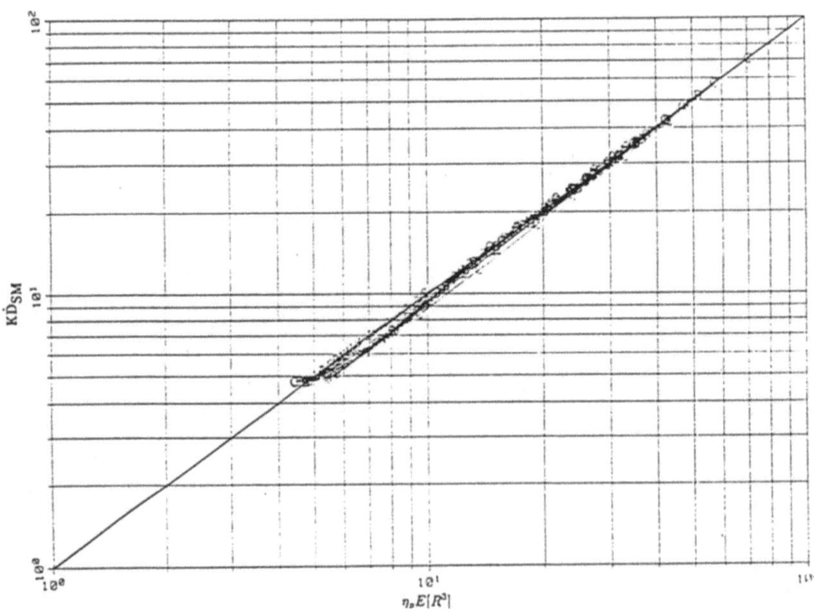

Figure 5(d) SM Damage Rate vs. Simulation Result Damage Rate, Two-block psd's, m=3

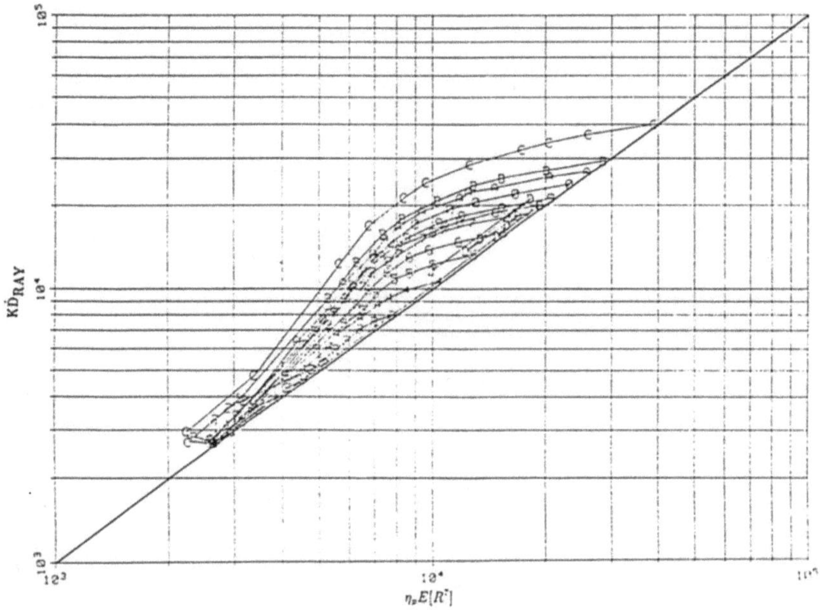

Figure 6(a) Rayleigh Damage Rate vs. Simulation Result Damage Rate, Two-block psd's, m=7

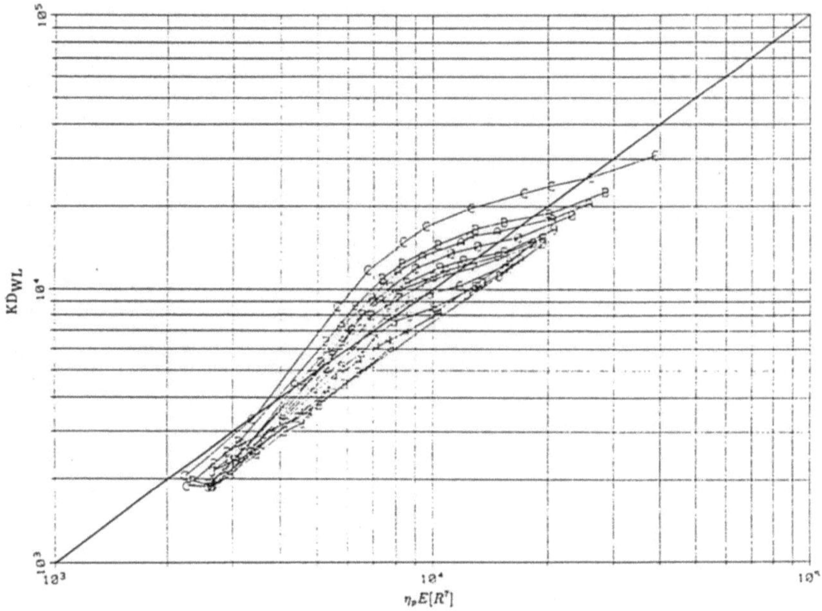

Figure 6(b) Wirsching & Light Damage Rate vs. Simulation Result Damage Rate, Two-block psd's, m=7

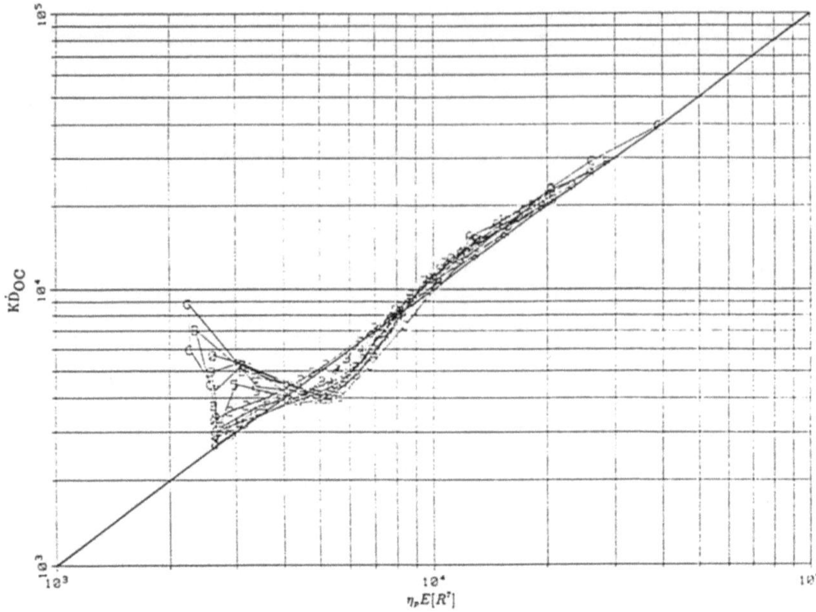

Figure 6(c) Ortiz & Chen Damage Rate vs. Simulation Result Damage Rate, Two-block psd's, m=7

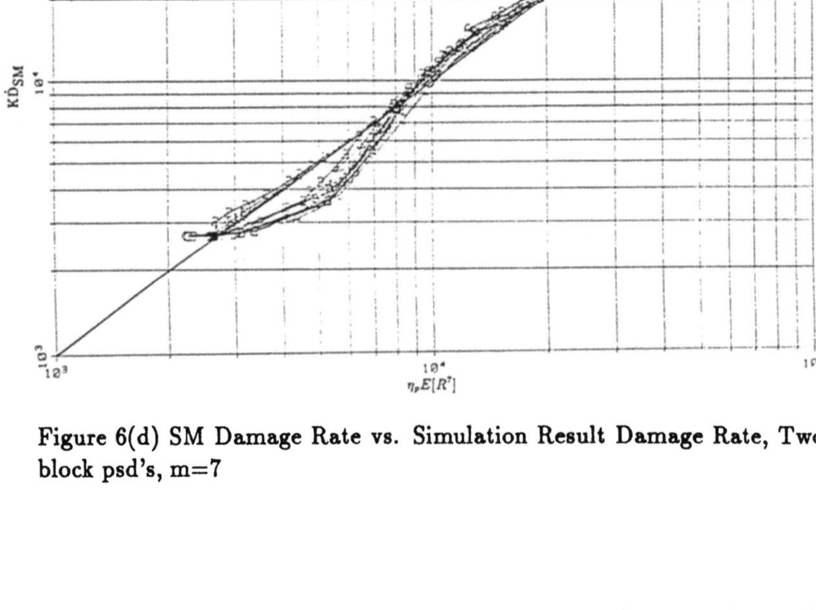

Figure 6(d) SM Damage Rate vs. Simulation Result Damage Rate, Two-block psd's, m=7

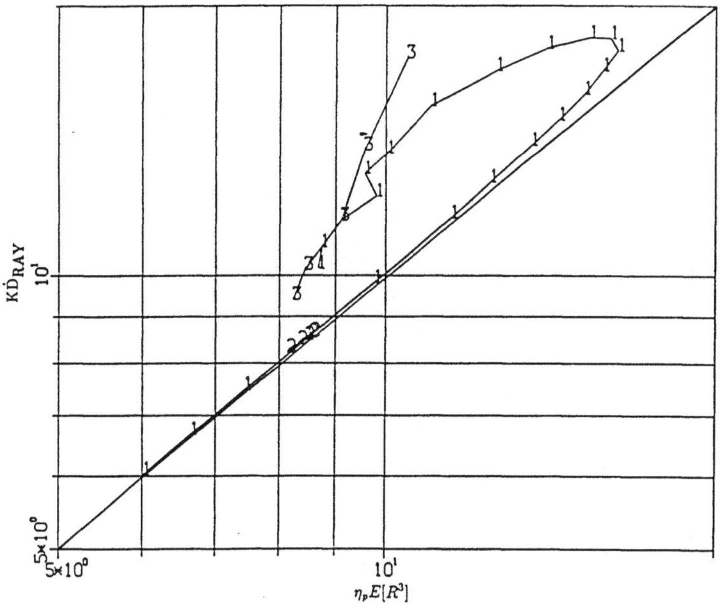

Figure 7(a) Rayleigh Damage Rate vs. Simulation Result Damage Rate, Smooth bimodal psd's, m=3

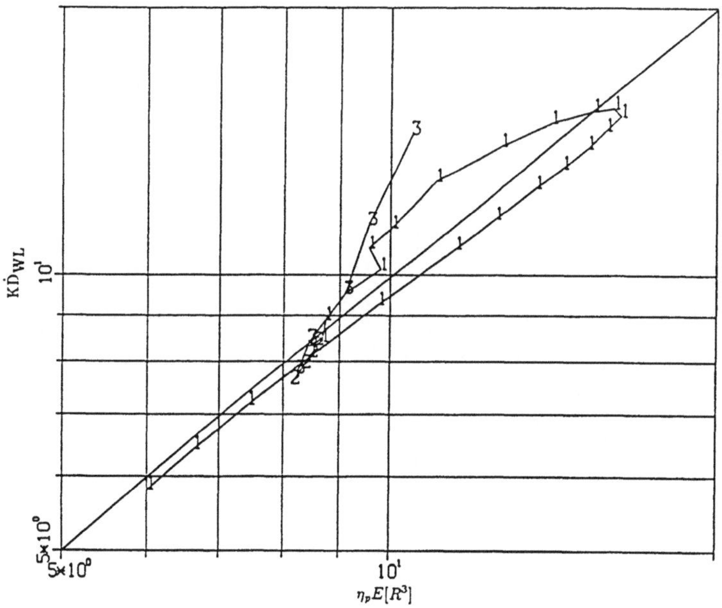

Figure 7(b) Wirsching & Light Damage Rate vs. Simulation Result Damage Rate, Smooth bimodal psd's, m=3

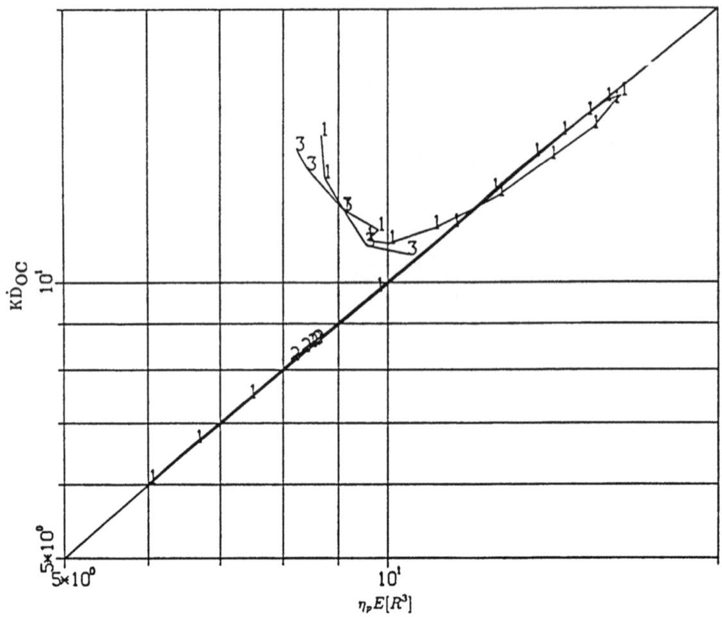

Figure 7(c) Ortiz & Chen Damage Rate vs. Simulation Result Damage Rate, Smooth bimodal psd's, m=3

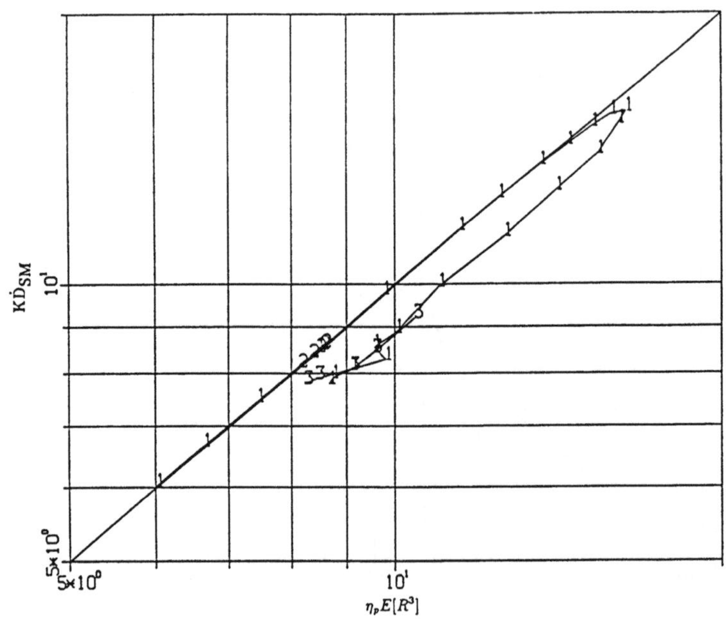

Figure 7(d) SM Damage Rate vs. Simulation Result Damage Rate, Smooth bimodal psd's, m=3

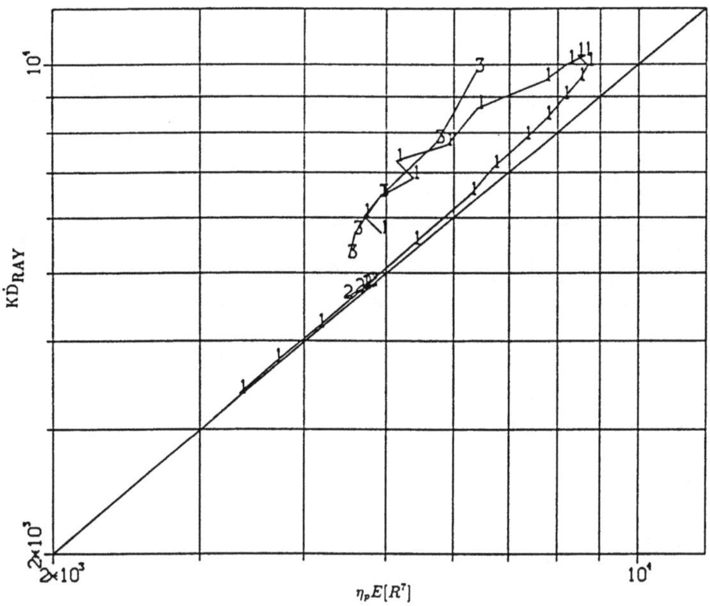

Figure 8(a) Rayleigh Damage Rate vs. Simulation Result Damage Rate, Smooth bimodal psd's, m=7

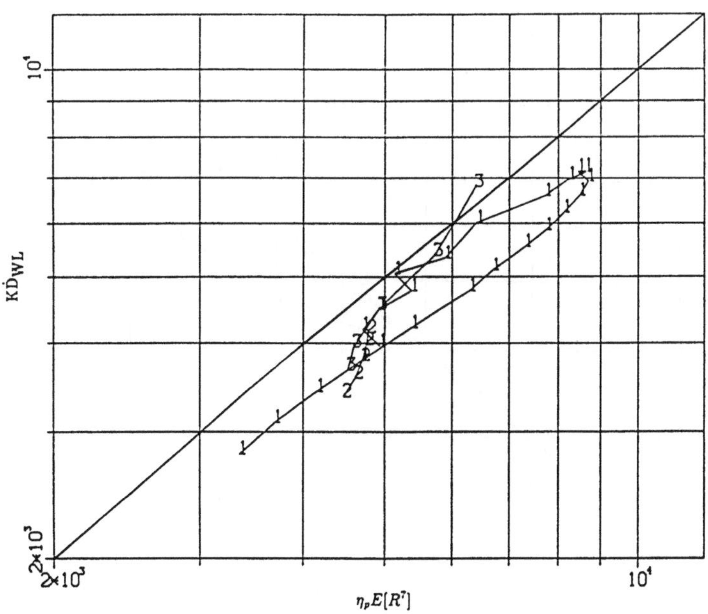

Figure 8(b) Wirsching & Light Damage Rate vs. Simulation Result Damage Rate, Smooth bimodal psd's, m=7

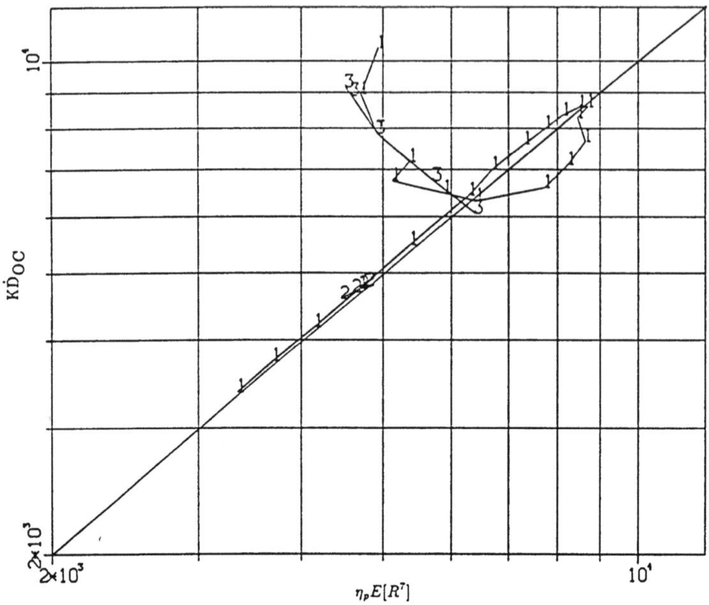

Figure 8(c) Ortiz & Chen Damage Rate vs. Simulation Result Damage Rate, Smooth bimodal psd's, m=7

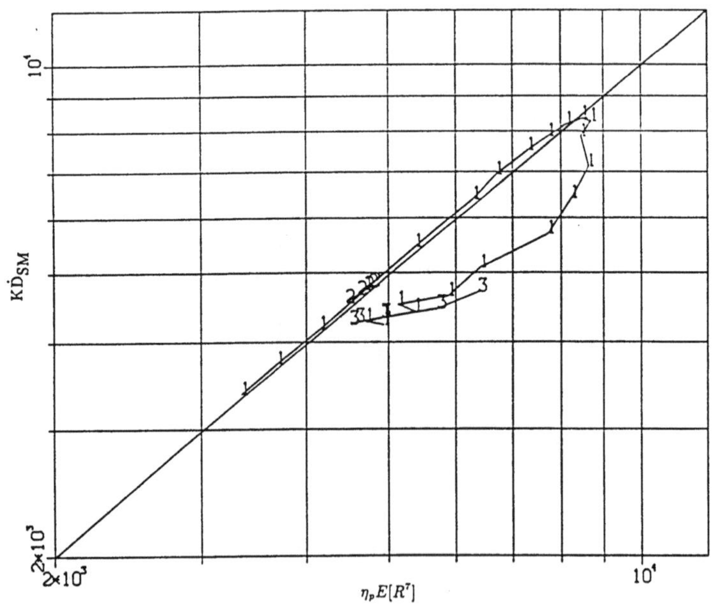

Figure 8(d) SM Damage Rate vs. Simulation Result Damage Rate, Smooth bimodal psd's, m=7

On the Computation of the Stochastic Response of Highly Nonlinear Large MDOF-Systems Modeled by Finite Elements

H. J. PRADLWARTER and Wenlung LI*

Institute of Engineering Mechanics
University of Innsbruck, Austria

* On leave from Dept. Mechanical Eng'g, NTIT,
Taipei, Taiwan 10626, Rep. Of China

Summary

A computational procedure to evaluate the stochastic response of a general non-linear finite element model is presented. Special attention is given to the storage of large systems and to computational efficiency by use of complex modal analysis. The stochastic equivalent linearization method is utilized in the procedure. It also allows one to use all available "deterministic" nonlinear elements. Adaptive Monte Carlo simulation is suggested to determine the stochastic equivalent linear incremental properties of the non-linear finite element.

1. INTRODUCTION

For pratical applications of non-linear stochastic dynamics on engineering structures, the following requirements should be met: (a) Nonlinear structural model, especially finite element (FE) models, used for deterministic analysis should also be applicable for a stochastic analysis including random excitation and random structural parameters; (b) The method of solution should not be restricted to weakly nonlinear problems and special types of nonlinearities; and (c) The computational effort of the method must be considerably less than that of the crude Monte Carlo simulation. These requirements can presently only be partially met. Most advanced approaches in stochastic dynamics such as methods based on the the Fokker–Planck equation, stochastic averaging methods, the utilization of moment equations, and other approaches give valuable insight for the stochastic response of minimum size problems. However, these methods can not be applied for systems with a large number of degree-of-freedom (DOF). Alternative approaches with less difficulties to cover large MDOF systems are the perturbation method [1] and methods of stochastic equivalent linearization (EQL). The accuracy in both approaches might not always be satisfactory. For example, the perturbation method is restricted to weakly nonlinear problems and will not be discussed further in the present paper.

I. Elishakoff · Y. K. Lin (Eds.)
Stochastic Structural Dynamics 2
New Practical Applications
2nd International Conference on Stochastic Structural Dynamics
May 9-11, 1990, Boca Raton, Florida

The method of equivalent linearization, on the contrary, is also applicable for systems with strong nonlinearities. Its limited accuracy stems from the necessity to assume a priori the type of distribution of the stochastic response [2].

Recognizing the need for a stochastic dynamic analysis on large MDOF systems, considerable research efforts have been spent in the last decade to generalize the method of equivalent linearization and to make it computationally more efficient. Based on the pioneer work by Wen [3], Bouc's endochronic model has been generalized to treat by EQL two- and three- dimension frames subjected to random loads [4]. Mochio et al. [5] were one the first to combine the FE-method with EQL. This idea has been generalized further by Simulescu et al. [6]. The statistical linearization of Bouc's model has been carried further to model 2-D continua [7] and the yielding interaction between the stress components [8]. Chang et al. [9] discussed complex modal analysis to determine the stochastic response of the time variant linearized system in an efficient manner.

The present paper is focused on the utilization of arbitrary finite elements in statistical EQL and on computational aspects arising from a large number of DOF. Adaptive Monte Carlo simulation is suggested to determine the statistically equivalent incremental properties of general nonlinear finite elements. Special attention is given to complex modal analysis for large systems.

2. NONLINEAR EQUATIONS OF MOTION

The most general way to describe the nonlinear behavior of a structural system is the finite element method (FEM). This is mainly because the FEM can model the dynamical responses quite flexible to any required accuracy. When applying the FEM, the equations of motion for the structural system can be expressed in matrix form

$$\underset{\sim}{M} \cdot \ddot{\underset{\sim}{u}} + \underset{\sim}{C} \cdot \dot{\underset{\sim}{u}} + \underset{\sim}{r}(\underset{\sim}{u}, \dot{\underset{\sim}{u}}, \underset{\sim}{q}, \underset{\sim}{w}; t) = \underset{\sim}{f}(t), \tag{1}$$

where $\underset{\sim}{M}$ and $\underset{\sim}{C}$ are mass and damping matrices, respectively; the vector $\underset{\sim}{r}(t)$ contains nonlinear restoring forces; $\underset{\sim}{f}(t)$ stands for the stochastic excitation vector; $\underset{\sim}{u}$ and $\dot{\underset{\sim}{u}}$ are the displacement and its time derivative vectors, $\underset{\sim}{q}$ the nodal force vector, and $\underset{\sim}{w}$ other time-variant quantities. Note that nonlinear restoring forces $\underset{\sim}{r}(t)$ can be represented as a linear function of element force vector $\underset{\sim}{q}(t)$

$$\underset{\sim}{r}(t) = \underset{\sim}{B}\,\underset{\sim}{q}(t), \tag{2}$$

where $\underset{\sim}{B}$ is a constant matrix. All types of nonlinearities including hysteresis can be formulated by the FEM as

$$\dot{q}(t) = H(t) \cdot \dot{u}(t), \tag{3}$$

where H is the tangent stiffness matrix. In general, H is function of u,$\dot{u}$,q, and w, i.e., $H(t) = H(u, \dot{u}, q, w;\ t)$.

The excitation term, $f(t)$, is conveniently split into two parts

$$f(t) = E[f(t)] + e(t), \tag{4}$$

where $E[\cdot]$ denotes the expectation values and $e(t) = (f(t) - E[f(t)])$ represents the random part of the excitation. Generally, $e(t)$ can be modeled by a stochastic process with certain spectral properties. For the sake of computational efficiency, it may be expressed as filtered modulated white noise

$$e(t) = -B \cdot z(t), \tag{5}$$

in which B is a constant matrix and $z(t)$ follows the linear differential equation

$$\dot{z}(t) = Z \cdot z(t) + \{n(t)\ m(t)\}, \tag{6}$$

where the matrix Z is defined by the filter properties, $n(t)$ denotes stationary white noise with unit spectral density, and $m(t)$ is a vector of deterministic modulating functions.

Some other time-dependent quantities $w(t)$, such as degradation, may also involve in the equations of motion. These parameters can be formulated in the form of a differential equation

$$\dot{w}(t) = g(u, \dot{u}, q, w; t). \tag{7}$$

All together, one can thus write equation (1) in the explicit form in terms of the first order differential equation as

$$\mathcal{M}\left(\frac{d}{dt}y(t)\right) = h\left(y(t)\right) + \check{b}(t) + c(t), \tag{8}$$

where $y(t)$ denotes the state vector, $\mathcal{M}$ the generalized mass matrix, $\check{b}(t)$ the mean of excitations, and $c(t)$ the modulated zero-mean white noise. These quantities are defined as

$$\mathcal{M} = \begin{pmatrix} I & 0 & 0 & 0 & 0 \\ 0 & M & 0 & 0 & 0 \\ 0 & 0 & I & 0 & 0 \\ 0 & 0 & 0 & I & 0 \\ 0 & 0 & 0 & 0 & I \end{pmatrix}, \tag{9a}$$

$$y(t) = \{u, \dot{u}, z, q, w\}^T, \tag{9b}$$

$$\check{b}(t) = \{0, E[f(t)], 0, 0, 0\}^T, \tag{9c}$$

$$c(t) = \{0, 0, \{n(t)\ m(t)\}, 0, 0\}^T, \tag{9d}$$

and the nonlinear vector function $\underset{\sim}{h}(t)$ reads

$$\underset{\sim}{h}(t) = \begin{cases} \dot{\underset{\sim}{u}} \\ \underline{C} \cdot \dot{\underset{\sim}{u}} + \underline{B} \cdot \underset{\sim}{z} + \underline{R} \cdot \underset{\sim}{q} \\ \underline{Z} \cdot \underset{\sim}{z} \\ \underline{H}(t) \cdot \dot{\underset{\sim}{u}} \\ \underset{\sim}{g}(t) \end{cases} \tag{9e}$$

Up to the authours' knowledge, there is no technique available to obtain the exact response for general nonlinear systems in equation (8). Especially, when the number of the DOF is large, the problem becomes much more involved. This can be handled by linearizing the nonlinear system to a statistically equivalent linear system with satisfactory accuracy. Therefore, in the next sections, the attention is focused on the efficient linearization of a nonlinear system when the FEM is applied.

3. STATISTICAL EQUIVALENT LINEARIZATION

Basically, the EQL method replaces the nonlinear system equations of motion (8) by a set of statistical equivalent linear ones

$$\underline{M}\left(\frac{d}{dt}\underset{\sim}{x}(t)\right) = \underline{K}(t)\underset{\sim}{x} + \underset{\sim}{b}(t) + \underset{\sim}{c}(t), \tag{10}$$

in which $\underset{\sim}{x}(t)$ is the state vector of the EQL system, and $\underline{K}(t)$ and $\underset{\sim}{b}(t)$ are defined by the following relations [10, 11]

$$\underline{K}(t) = \Big\{E[\underset{\sim}{h}(\underset{\sim}{y}(t)) \cdot \underset{\sim}{y}^T(t)] - E[\underset{\sim}{h}(\underset{\sim}{y}(t))] \cdot E[\underset{\sim}{y}^T(t)]\Big\} \cdot \underline{S}^{-1}(t), \tag{11a}$$

and

$$\underset{\sim}{b}(t) = \check{\underset{\sim}{b}}(t) + E[\underset{\sim}{h}(\underset{\sim}{y}(t))] - \underline{K} \cdot E[\underset{\sim}{y}(t)], \tag{11b}$$

where $\underline{S}(t)$ denotes the covariance matrix

$$\underline{S}(t) = E[\underset{\sim}{y}(t) \cdot \underset{\sim}{y}^T(t)] - E[\underset{\sim}{y}(t)] \cdot E[\underset{\sim}{y}^T(t)]. \tag{11c}$$

The equivalent relation of the two systems is considered under conditions that the first two statistical moments of the target nonlinear system and those of the EQL system are equal

$$E[\underset{\sim}{x}(t)] = E[\underset{\sim}{y}(t)], \tag{12a}$$

$$E[\underset{\sim}{x}(t) \cdot \underset{\sim}{x}^T(t)] = E[\underset{\sim}{y}(t) \cdot \underset{\sim}{y}^T(t)]. \tag{12b}$$

It is also noteworhty to mention that the parameters of the linearized system in equation (11) are function of time t, the EQL procedure has to be carried out in each time step. As the consequence, the procedure is capable to predict the nonlinear responses subjected to non-stationary excitations. Furthermore, one may notice from equation (11) that $\underset{\sim}{b}(t)$ and $\underline{K}(t)$

can be evaluted only after the joint distribution $f_{\underset{\sim}{y}}(\underset{\sim}{y})$ of the nonlinear system is known a priori. However, the joint distribution of $\underset{\sim}{y}(t)$ is not available. The way to resolve this is to apply an iteration technique which will be discussed in detail in Section 4.

Comparing equation (10) with equation (8), one concludes that matrix $\underset{\sim}{K}(t)$ must have the form

$$\underset{\sim}{K}(t) = \begin{pmatrix} 0 & \underline{I} & 0 & 0 & 0 \\ 0 & \underline{C} & \underline{B} & \underline{B} & 0 \\ 0 & 0 & \underline{Z} & 0 & 0 \\ & & \underline{Q}(t) & & \\ & & \underline{G}(t) & & \end{pmatrix}, \tag{13}$$

where $\underline{I}$ is an identity matrix and sub-matrices $\underline{Q}$ and $\underline{G}$ contain the linearization coefficients resulting from the nonlinear function $\underset{\sim}{g}(t)$ and $\dot{\underset{\sim}{q}}(t)$equation (3), namely,

$$\begin{aligned} &\underline{S}(t) \cdot \underline{Q}^T(t) = \\ &\quad E[\dot{\underset{\sim}{q}}^T(t) \cdot \underset{\sim}{y}(t)] - E[\dot{\underset{\sim}{q}}^T(t)] \cdot E[\underset{\sim}{y}(t)], \end{aligned} \tag{14a}$$

$$\begin{aligned} &\underline{S}(t) \cdot \underline{G}^T(t) = \\ &\quad E[\underset{\sim}{g}^T(t) \cdot \underset{\sim}{y}(t)] - E[\underset{\sim}{g}^T(t)] \cdot E[\underset{\sim}{y}(t)]. \end{aligned} \tag{14b}$$

Obvioully, most parts of the linearization coefficient matrix $\underset{\sim}{K}(t)$ remain unchanged except those parts related with nonlinear functions during the linearization procedure. In addition, this is also true from a physical point of view since the nonlinear effects are generally localized or isolated within a small portion of the whole system. For example, the most serious hysteretic effect often occurs at the lower levels of the building when it is under seismic loads.

Assuming for time t_0 that both $E[\underset{\sim}{y}(t_0)]$ and $E[\underset{\sim}{y}(t_0) \cdot \underset{\sim}{y}^T(t_0)]$ are known as initial conditions, the constraint conditions given in equation (12) ensure one to have the EQL matrix $\underset{\sim}{K}(t_0)$ together with $E[\underset{\sim}{x}(t_0)]$ and $E[\underset{\sim}{x}(t_0) \cdot \underset{\sim}{x}^T(t_0)]$. As time goes by, say at t_1, parts of $\underset{\sim}{K}(t_1)$ must be updated due to nonlinear effects. Thus, to determine the equivalent linearization coefficients from equation (14), one has to construct the joint distribution for the state vector $\underset{\sim}{y}$ at time t_1. Generally speaking, each component of $\underset{\sim}{y}(\underset{\sim}{x})$ is function of all components of the linear state vector $\underset{\sim}{x}$. This makes the problem quite involved. The easiest approach is to assume both the nonlinear and linear systems have the same jointly Gaussian distribution, $f_{\underset{\sim}{y}}(\underset{\sim}{y}) = f_{\underset{\sim}{x}}(\underset{\sim}{y})$. Subsequently, the state vector $\underset{\sim}{y}$ can be evaluated.

However, the assumption of jointly normal distribution for the nonlinear system is obviously inadequate in cases where strong nonlinear behavior occurs. It is clear when the system is nonlinear, the system response is not Gaussian any more even though the excitation is normal (with very few exceptions [12]). As the consequence, the identity relations between the target nonlinear and the EQL systems must be modified in such a way that non-Gaussain effects can be taken into account. Since equation (12) holds, the distribution function of the nonlinear state vector $\underset{\sim}{y}$ can be established as a nonlinear function of $E[\underset{\sim}{x}]$ and $E[\underset{\sim}{x}\underset{\sim}{x}^T]$. This nonlinear transformation from $\underset{\sim}{x}$ to $\underset{\sim}{y}$ must not change the constraint conditions in equation (12). Otherwise, the transformation cannot be used. As the following step, a more practicable simplification is to construct the joint distributions of the nonlinear system by its marginals $f_{\underset{\sim}{y}}(y_i)$ and the covariance matrix.

Based on the authors' knowledge, the best representation for the multivariate distribution is Nataf's model [13] in case the marginals and covariance matrix are known. The model is applicable for an arbitrary number of random variables. It is also clear that the correlation coefficient $\rho_{y_i y_j}$ is not equal to $\rho_{x_i x_j}$ any more because of the equality condition $E[y_i(x_i)\, y_j(x_j)] = E[x_i\, x_j]$ if the nonlinear transformation is adopted. Furthermore, one can modify the correlation coefficients ρ'_{ij} by the following integral equation [13,14] by use of nonlinear transformation, namely

$$E[x_i\, x_j] = \int_{-\infty}^{-\infty} \int_{-\infty}^{-\infty} y_i(x_i)\, y_j(x_j)\, f_{x_i x_j}\left(x_i, x_j; \rho'_{ij}\right) dx_i\, dx_j, \tag{15}$$

where $f_{x_i x_j}(\cdot)$ represents a two-dimensional normal probability density function. Or, equivalently, the joint distribution of the state vector $\underset{\sim}{y}$ can be written as a function of $f_{\underset{\sim}{x}}(\underset{\sim}{x})$ [13] which is jointly Gaussian

$$f_{\underset{\sim}{y}}(\underset{\sim}{y}) = f_{\underset{\sim}{x}}(\underset{\sim}{x}; \underset{\sim}{R}') \prod_{k=1}^{N} \left(\frac{\partial y_k}{\partial x_k}\right)^{-1}, \tag{16}$$

and where the correlation matrix $\underset{\sim}{R}'$ contains all modified correlation coefficients obtained from equation (15).

4. COMPLEX MODAL ANALYSIS FOR MANY-DOF SYSTEMS

A large number of DOF is common when applying the FE formulation in the structural analysis. Tens of hundreds of DOF may be needed for a practical system if detailed response quantities are required. Under this circumstance, the technique of mode superposition may be adopted to the EQL procedure. The technique simply involves a coordinate transformation from the finite element state vector $\underset{\sim}{x}$ to the modal coordinates $\underset{\sim}{\xi}$ by

$$\underset{\sim}{x} = \underline{\Phi}\, \underset{\sim}{\xi}, \tag{17}$$

where $\underline{\Phi}$ is obtained by solving eigenvalue problems

$$\underline{K}\,\underline{\Phi} = \underline{M}\,\underline{\Phi}\,\underline{\Lambda}, \tag{18a}$$

and

$$\underset{\sim}{K}^T \underline{\Psi} = \underset{\sim}{M}^T \underline{\Psi}\, \underline{\Lambda}, \tag{18b}$$

with

$$\underline{\Lambda} = \underline{\Psi}^T \underset{\sim}{K} \underline{\Phi}, \quad \text{and} \quad \underset{\sim}{I} = \underline{\Psi}^T \underset{\sim}{M}\, \underline{\Phi}, \tag{18c}$$

in which $\underline{\Lambda}$ is a diagonal matrix containing eigenvalues λ_i in its diagonal, $\underline{\Phi}$ and $\underline{\Psi}$ are the RHS and LHS modal matrices containing the RHS and LHS normalized eigenvectors, respectively. The first order system equation in the modal coordinates becomes

$$\dot{\underset{\sim}{\xi}} = \underline{\Lambda} \cdot \underset{\sim}{\xi} + \underset{\sim}{p}(t), \tag{19a}$$

where

$$\underset{\sim}{p}(t) = \underline{\Psi}^T \cdot \left[\underset{\sim}{b}(t) + \underset{\sim}{c}(t) \right]. \tag{19b}$$

Since the coefficient matrix $\underset{\sim}{K}$ in equation (10) is generally nonsymmetric, the eigenvalues associated with it are in general complex, and so are eigenvectors.

It is also clear that the complex eigenvalues are physically related with the frequencies and dampings of vibrations. Usually, the frequency range of interest to the dynamical analysis is limited in a few lower frequency modes for most structural systems. In addition, the responses of high frequencies ($|\Re(\lambda_i)| \gg 1$) are damped away quickly. However, these eigenvalues and eigenvectors tend to be unreliable. Consequently, it would not be economical to calculate all eigenvalues and eigenvectors. For example, let the order of $\underset{\sim}{K}$ be n, and the number of modes of interest to seek be m, where $m \leq n$. The dimension of $\underline{\Phi}$ is then $n \times m$ and so is $\underline{\Psi}$. This implies $\underset{\sim}{\xi}$ is an m dimensional array while $\underset{\sim}{x}$ has dimensions of n. One can clearly see that the storages and computation effort can be drastically reduced if m is chosen such that $m \ll n$. Note that, the choice of these m characteristic values will not lose the information of interest, or only negligible. Therefore, without loss of generality, notations $\underline{\Phi}$ and $\underline{\Psi}$ are used hereafter to denote the reduced dimension eigenvector matrices with dimensions $n \times m$, and $\underline{\Lambda}$ the reduced eigenvalue matrix of dimension $m \times m$.

Using the coordinate transformation defined by equation (18) in the complex space, the relations between the covariance matrix in the modal coordinates $\underset{\sim}{S}(\underset{\sim}{\xi}(t))$ and that in the displacement coordinates $\underset{\sim}{S}(\underset{\sim}{x}(t))$ are

$$E[\underset{\sim}{x}] = \underline{\Phi}\, E[\underset{\sim}{\xi}], \tag{20a}$$

and

$$\underline{S}(\underset{\sim}{x}) = \underline{\Phi}\, \underline{S}(\underset{\sim}{\xi})\, \underline{\Phi}^T. \tag{20b}$$

The above relations ensure one to transform the first two moments back and forth from the real space $\underset{\sim}{x}$ to the complex space $\underset{\sim}{\xi}$ at any time t with less computation. This in turn, necessitates the storage of Φ and Ψ^T all the time. Furthermore, the modal responses $\xi_k, k = 1, 2, \cdots, m$ at time $t_1 = t_0 + \tau$ can be evaluated, if the EQL system is considered as constant within the time increment τ, from the given initial conditions at time t_0 [9] by

$$E[\xi_k(t_1)] = e^{\lambda_k \tau} E\Big[\xi_k(t_0)\Big] + \int_0^\tau p_k(t_0 + s)\, e^{-\lambda_k s}\, ds, \tag{21}$$

where $p_k(t) = \sum_{j=1}^n \psi_{jk} b_j(t)$ is the mean of modal excitation associated with the k-th eigenvalue. It is assumed to vary linearly within $[t_0, t_1]$. The covariance in the modal coordinates $\underset{\sim}{S}(\xi)$ is easily integrated. For the zero-mean white noise excitation

$$S_{\ell j}(t_0 + \tau) = e^{(\lambda_\ell + \lambda_j^*)} \cdot \left[S_{\ell j}(t_0) + \int_0^\tau \mathcal{B}_{\ell j}(t_0 + s)\, e^{-(\lambda_\ell + \lambda_j^*) s} ds \right], \tag{22}$$

in which the super asterisk denotes the complex conjugate value, and $\mathcal{B}_{\ell j}$ is the modulated white noise intensity after being transformed to the modal coordinates, i.e., $\mathcal{B} = \Psi^T\, E[\underset{\sim}{c}(t) \underset{\sim}{c}^T(t)]\, \Psi$.

However, the evaluation of $\xi_k(t_1)$ from equation (21) involves the values of $\lambda_k(t_0)$, $\Phi(t_0)$ and $\Psi(t_0)$. They in turn are functions of the coefficient matrix $\mathcal{K}(t_0)$ which contains the linearization coefficients in the sub-matrices Q and G. Because of the assumption that the system matrix remains constant within time interval $[t_0, t_1]$, the values of $Q(t_0)$ and $G(t_0)$ are initially unknown. That is, assuming the computation starts at $t_0 = 0$, the representation of $\mathcal{K}(t_0)$ is actually located at the mid-point of interval $[0, \tau]$, refer to fig. 1. Thus, the evaluation of the EQL system at t_0 is firstly done by computing the system at t_1^- while the final representation for $\mathcal{K}(t_0)$ is the average of the two. In a general recurrsive form, $\mathcal{K}(t_\imath) = \frac{1}{2}[\mathcal{K}(t_i) + \mathcal{K}(t_{i+1})]$, where $t_\imath$ denotes for the system representation at state $\imath$. As the consequence, a self-correcting iteration process is needed, even for the first step, until convergence is reached. Convergence can be generally obtained with much less computation effort if the immediate previous values is utilized as the starting values. In addition, the state vector $\underset{\sim}{x}$ must be continuous all the time. In other words, the condition that $\underset{\sim}{x}(t_1^-) = \underset{\sim}{x}(t_1^+)$ must be satisfied at the adjacent point, shown as double vertical lines in figure 1. This continuity condition can be ensured by introducing the following steps. Again refer to fig. 1, let Φ_0, Ψ_0^T, Φ_1 and Ψ_1^T represent the modal matrices of two different time states t_0 and t_1, respectively. From the relation in equation (18) one has

$$\underset{\sim}{x}_0 = \Phi_0 \cdot \underset{\sim}{\xi}_0 \qquad \text{and} \qquad \underset{\sim}{x}_1 = \Phi_1 \cdot \underset{\sim}{\xi}_1, \tag{23}$$

which in turn, give the relations that guarantee the continuity as

$$E[\underset{\sim}{\xi}_1] = \underline{\Psi}_1^T \underline{M} \underline{\Phi}_0 \cdot E[\underset{\sim}{\xi}_0], \quad \text{and} \tag{24a}$$

$$\underline{S}_1(\underset{\sim}{\xi},t) = (\underline{\Psi}_1^T \underline{M} \underline{\Phi}_0) \cdot \underline{S}_0 \cdot (\underline{\Psi}_1^T \underline{M} \underline{\Phi}_0)^T. \tag{24b}$$

Of course, all $\underline{\Lambda}$, $\underline{\Phi}$ and $\underline{\Psi}$, or $\underline{K}$ must be updated for each iteration. The stop criterion would be a small acceptable relative error between two consecutive iterations.

The matrix $\underline{K}(t)$ is sparse but not symmetric. In order to make the present EQL procedure practical for systems of several hundred DOF, the eigenvalue problem defined in equation (18) must be solved efficiently. In addition, the storage of matrices $\underline{K}$ and $\underline{M}$ becomes a very important aspect since it is often impossible to store a full quadratic matrix of high order, say $n > 500$. Of course, the storage depends on the solution technique. In the present approach, matrices $\underline{K}$ and $\underline{M}$ are only required to determine the eigenvalues $\underline{\Lambda}$ and modal matrices $\underline{\Phi}$ and $\underline{\Psi}$. Therefore, the authors suggest to use a solution technique where only matrix multiplication but not matrix inversions are required. As the consequence, the following simple and compact storage can be used. The matrix $\underline{K}$, for instance, is stored in three arrays: $I\mathcal{K}(i)$, $J\mathcal{K}(i)$ and $C\mathcal{K}(i)$. The first two arrays contain the location information (in row and column of $\underline{K}$) where the elements are not zero, while their values are kept in array $C\mathcal{K}(i)$. A similar storage is recommended for matrix $\underline{M}$ as well.

Actually, the eigenvalues and eigenvectors can be found by inverse vector iteration together with an additional use of shifts [15]. The inverse vector iteration, however, requires the inversion of the matrix $(\underline{K} - \mu\underline{M})^{-1}$ where μ is a complex valued shift. Instead of the inversion, the utilization of the conjugate gradient [16] method is suggested, in which only matrix multiplcations $(\underline{K} - \mu\underline{M}) \cdot \underset{\sim}{v}$ are needed. After some simple re-ordering, the suggested storage mentioned earlier is especially suitable for vector computers.

In addition, one should be aware that the modal matrices $\underline{\Phi}$, $\underline{\Psi}$ and the eigenvalues must be evaluated in each iteration and time step. Once these matrices are found in the beginning of the EQL procedure, the eigenvalues and the associated eigenvectors change only slightly in each iteration. By use of the previous eigenvalues as the shift for inverse vector iteration, actual eigenvalues and eigenvectors can be computed economically by additional use of subspace iteration.

5. APPLICATIONS OF EQL ON FINITE ELEMENTS

For the practical application of the EQL procedure, it is important to mention that equation (14) is also applicable to individual elements. This enables one to write equation (14) for each nonlinear element

$$^{\ell}\underline{S}(t) \cdot {}^{\ell}\underline{Q}^T(t) = $$
$$E[{}^{\ell}\dot{\underset{\sim}{q}}^T(t) \cdot {}^{\ell}\underset{\sim}{y}(t)] - E[{}^{\ell}\dot{\underset{\sim}{q}}^T(t)] \cdot E[{}^{\ell}\underset{\sim}{y}(t)], \tag{25a}$$
$$^{\ell}\underline{S}(t) \cdot {}^{\ell}\underline{G}^T(t) = $$
$$E[{}^{\ell}\underset{\sim}{g}^T(t) \cdot {}^{\ell}\underset{\sim}{y}(t)] - E[{}^{\ell}\underset{\sim}{g}^T(t)] \cdot E[{}^{\ell}\underset{\sim}{y}(t)], \tag{25b}$$

where ${}^{\ell}\underset{\sim}{y}(t)$ is the state vector of the element

$$^{\ell}\underset{\sim}{y}(t) = \{{}^{\ell}\underset{\sim}{u}, {}^{\ell}\dot{\underset{\sim}{u}}, {}^{\ell}\underset{\sim}{q}, {}^{\ell}\underset{\sim}{w}\}^T, \tag{25c}$$

and notations with superscript ℓ denote the elemental parameters of the ℓ-th element. Note that the elemental state vector ${}^{\ell}\underset{\sim}{y}$ shall at least comprise all variables of the ℓ-th moments as shown in equation (25c).

It is clear that the covariance matrix $\underset{\sim}{S}(\xi)$ contains all information concerning the correlation of the state vector $\underset{\sim}{x}$ in the modal coordinates. During each iteration, the updated information at time step t is kept in the new covariance matrix. However, it is also inconvenient since all elements of $\underset{\sim}{S}(\xi)$ are complex. Thus, before proceeding further, it may prove useful to get the relevant information back into the real space. This can be done again by solving an eigenvalue problem of $\underset{\sim}{S}(\xi)$, i.e.,

$$\underset{\sim}{S}(\xi) = \hat{\underline{\aleph}}\, \underline{\Sigma}\, \hat{\underline{\aleph}}^T, \tag{26}$$

where $\underline{\Sigma}$ is a diagonal matrix containing the eigenvalues σ^2 of $\underset{\sim}{S}(\xi)$, and $\hat{\underline{\aleph}}$ stands for its eigen matrix. It is apparent that all σ^2's in $\underline{\Sigma}$ are real since covariance matrix $\underset{\sim}{S}(\xi)$ is Hermitian. Substituting equation (26) into (20), one concludes that $\underline{\Phi}\, \hat{\underline{\aleph}}$ must be real valued. Therefore, it is reasonable to express $\underline{S}(\underset{\sim}{x})$ in terms of

$$\underline{S}(\underset{\sim}{x}) = \underline{\aleph}\, \underline{\aleph}^T, \tag{27a}$$

where

$$\underline{\aleph} = \underline{\Phi}\, \hat{\underline{\aleph}}\, [\sigma], \tag{27b}$$

is the operator with dimensions $n \times m$ for transforming the uncorrelated bases into correlated ones. Actually, the matrix $\underline{\aleph}$ contains "correlation modes" which describe the correlation of the state vector $\underset{\sim}{x}$. The number of nonzero columns in $\underline{\aleph}$ also gives the rank of the covariance matrix $\underline{S}(\underset{\sim}{x})$. This expression also substantiates that there is no need to include all n components of $\underset{\sim}{x}$ during the linearization procedure for the ℓ-th element.

In order to obtain the linearization coefficients, the Monte Carlo (MC) simulation technique has been used in the EQL procedure. One might question about the use of this simulation technique for obtaining the linearization coefficients. The adoption of MC simulation, however, has no limitation for the type of nonlinear elements. It works for all nonlinear, determinisitic elements as long as the relations among the element state are correctly specified. In addition, for most applications, only a few hundred simulations are usually sufficient. The number of simulations can be even further reduced if the concept of importance sampling [17] is utilized.

Applying equation (27), one can independently generate a set random numbers in the uncorrelated space with m components, say $\overset{k}{\underset{\sim}{z}}$, while the corresponding state vector $\overset{k}{\underset{\sim}{x}}$ can be readily computed

$$\overset{k}{\underset{\sim}{x}}_i,\ i \in \mathcal{I}^\ell = \sum_{j=1}^{m} \aleph_{ij} \overset{k}{z}_j, \quad \text{and} \quad k = 1, 2, \cdots, N_s, \tag{28}$$

where $\mathcal{I}^\ell$ denotes the index set for those components of the ℓ-th element, over-script k simulation index number runs from 1 to the total number of simulations N_s. As soon as the state vector $\underset{\sim}{x}$ is obtained from equation (28), $\underset{\sim}{y}$ can be subsequently evaluated in accordance with a suitable nonlinear transformation function. Assuming further that the correlation coefficients ρ'_{ij} have been also modified by using equation (15), one thus has no difficulty to evaluate the equivalent linearization coefficients for the indicated element from equation (25) and the iteration technique given above.

Practically, however, equation (25) can be also formulated as a least-square problem, take (25b) as an example for ℓ-th element,

$$\sum_j \left[\overset{k}{y}_j - E[y_j]\right] G_{\ell j} = g_\ell(\overset{k}{\underset{\sim}{y}}) - E[g_\ell], \tag{29}$$

where $j = 1, 2, \cdots, N$, N total number of linearization coefficients in the ℓ-th element, i.e., $\{y_j | j \in \mathcal{I}^\ell\}$, and $k = 1, 2, \cdots, N_s$, the over-script k denotes the k-th simulated state vector. In case $N_s \to \infty$, the equivalence of equation (29) and (25) can be proved. It may be noteworthy to note that this number of simulation (N_s) required to obtain certain accuracy strongly depends upon the system nonlinearities of the element like ${}^\ell\underset{\sim}{\dot{q}}(t)$ and ${}^\ell\underset{\sim}{g}(t)$ in equation (25). This can be seen from the case where $g_\ell({}^\ell\underset{\sim}{y})$ is linear, then $N_s = N + 1$ will be enough to obtain the exact solution. In other words, the simulated covariance matrix $\underset{\sim}{S}^{\text{sim}}$ may be considerably different from the exact $\underset{\sim}{S}$, the exact solution for the linearization coefficients can still be obtained if $g_\ell({}^\ell\underset{\sim}{y})$ is linear. Therefore, it is then resonable to compute the linearization coefficients with N_s small for a weakly nonlinear function, but N_s large for strongly nonlinear functions. Let the total number of simlutions $N_s = N'_s \times M_s$, where N'_s is selected in such a way that it is suitable for a vector computer, e.g., $N'_s = 128, 256$, etc., and M_s is the total sets of N'_s required. Next, a stop criterion based on the information how the computed linearization coefficients spread around the mean is set. Starting from set number $j = 1$, the criterion is checked after each set is completed. In case the stop criterion can be satisfied before $j = M_s$ is reached, then stop. The final linearization coefficients are then taken from the average of all j sets data. However, $j \geq 3$ is recommended in the real application.

Solving for the equivalent linearization coefficients either from equation (25) or from equation (29), one has to take care that an ill-conditioned matrix may be obtained. In this case, those techniques like Gaussian elimination or decomposition techniques such as the Cholesky method could cause some difficulties. Using the mentioned conjugate gradient method [18], this problem can be resolved. The method allows one to find a solution close to the previous linearization coefficients in case an ill-conditioned covariance matrix of the element ${}^{\ell}\underset{\sim}{S}$ is obtained.

6. A NUMERICAL EXAMPLE

The stochastic responses of a bridge, shown in fig. 2, subjected to dead loads, and random horizontal and vertical ground motions is investigated. The bridge is modeled by seven horizontal and two vertical 2-D beam elements. The cross section of the horizontal part consists of a concrete plate at the top and two steel girders at the bottom, while the whole horizontal part is supported by two steel columns. In the present investigation, the following structural parameters have been applied for both the horizotal (index h) and vertical (index v) beam elements

m_h= 44,000 kg	m_v= 123.0 kg
EA_h= 39289.9 MN	EA_v= 2520.0 MN
GA_h= 808.0 MN	GA_v= 133.4 MN
EI_h= 6124.7 MNm^2	EI_v= 622.2 MNm^2

where m_h & m_v denote the mass per unit length, E Young's modulus, G the modulus of rigidity, A the cross-sectional area, and I the area moment inertia.

In case of small deflections where the bridge reacts linearly, the first ten eigen-frequencies (rad/sec) have been as

$$\omega_j = \{6.77,\ 11.2,\ 19.4,\ 27.8,\ 32.2,$$
$$42.0,\ 44.5,\ 70.7,\ 129.,\ 132.\}, \quad 1 \leq j \leq 10\,.$$

Since the shear stiffness GA_h is relatively small, the bending behavior of the horizontal beams are strongly affected by the shear deformation. The horizontal beams react linearly within the range $[-3.6\,, +4.2]$ MNm of bending moment. The ultimate yielding moments are -4.1 and +5.0 MNm, where the positive bending moment causes tension in the lower part of steel girders. The loss in stiffness between ultimate linear and ultimate yielding moment is assumed to vary linearly.

For the evaluation of the nonlinear dynamic response, the stress state due to dead loads is important. Because of the composite method of construction, the stress state in the steel girder is influenced by the construction process. For this reason, a static analysis of progressive construction states is required prior to the dynamic analysis can be performed. The cross-sectional properties shown above actually represent the parameters for the state after completion of construction. In the present report, the horizontal and vertical ground motions are modeled as filtered white noise with uncertainties in the filter properties. The ground accelerations e_h and e_v, respectively, are modeled by

$$e_h = (\dot{z}_1 + \dot{z}_2)\frac{1}{sec}, \tag{30a}$$

$$e_v = (\dot{z}_3 + \dot{z}_4)\frac{1}{sec}, \tag{30b}$$

where the filter equations read

$$\ddot{z}_i + 2\zeta_i\Omega_i\dot{z}_i + \Omega_i^2 z_i = n_i(t)m_i(t), \quad i = 1,2,3,4. \tag{31}$$

These filter parameters are assumed to be uncertain and are therefore represented as time-invariant random variables. The following means and standard deviations have been assumed

	μ	σ		μ	σ
Ω_1 :	(6.00,	1.13)	ζ_1 :	(0.30,	0.060),
Ω_2 :	(12.0,	2.12)	ζ_2 :	(0.50,	0.173),
Ω_3 :	(10.0,	1.41)	ζ_3 :	(0.25,	0.050),
Ω_4 :	(20.0,	2.83)	ζ_4 :	(0.60,	0.141).

The modulating functions are all assumed to follow the form [18]

$$m(t) = \frac{[e^{-a_i t} - e^{-b_i t}]}{c_i}, \tag{32}$$

where $i = 1,2,3,4$, c_i are normalizing constants so that the maximum value of $m(t)$ is unity, and $a_i = \{0.3, 0.7, 0.5, 0.9\}$ and $b_i = \{1.0, 1.0, 0.6, 1.0\}$. It is further assumed that all white noise components are independent and their intensities are given by $I_i = \{10, 10, 5, 5\}\ m^2/\sec^3$. The data lead to quite high horizontal and vertical ground motions. The largest vertical standard deviation of the ground acceleration occurs at time $t = 1.9$ sec with σ_{e_v}=0.748 $m/\sec^2$ and for the horizontal acceleration at $t = 2.5$ sec with $\sigma_{e_h} = 1.225\ m/\sec^2$. The horizontal ground motion causes significant horizontal forces in the upper left support (±4.2 MN), but has little influence on the bending moments and on the hysteretic reaction of the bridge. The results of the stochastic response are summarized in figs. 3 and 4. Figure 3 shows the vertical displacements of the bridge at time t =3 sec, when the maximum standard deviation took place. In addition to the nonlinear response, the static and the linear results are shown for reasons of comparison. For the linear case, the mean values for the filter have been set to be fixed. It can be recognized that the standard deviations of the nonlinear response are considerably smaller that those of the linear one. The difference would be even more pronounced if the filter parameters were assumed to be fixed. The bending moments in the horizontal elements at t = 3 sec are shown in fig. 4.

7. CONCLUSIONS

Based on the developments presented herein, the following conclusions are drawn:

(1) The stochastic dynamic response of an arbitrary nonlinear FE–model can be determined by the suggested numerical procedure. The procedure is based on stochastic equivalent linearization and on complex modal analysis.

(2) Using adaptive Monte Carlo simulation to evaluate the statistically equivalent linear incremental properties of the nonlinear element, all available nonlinear elements can be utilized in the model including "black box" routines.

(3) The presented procedure allows a compact storage of the sparse matrices of the linearized system and of the covariance matrix. Such compact storage becomes indispensable for large system with several hundred degrees of freedom.

(4) The suggested approach is computationally highly efficient in comparison with straightforward Monte Carlo simulation.

ACKNOWLEDGEMENTS

This research is supported by the Austrian Industrial Research Promotion Fund under Project No. 6/546/B1 which is gratefully acknowledged by the authors.

REFERENCES

1. Benaroya, H. and Rehak, M., Finite element methods in probabilistic structural analysis: A selective review, *Applied Mechanics Review*, **41**(5), 1988, 201–213.

2. Pradlwarter, H.J., Non-Gaussian linearization – An efficient tool to analyze nonlinear MDOF systems, *Proc. SMiRT 10*, Anaheim, California, 1989, **M**, 19–30.

3. Wen, Y.K., Equivalent linearization for hysteretic systems under random excitation, *Journal of Applied Mechanics*, **47**, 1980, 150–154.

4. Casciati, F. and Faravelli, L., Stochastic linearization for 3–D frames, *Journal of Engineering Mechanics*, ASCE, **114**(10), 1988.

5. Mochio, T., Samaras, E. and Shinozuka, M., Stochastic equivalent linearization for finite element based on reliability analysis, *Proc. of ICOSSAR, the 4th International Conf. on Structural Safety and Relability*, Kobe, 1985, **I**, 375–384.

6. Simulescu, I., Mochio, T. and Shinozuka, M., Equivalent linearization method in nonlinear FEM, *Journal of Engineering Mechanics*, ASCE, **115**(3), 1989, 475–492.

7. Casciati, F. and Faravelli, L., Stochastic equivalent linearization for dynamic analysis of continuous structures, *Proc. ASME/SES for Meeting Computational Probabilistic Methods*, 1988, 205–210.

8. Wen, Y.K. & Yeh, C.H., Biaxial and torsional response of inelastic structures under random excitation, *Proc. of the Symposium of Stochastic structural dynamics*, Univ. Illinois Urbana–Champaign, 1988, 391–411.

9. Chang, T-P., Mochio, T. and Samaras, E., Seismic response analysis of nonlinear structures, *Probabilistic Engineering Mechanics*, **1**(3), 157–166.

10. Kozin, F., The method of statistical linearization for nonlinear stochastic vibrations, *Proc. of IUTAM Symposium*, Innsbruck/Igls, 1987, F. Ziegler and G.I. Schuëller (Eds.), Springer Verlag, 1988, 455-466.

11. Pradlwarter, H.J., On existence of true equivalent linear systems for the evaluation of the nonlinear stochastic response due to non-stationary Gaussian excitation, (submitted for publication), 1990.

12. Lin, Y.K., Recent advances in nonlinear random vibration, *Proc. 3rd Int. Conf. on Recent Advances in Structural Dynamics*, 1988, Southampton, M. Petyt, H. Wolfe and C. Mei (Eds), **II**, 717–736.

13. Nataf, A., Determination des distribution dont les marges sont donnees, *Comptes Rendus de l'Academie des Sciences*, Paris, 1962, **225**, 42–43.

14. Liu, P-L. & Der Kiureghian, A., Multivariate distribution models with prescribed marginals and covariances, *Probabilistic Engineering Mechanics*, **1**(2), 105–112.

15. Wilkinson, J.H. and Reinsch, C., *Handbook for Automatic Computation*, Vol. **II**, Linear Algebra, 1971, 339–358 and 359–371.

16. Hestenes, M.R., The conjugate–gradient method for solving linear systems, *Proceedings of Symposia in Applied Mathematics*, **VI**, McGraw-Hill, NY, 1956, 83–102.

17. Schuëller, G.I. & Stix, R. A critical appraisal of method to determine failure probabilities, *Structural Safety*, 1987, **4**(4), 293–309.

18. Shinozuka, M. and Sato, Y. Simulation of nonsationary random process, *Journal of the Engineering Mechanics Division*, ASCE, 1967, **93**, No. EM1, 11–40.

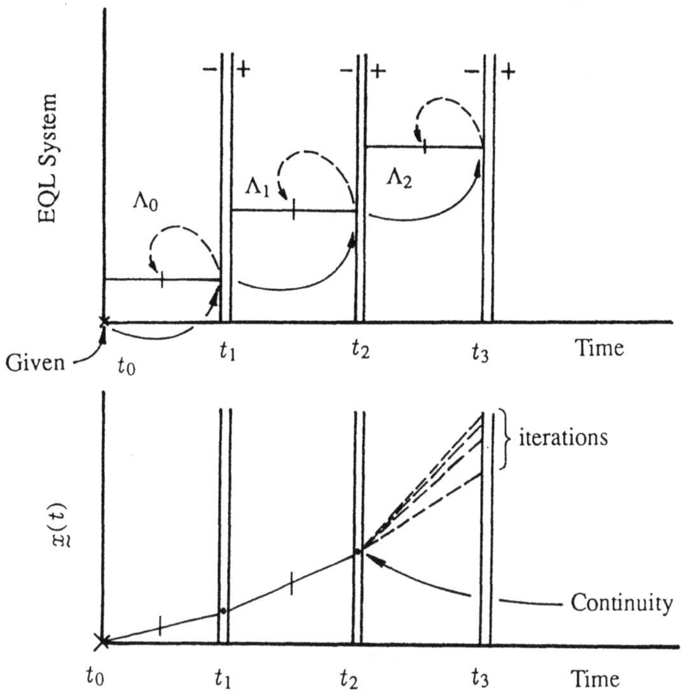

Fig. 1 EQL Iterations (solid lines: forward computation, dotted lines: backward averaging).

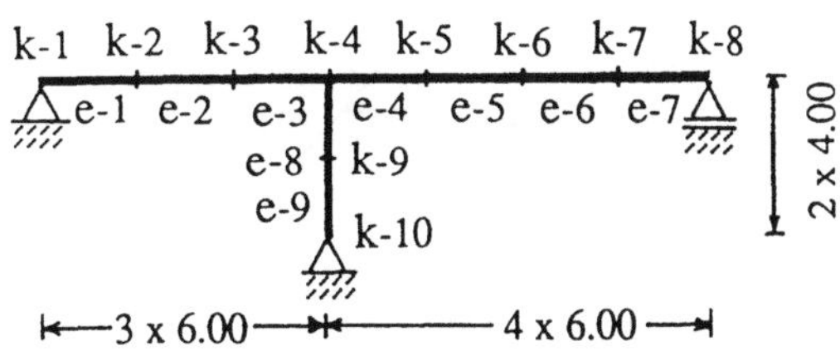

Fig. 2 FE–Model of a Bridge.

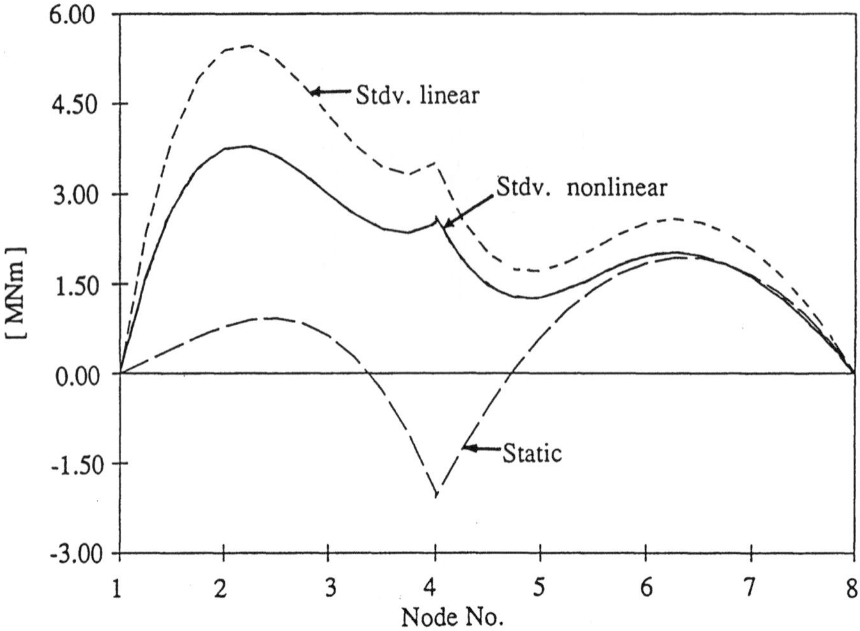

Fig. 3 Vertical Displacements of the Bridge.

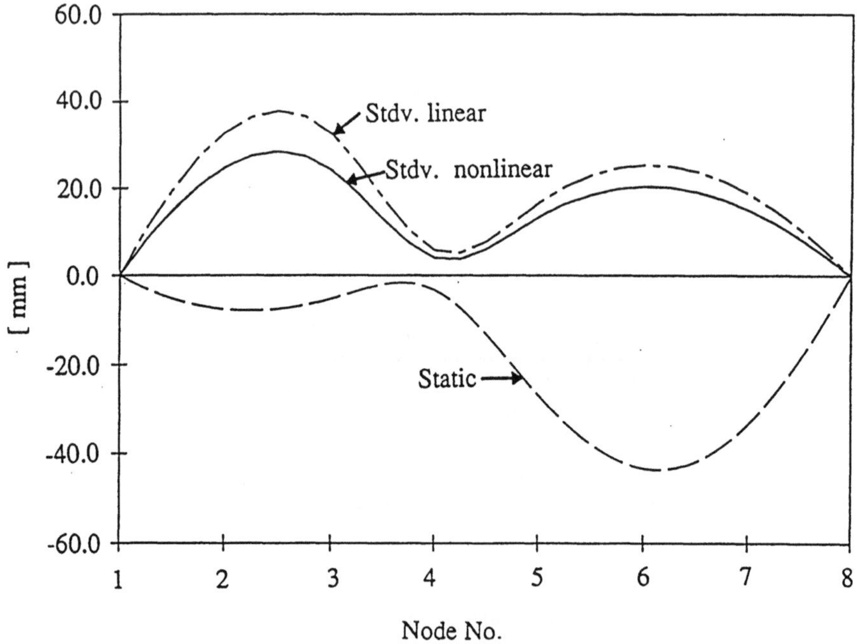

Fig. 4 Bending Moments of the Bridge.

Reliability Estimate of Linear Oscillator with Uncertain Input Parameters

S. T. QUEK, Y. P. TEO and T. BALENDRA

Department of Civil Engineering
National University of Singapore
Republic of Singapore

Abstract

The first passage probability of a linear oscillator subjected to non-stationary uniformly-modulated seismic excitation is estimated considering the uncertainties of the parameters in the structure and input model. A geophysical model for the input power spectral density of the seismic excitation is used and the probability distributions of the assumed time-invariant parameters are estimated from regression analyses using actual accelerogram records. The statistics of the structural parameters are adopted from current literature. The time-variant reliability of the oscillator are estimated using a Markovian extreme point process model and two different methods of incorporating the uncertain input parameters, namely, the method of moments and the Advanced First-Order Second-Moment (AFOSM) method are presented. Using a numerical example it is shown that uncertainties in the input parameters affect the reliability significantly. The AFOSM is preferred and should be used when the probability distributions of the uncertain parameters can be reasonably approximated whereas the method of moments albeit simple is sensitive to the assumed distribution of the maximum peak of the response, S. It is further noted that the variabilities of geophysical model parameters contribute significantly to the variance of S compared to that of the structural parameters.

1. Introduction

In the computation of the first-passage probability of a single degree of freedom (SDOF) oscillator, the excitations are usually characterized by means of power spectral density (PSD) functions in which the parameters of the input PSD and the structural parameters are treated as deterministic [1]. In general, this may not be always appropriate and a more realistic estimate of the reliability should therefore account for the uncertainties of these parameters.

In this paper, the uncertainties of the structural and input parameters are considered in the reliability estimate of a SDOF linear oscillator subjected to a nonstationary uniformly-modulated seismic excitation in which a geophysical model [2] is used. The parameters are modeled as time-invariant random variables and the statistics of the input model parameters are estimated from regression analyses using 54 accelerogram records

I. Elishakoff · Y. K. Lin (Eds.)
Stochastic Structural Dynamics 2
New Practical Applications
2nd International Conference on Stochastic Structural Dynamics
May 9–11, 1990, Boca Raton, Florida

whereas the statistics of the structural parameters are adopted from available literature. These are incorporated in the reliability analysis using either the method of moments [3] or the AFOSM method [4] where the distribution of the maximum peak of the response is considered.

2. Evolutionary Response of SDOF Oscillator

For a linear SDOF structure subjected to a nonstationary ground acceleration, $a(t)$, the relative displacement, $y(t)$, of the structure is governed by

$$\frac{d^2y(t)}{dt^2} + 2\xi\omega_o\frac{dy(t)}{dt} + \omega_o^2y(t) = -a(t) \tag{1}$$

where ξ is the damping coefficient and ω_o the natural circular frequency of the structure. For a system initially at rest, if $a(t)$ is a uniformly modulated Gaussian random process admitting an evolutionary spectral representation [2]

$$a(t) = \psi(t)\tilde{a}(t) = \int_{-\infty}^{\infty} \psi(t)e^{i\omega t}d\tilde{Z}(\omega) \tag{2}$$

where $\psi(t)$ is a time modulating function and $\tilde{a}(t)$ is a zero-mean unit-variance stationary process, it follows that $y(t)$ is also a Gaussian random process given by

$$y(t) = \int_{-\infty}^{\infty} M(t,\omega)e^{i\omega t}d\tilde{Z}(\omega) \tag{3}$$

$$\text{where } M(t,\omega) = -\int_0^t \frac{e^{-\xi\omega_o\tau}}{\omega_o\sqrt{(1-\xi^2)}}\sin[\tau\omega_o\sqrt{(1-\xi^2)}]\ \psi(t-\tau)e^{-i\omega\tau}d\tau \tag{4}$$

The evolutionary PSD of the response, $S_y(t,\omega)$, and the variance of the response, $\sigma_y^2(t)$, are then given by

$$S_y(t,\omega) = |M(t,\omega)|^2S_{\tilde{a}}(\omega) \tag{5}$$

$$\sigma_y^2(t) = 2\int_0^{\infty} S_y(t,\omega)d\omega \tag{6}$$

where $S_{\tilde{a}}(\omega)$ is the two-sided PSD of $\tilde{a}(t)$. Defining the nonstationary envelope process, $V(t)$, of $y(t)$ as $V(t)=\sqrt{[y^2(t)+\hat{y}^2(t)]}$, where $\hat{y}(t)$ is the Hilbert transform of $y(t)$, the probability density function (PDF) of $v(t)$, $f_V(v,t)$, and the joint PDF of $v(t_1)$ and $v(t_2)$, $f_{VV}(v_1,v_2;t_1,t_2)$, are given by [1]:

$$f_V(v;t) = \frac{v}{\sigma^2}\exp\left(-\frac{v^2}{2\sigma^2}\right) \tag{7a}$$

$$f_{VV}(v_1,v_2;t_1,t_2) = \frac{v_1v_2}{\sigma_1^2\sigma_2^2(1-r_{1,2}^2)}\exp\left\{-\frac{(\sigma_2^2v_1^2+\sigma_1^2v_2^2)}{2\sigma_1^2\sigma_2^2(1-r_{1,2}^2)}\right\}\mathrm{I}_o\left[\frac{v_1v_2r_{1,2}}{\sigma_1\sigma_2(1-r_{1,2}^2)}\right] \tag{7b}$$

where $\sigma = \sigma_y(t)$; $\sigma_1 = \sigma_y(t_1)$; $\sigma_2 = \sigma_y(t_2)$;

$$r_{1,2}^2 = \rho^2(t_1,t_2)+\lambda^2(t_1,t_2);$$

$$\rho(t_1,t_2) = \frac{2}{\sigma_1\sigma_2}\int_0^\infty |M(t_1,\omega)||M(t_2,\omega)|\cos[\omega(t_1-t_2)+\delta(t_1,\omega)-\delta(t_2,\omega)]S_{\tilde{a}}(\omega)d\omega;$$

$$\lambda(t_1,t_2) = \frac{2}{\sigma_1\sigma_2}\int_0^\infty |M(t_1,\omega)||M(t_2,\omega)|\sin[\omega(t_1-t_2)+\delta(t_1,\omega)-\delta(t_2,\omega)]S_{\tilde{a}}(\omega)d\omega;$$

$$\delta(t,\omega) = \tan^{-1}\{\mathrm{Im}[M(t,\omega)]/\mathrm{Re}[M(t,\omega)]\};$$

and $\mathrm{I}_o[z] = \frac{1}{2\pi}\int_0^{2\pi} e^{z\cos\theta}\,d\theta$ which is the zero-order modified Bessel function of the first kind. Equation 7 may then be used to compute the first passage probability of the oscillator.

3. Nonstationary Ground Excitation Model

Generalizing Vanmarcke and Lai's relation for the PSD of a stationary process [5] to the nonstationary case, the two-sided evolutionary PSD of $a(t)$ can be written as

$$S_a(t,\omega) = \psi^2(t)S_{\tilde{a}}(\omega) = \psi^2(t)\bar{A}_e^2(\omega)/(2\pi I) \tag{8}$$

where $I = \int_0^\infty \psi^2(t)dt$ (9)

and $\bar{A}_e(\omega)$ is the equivalent-stationary averaged Fourier amplitude (FA) of $a(t)$. The characterization of the input PSD will therefore involve the characterization of $\bar{A}_e(\omega)$ in the frequency domain and $\psi(t)$ in the time domain.

3.1 Geophysical Model

The equivalent-stationary averaged FA, $\bar{A}_e(\omega)$, of nonstationary ground accelerations at rock sites due to seismic excitation can be modeled based on geophysical considerations [6,7] as

$$\bar{A}_e(\omega) = \left(\frac{R_\phi \mathcal{F}_s P_r}{4\pi\rho\beta_w^3}\right)\frac{m_o}{R}\,\frac{\omega^2}{1 + (\omega/\omega_c)^2}\,e^{-\kappa|\omega|}, \qquad -\infty < \omega < \infty. \tag{10}$$

where R_ϕ accounts for the radiation pattern of the seismic wave, $\mathcal{F}_s$ for free surface effect, P_r for partition of energy into horizontal components, ρ is the density of the medium, β_w the seismic wave velocity, R the source-to-site distance and m_o the seismic moment. m_o (in Nm) is related to the recorded earthquake magnitude, M_{rec}, by [6]

$$m_o = 10^{(1.5\gamma_M M_{rec} + 9.05)} \tag{11}$$

in which γ_M is a bias factor to account for the uncertainty in M_{rec}. ω_c is the corner circular frequency above which the spectral amplitude of the ground displacement diminishes rapidly and can be computed from [6]

$$\omega_c = 2\pi(0.49\beta_w)(\Delta\sigma/m_o)^{1/3} \tag{12}$$

where $\Delta\sigma$ is the drop in the stress built-up along the fault surface. κ characterizes the attenuation of seismic wave and is related to R by [7]:

$$\kappa = \kappa_R R + \kappa_c \tag{13}$$

where κ_R and κ_c are both the attenuation constants.

3.2 Time Modulating Function

The time modulating function, $\psi(t)$, is modeled by [8]

$$\psi(t) = A_o(e^{-bt} - e^{-\alpha bt}), \qquad \alpha > 1 \tag{14}$$

where A_o is a scaling factor; b and α can be computed for any earthquake record through the following equations [2]:

$$T_o = t_{95} - t_5 \tag{15}$$

$$\varepsilon = t_m/t_{95} \tag{16}$$

where

$$\int_0^{t_5}\psi^2(t)dt = 0.05\int_0^{\infty}\psi^2(t)dt \tag{17a}$$

$$\int_0^{t_{95}}\psi^2(t)dt = 0.95\int_0^{\infty}\psi^2(t)dt \tag{17b}$$

ε is the rise time fraction, in which t_m is the time at which $\psi(t)$ attains its maximum value, and T_o is the strong motion duration [9]. T_o is assumed as dependent on M_{rec} and R [9]:

$$T_o = \tau_M M_{rec} + \tau_R R + \tau_c \tag{18}$$

where τ_M, τ_R and τ_c are the duration constants.

4. Statistics of Input Parameters

4.1 Selection of Random Variables

For a particular input excitation specified in terms of M_{rec} and R, $\bar{A}_e(\omega)$ is completely defined if the values of the parameters R_ϕ, $\mathcal{F}_s$, P_r, ρ, β_w, γ_M, $\Delta\sigma$, κ_R, κ_c and κ_o are known. M_{rec} is taken as the local magnitude, M_L, and R is taken as the epicentral distance, R_{epi}. Since a study by McGuire [10] showed that the FA of the ground acceleration is lognormally distributed about its median value, a regression analysis using actual accelerograms can be performed on the logarithmic version of eq. 10,

$$\ln[\bar{A}_e(\omega)] = \ln\left(\frac{R_\phi \mathcal{F}_s P_r m_o}{4\pi\rho\beta_w^3 R}\right) + 2\ln(\omega) - \ln[1+(\omega/\omega_c)^2] - \kappa\omega + \epsilon, \qquad \omega > 0 \tag{19}$$

in which ϵ is the multiplicative residual. It should be noted that a regression analysis of $\ln[\bar{A}_e(\omega)]$ against ω implies that (a) the equivalent-stationary FA is averaged by taking its median value in the frequency domain, and (b) only three coefficients will be obtained which comprise the first term on the right hand side of eq. 19, ω_c and κ.

For the first coefficient, m_o is considered most uncertain and through eq. 11, γ_M will be selected as a random variable. Similarly, using the regressed ω_c and through eq. 12, $\Delta\sigma$ is chosen as the second random variable. Based on a study by the authors which shows that the variations of κ along the regressed line on the plot of κ against R_{epi} are quite constant, eq. 13 may be modified to account for the variability of κ as follows:

$$\kappa = \kappa_R R_{epi} + \kappa_c + \kappa_o \tag{20}$$

where κ_o is introduced as the third random variable with zero mean; κ_R and κ_c are the regression constants.

The remaining parameters of the geophysical model are treated as deterministic constants, and for the earthquake region under consideration in this study, take the following values [6]: ρ = 2700 kg/m^3, β_w = 3200 m/sec, $R_\phi = 0.63$, $P_r = 0.71$ and $\mathcal{F}_s = 2.0$.

The modulating function, $\psi(t)$, is defined if T_o and ε are known. The uncertainty in T_o may be accounted for by introducing a zero-mean random variable, τ_o, in eq. 18 as

$$T_o = \tau_M M_L + \tau_R R_{epi} + \tau_c + \tau_o \tag{21}$$

where τ_M, τ_R and τ_c are determined from regression analysis. The rise time fraction, ε, is modeled as a uniformly distributed random variable in the range (0,0.317).

The vector of basic random variables, **X**, defining the randomness of the complete oscillator system is given thus by

$$\mathbf{X} = \langle \gamma_M,\ \Delta\sigma,\ \kappa_o,\ \tau_o,\ \varepsilon,\ T_n,\ \xi \rangle^T \tag{22}$$

where T_n and ξ are the natural period and damping coefficient of the linear oscillator.

4.2 Statistics of Random Variables

For the purpose of this study, 54 earthquake accelerograms (all of which are in the western region of USA) are selected where $4.5 < M_L < 7.2$, $7\text{x}10^3\text{m} < R_{epi} < 135\text{x}10^3\text{m}$ and the recording sites are on rock or soft rock. The digitized ground accelerations are extracted from Caltech California corrected accelerograms (Vol II, 1933-1971), resulting in a data base of 54 acceleration records from 8 earthquakes. The analysis is carried out using the IMSL routine RNLIN [11] which minimizes the summation of the square of residuals. This results in a set of 54 coefficients for each of the parameters γ_M, $\Delta\sigma$ and κ. The values of κ (in sec/rad) are first regressed against R_{epi} (in m), giving the relation:

$$\kappa = 5.007\text{x}10^{-8} R_{epi} + 0.0293 + \kappa_o \tag{23}$$

for $7\text{x}10^3\text{m} < R_{epi} < 135\text{x}10^3\text{m}$, from which a sample set of 54 values of the residuals, κ_o, may be obtained. Similarly, the strong motion duration, T_o (in sec), for each record can be computed in the time domain and linear-regressed against M_L and R_{epi} (in m) to yield

$$T_o = 6.3 M_L + 0.118\text{x}10^{-3} R_{epi} - 31.06 + \tau_o \tag{24}$$

and a corresponding set of τ_o. The probability distributions of γ_M, $\Delta\sigma$, κ_o

and τ_o are selected based on their probability plots, where the empirical distributions are chosen based on their correlation coefficients. They all satisfy the Kolmogorov-Smirnov goodness-of-fit test at a significance level of 0.10. The results are summarized in Table 1.

Table 1 - Sample Statistics of **X**

i	x_i	$\bar{x}_i$	δ_{x_i}	$s(x_i)$	Distribution	ρ
1	γ_M	0.969	0.08	0.076	Lognormal	0.990
2	$\Delta\sigma$	$1.413\text{x}10^8 (N/m^2)$	1.89	$2.676\text{x}10^8$	Lognormal†	0.988
3	κ_o	0.0 (sec/rad)		0.0065	Normal	0.984
4	τ_o	0.0 (sec)		4.31	Normal	0.973
5	ε	0.159	0.58	0.092	Uniform	
6	T_n		0.10		Lognormal	
7	ξ		0.65		Lognormal	

$\bar{x}_i$ = sample mean of x_i

δ_{x_i} = sample coefficient of variation of x_i

$s(x_i)$ = sample standard deviation x_i

ρ = correlation coefficient on probability plot

†x_{mode} = $1.44\text{x}10^7$ Nm (144 bar)

Following O'Connor and Ellingwood [12], T_n and ξ are taken to be lognormally distributed and the coefficients of variation (c.o.v.) of 0.10 and 0.65 are adopted based on a study by Sues, Wen and Ang [13] for steel structures in which the stiffness and mass are assumed as statistically independent.

5. Reliability of Linear Oscillator

For a linear oscillator having a mean natural period, μ_{T_n}, and mean damping ratio, μ_ξ, subjected to a uniformly modulated ground acceleration, prescribed in terms of M_L and R_{epi}, the relative displacement, $y(t)$, is a non-stationary random process with evolutionary PSD. If the failure criterion is such that the oscillator is deemed to have failed when the absolute relative displacement exceeds some critical level, η, during an observation period $(0,T_{obs})$, then the probability of failure conditioned on a specific set of input model and structural parameters, $\mathbf{X} = \mathbf{x}$, is given by

$$P_f(\eta|\mathbf{X}=\mathbf{x}) = 1 - F_{S|\mathbf{X}}(\eta|\mathbf{X}=\mathbf{x}) \tag{25}$$

where S is the maximum peak of the absolute response, and the conditional cumulative distribution of S is approximated as [1,2]

$$F_{S|\mathbf{X}}(\eta|\mathbf{X}=\mathbf{x}) \simeq \exp\left[-\sum_{i=1}^{N} h(i,\eta)\Big|_{\mathbf{X}=\mathbf{x}}\right] \tag{26}$$

N is the number of half-cycles during $(0,T_{obs})$, and for a lightly damped system can be approximated by $N \simeq 2T_{obs}/T_n$; $h(i,\eta)$ is the probability that $|y(t)|$ will exit the barrier, $|y(t)| = \eta$, in the ith half-cycle given no prior crossing. By considering the peaks as constituting a Markovian point process [1], $h(i,\eta)$ can be written as

$$h(1,\eta) \simeq \bar{f}_1(1,\eta) \tag{27}$$

$$h(i,\eta) \simeq \bar{f}_2^*(i-1,i;\eta)/[\ 1 - \bar{f}_1(i-1,\eta)\ ], \qquad i\geq 2 \tag{28}$$

where $\bar{f}_1(i,\eta)$ and $\bar{f}_2^*(i,j;\eta)$ are approximated by those of the nonstationary envelope process, $V(t)$, by

$$\bar{f}_1(i,\eta) \simeq \int_\eta^\infty f_V(v;iT_n/2)\mathrm{d}v \tag{29}$$

$$\bar{f}_2^*(i,j;\eta) \simeq \int_0^\eta\int_\eta^\infty f_{VV}(v_i,v_j;\ iT_n/2,\ jT_n/2)\mathrm{d}v_j\mathrm{d}v_i \tag{30}$$

in which $f_V(v,t)$ and $f_{VV}(v_1,v_2;t_1,t_2)$ are given in eq. 7. $\bar{f}_1(i,\eta)$ and $\bar{f}_2^*(i,j;\eta)$ have been solved analytically for the case of Gaussian response process [1]. The overall probability of failure, treating $\mathbf{X}$ as a vector of random variables, is given by

$$P_f(\eta) = \int_{\mathbf{X}} P_f(\eta|\mathbf{X}=\mathbf{x})\ f_{\mathbf{X}}(\mathbf{x})\mathrm{d}\mathbf{x} \tag{31}$$

5.1 Solution by Method of Moments

Based on the second-moment method, the expected value of S, $E(S)$, and the variance of S, $Var(S)$, can be written in terms of the first two moments of $\mathbf{X}$ through first-order Taylor series expansion about the mean vector, $\mu_{\mathbf{X}}$, giving [3]

$$E(S) \simeq E(S|\mathbf{X}=\mu_{\mathbf{X}}) \tag{32}$$

$$\mathrm{Var}(S) \simeq \mathrm{Var}(S|\mathbf{X}=\mu_{\mathbf{X}}) + \sum_{i=1}^{n}\sum_{j=1}^{n} \nabla E(S|\mathbf{X}),_{x_i}\Big|_{\mathbf{X}=\mu_{\mathbf{X}}} . \nabla E(S|\mathbf{X}),_{x_j}\Big|_{\mathbf{X}=\mu_{\mathbf{X}}} . \mathrm{Cov}(x_i,x_j) \quad (33)$$

in which n is the size of vector $\mathbf{X}$, $\nabla E(S|\mathbf{X}),_{x_i}$ the partial derivative of $E(S|\mathbf{X})$ with respect to x_i and $\mathrm{Cov}(x_i,x_j)$ the covariance of (x_i,x_j). The conditional mean and variance of S in eqs. 32 and 33 can be evaluated using its conditional distribution given in eq. 26 as follows:

$$E(S|\mathbf{X}=\mu_{\mathbf{X}}) = \lim_{\eta_o\to\infty} \int_0^{\eta_o} [1-F_{S|\mathbf{X}}(\eta|\mathbf{X}=\mu_{\mathbf{X}})]d\eta \quad (34)$$

$$\mathrm{Var}(S|\mathbf{X}=\mu_{\mathbf{X}}) = \lim_{\eta_o\to\infty} 2\int_0^{\eta_o} \eta\, [1-F_{S|\mathbf{X}}(\eta|\mathbf{X}=\mu_{\mathbf{X}})]d\eta - [E(S|\mathbf{X}=\mu_{\mathbf{X}})]^2 \quad (35)$$

To estimate the failure probability of the system using the results of eqs. 32 and 33, the distribution of S must be assumed, and the problem of tail sensitivity will be encountered if the probability is small.

5.2 Solution by AFOSM Method

An alternative solution is by the AFOSM method [4]. For the case of barrier crossing at a constant level η considered, the limit-state function can be written as

$$G(\mathbf{X},S) = \eta - S(\mathbf{X}) \quad (36)$$

where the failure domain is defined by the region $\{ G(\mathbf{X},S) < 0 \}$ and S is included as one of the basic variables. The basic variables can be transformed into a set of n+1 independent standard normal variates, $\mathbf{U}$, giving

$$G(\mathbf{X},S) = G(\mathbf{T}^{-1}(\mathbf{U})) = g(\mathbf{U}) \quad (37)$$

where $\mathbf{T}$: $x_1 \longrightarrow u_1 = \Phi^{-1}[F_{X_1}(x_1)]$

$$x_i \longrightarrow u_i = \Phi^{-1}[F_{X_i|X_1,\ldots,X_{i-1}}(x_i|x_1,\ldots,x_{i-1})], \qquad i = 2 \text{ to } n$$

$$s \longrightarrow u_s = \Phi^{-1}[F_{S|X}(s|\mathbf{x})] \quad (38)$$

in which Φ is the standard normal distribution and $F_{X_i|X_1..X_{i-1}}(x_i|x_1..x_{i-1})$ the conditional distribution of X_i. The shortest distance from the origin to the failure surface in the transformed space is given by

$$\beta = [g(\mathbf{u}^*) - \nabla g^T(\mathbf{u}^*).\mathbf{u}^*]/|\nabla g(\mathbf{u}^*)| \quad (39)$$

where $\mathbf{u}^*$ is the corresponding point on the failure surface, generally known as the design point. $\nabla g(\mathbf{U})$ is a vector of the derivatives of $g(\mathbf{U})$ with respect to $\mathbf{U}$ and is related to the vector of the derivatives of $G(\mathbf{X},S)$ with respect to $\mathbf{X}$ and S, $\nabla G(\mathbf{X},S)$, by

$$\nabla g(\mathbf{U}) = (\mathbf{J}^{-1})^T \nabla G(\mathbf{X},S) \tag{40}$$

where $\nabla G(\mathbf{X},S) = \langle \partial G/\partial x_1, \ldots, \partial G/\partial s\rangle^T = \langle 0,0,0,\ldots,-1\rangle^T$ and $\mathbf{J}$ is the $(n+1)\mathrm{x}(n+1)$ Jacobian matrix with elements

$$J_{ij} = \partial u_i/\partial x_j = \begin{cases} 0 & , \; j > i \\ \nabla F_{X_i|X_1\ldots X_{i-1}}(x_i|x_1\ldots x_{i-1}),_{x_j}/\phi(u_i) & , \; j \le i \end{cases}$$

$$\text{for } i = 1 \text{ to } n, \; j = 1 \text{ to } n+1$$

$$J_{n+1,j} = \partial u_s/\partial x_j = \nabla F_{S|\mathbf{X}}(s|\mathbf{x}),_{x_j}/\phi(u_s) \qquad \text{for } j = 1 \text{ to } n$$

$$J_{n+1,n+1} = \partial u_s/\partial s = f_{S|\mathbf{X}}(s|\mathbf{x})/\phi(u_s) \tag{41}$$

where $\nabla F(.),_{x_j}$ is the derivative of $F(.)$ with respect to x_j, $f_{S|\mathbf{X}}(s|\mathbf{x})$ the conditional PDF of S and ϕ the standard normal PDF. The probability of failure of the system can be approximated as

$$P_f(\eta) \simeq \Phi(-\beta) \tag{42}$$

The effect of changes in the random variables on the reliability at the design point is given by the vector of sensitivity factors

$$\alpha(\mathbf{u}^*) = \nabla\beta,_{\mathbf{u}^*} = -\nabla g(\mathbf{u}^*)/|\nabla g(\mathbf{u}^*)| \tag{43}$$

where $\nabla\beta,_{\mathbf{u}^*}$ denotes the vector of the derivatives of β with respect to $\mathbf{u}^*$. The corresponding vector of sensitivity factors with respect to $\mathbf{u}^*$ for the failure probability is

$$\nabla P_f,_{\mathbf{u}^*} = -\phi(\beta)\nabla\beta,_{\mathbf{u}^*} \tag{44}$$

A slightly different way of approaching the solution within the AFOSM method is to transform eq. 36 at the limit state, $G(\mathbf{X},S) = 0$, and using eq. 25 to lead to another limit-state function

$$\tilde{g}(\mathbf{U}) = -u_s - \Phi^{-1}\{P_f[\eta|\mathbf{T}^{-1}(\mathbf{u})]\} \tag{45}$$

where u_s is defined in eq. 38 and $P_f[\eta|\mathbf{T}^{-1}(\mathbf{u})]$ is the conditional probability of failure given the system parameters $\mathbf{X} = \mathbf{x} = \mathbf{T}^{-1}(\mathbf{u})$. Equation 45 is somewhat similar to the limit state proposed by Wen and Chen [14], with their $u_{n+1} = -u_s$. To evaluate β, the vector of the derivatives of $\tilde{g}(\mathbf{U})$ with respect to $\mathbf{U}$, $\nabla\tilde{g}(\mathbf{U})$, is needed and the vector of the first n elements is related to the vector of the derivatives of $P_f(\eta|\mathbf{x})$ with respect to $\mathbf{X}$, $\nabla P_f(\eta|\mathbf{x})$, by

$$\nabla\tilde{g}(\mathbf{U}) = -(\mathbf{J}^{-1})^T \nabla P_f(\eta|\mathbf{x})/\phi(-u_s) \tag{46}$$

and the $(n+1)$th element is $\partial\tilde{g}/\partial u_s = -1$. $\mathbf{J}$ is the nxn Jacobian matrix which is a subset of that given in eq. 41.

It can be shown that $\tilde{g}(\mathbf{U})$ of eq. 45 is related to $G(\mathbf{X},S)$ of eq. 36 by

$$\tilde{g}(\mathbf{U}) = \Phi^{-1}\left\{ F_{S|\mathbf{X}}\left[\left(G(\mathbf{X},S) + s\right)|\mathbf{X}=\mathbf{x}\right] \right\} - u_s \tag{47}$$

Taking the partial derivative of $\tilde{g}(\mathbf{U})$ in eq. 46 with respect to u_i, $i = 1$ to n,at the limit-state $G(\mathbf{X},S) = 0$ gives

$$\nabla\tilde{g},_{u_i}\Big|_{G=0} = (\partial u_s/\partial s)\ \nabla G,_{u_i}\Big|_{G=0} = \left[f_{S|\mathbf{X}}(s|\mathbf{x})/\phi(u_s)\right]\ \nabla G,_{u_i}\Big|_{G=0} \tag{48}$$

where $\nabla G,_{u_i}$ denotes the derivative of $G(\mathbf{X},S)$ with respect to u_i. From eq. 48, it is obvious that as long as the limit-state condition is satisfied for each iteration, the value of β calculated using eq. 39 will be identical for both limit-states $\tilde{g}(\mathbf{U})$ and $G(\mathbf{X},S)$.

The only advantage of using $\tilde{g}(\mathbf{U})$ instead of $G(\mathbf{X},S)$ is that $f_{S|\mathbf{X}}(s|\mathbf{x})$ needs not be evaluated during the construction of the Jacobian matrix, $\mathbf{J}$, as only nxn elements of $\mathbf{J}$ are needed (see eq. 41). The computational efforts for $\nabla F_{S|\mathbf{X}}(s|\mathbf{x}),_{x_j}$, $j = 1$ to n, needed in $\mathbf{J}$ when $G(\mathbf{X},S)$ is used, and $\nabla P_f(\eta|\mathbf{x}),_{x_j}$, in evaluating $\nabla\tilde{g}(\mathbf{U}),_{u_j}$ when $\tilde{g}(\mathbf{U})$ is used, are the same.

However, a direct result of using $\tilde{g}(\mathbf{U})$ is that since only nxn elements of $\mathbf{J}$ are evaluated, u_s must always be forced to take the value of $-\Phi^{-1}[P_f(\eta|\mathbf{x})]$. Experience shows that for a high barrier level η, an initial set of prescribed $\mathbf{X}$, e.g. mean of $\mathbf{X}$, can lead to a breakdown in the computation because $P_f(\eta|\mathbf{x})$ approaches zero. Thus, the initial values of $\mathbf{X}$ must be assumed based on judgment, or a physical feeling of the problem or obtained by trial and error before iterations can proceed. In contrast to this by using $G(\mathbf{X},S)$, an initial value of S, which is less than η, can be assigned, together with an initial set of $\mathbf{X}$, e.g. $\mu_{\mathbf{X}}$. After the first iteration, a new set of both $\mathbf{X}$ and S will be generated.

6. Example and Discussions

To illustrate the effect of the uncertainties in the structural and input model parameters on the response of a linear oscillator and to compare the reliability estimates using the method of moments and the AFOSM method, an oscillator with $\mu_{T_n} = 1$ sec. and $\mu_{\xi} = 0.05$ is assumed to be subjected to a seismic excitation of $M_L = 7.0$ and $R_{epi} = 10$ km. The set of random variables and their corresponding marginal distributions listed in Table 1 are adopted. The variables are assumed to be statistically independent. The observation period is taken to be $(0,\infty)$, and the constant barrier levels, η's, are prescribed as multiples of a reference response magnitude, σ_{st}, where

$$\sigma_{st}^2 = S_o \mu_{T_n}^3 / (16\pi^2 \mu_{\xi}) \tag{49}$$

$$S_o = \bar{A}_e^2 (2\pi/\mu_{T_n}) / (2\pi\mu_{T_o}) \tag{50}$$

σ_{st} is the (stationary) standard deviation of the relative displacement of the oscillator for a white noise excitation of intensity S_o, with the parameters taking on their mean values.

Three sets of probability values are obtained for different barrier level η's. The first set of results are obtained by assuming that $\mathbf{X}$ is deterministic with its value fixed at $\mu_{\mathbf{X}}$. The second set of results are obtained using the method of moments and the probabilities of failure for different η's assuming distribution of S to be lognormal, extreme Type I and normal are computed. The third set of results are obtained using the AFOSM method with $G(\mathbf{X},S)$ as the limit-state function. Using the limit-state function $\tilde{g}(\mathbf{U})$ instead of $G(\mathbf{X},S)$ with the appropriate initial values of $\mathbf{X}$ yields the same results. The three sets of results are summarized in Table 2.

The results clearly illustrate that the variability of $\mathbf{X}$ does contribute significantly towards the probability of failure. From the different distributions of S considered in the second moment method for this example, the lognormal distribution underestimates the probabilities of failure for

Table 2 - Probabilities of Failure
(σ_{st}=0.154m, M_L=7, R_{epi}=10km, μ_{T_n}=1sec, μ_ξ=0.05)

$\frac{\eta}{\sigma_{st}}$	η (m)	$P_f(\eta\|X=\mu_x)$	$P_f(\eta)$ (Method of Moments)			$P_f(\eta)$
			lognormal	extr I	normal	(AFOSM)
3	0.462	0.1898	0.2357	0.3652	0.4342	0.2659
4	0.616	0.0196	0.1595	0.2680	0.3231	0.1960
5	0.770	6.93×10^{-4}	0.1130	0.1928	0.2259	0.1436
6	0.924	1.78×10^{-6}	0.0829	0.1368	0.1479	0.1035
7	1.078	1.25×10^{-13}	0.0625	0.0961	0.0903	0.0723
8	1.232	0	0.0483	0.0670	0.0513	0.0483
9	1.386	0	0.0379	0.0465	0.0271	0.0300
10	1.540	0	0.0303	0.0322	0.0132	0.0165

the lower barrier portion and overestimates the higher barrier portion, while the extreme type I and normal distributions generally overestimates the probabilities of failure over the whole range of barriers considered. This illustrates the tail sensitivity problem, and the fact that all the results from the three distributions fail to match that from the AFOSM method indicates that the AFOSM method should be used when the distributions of the basic random variables can be reasonably approximated. The CPU time for each η using the AFOSM method is about 50 seconds on the IBM 3081 which is about four times of that of the mean value estimate, and about three times of that using the method of moments. The number of iterations needed for the convergence of β to within a tolerance of 0.001 seldom exceeds five.

The results of the sensitivity analysis at a barrier level of η/σ_{st} = 7 are tabulated in Table 3. It can be seen that for the system considered, changes in $\Delta\sigma$ and γ_M affect the probability of failure most, while changes in κ_o, τ_o, ε and the structural parameters have negligible effect implying that the latter parameters may be treated as deterministic. The high gradient of β with respect to $\Delta\sigma$ is due to the high variability of $\Delta\sigma$ (c.o.v. = 1.89) while that with respect to γ_M is due to the exponential relation of γ_M with $\bar{A}_e(\omega)$. As an illustration of the significance of the

Table 3 - Failure Point and Sensitivity Factors

(M_L=7, R_{epi}=10km, μ_{T_n}=1sec, μ_ξ=0.05, η/σ_{st}=7.0)

i	x_i	initial assumed x_i	u_i^*	x_i^*	$\nabla\beta,_{u_i^*}$	$\nabla P_f(\eta),_{u_i^*}$
1	γ_M	0.969	0.956	1.037	0.655	-0.0902
2	$\Delta\sigma$	1.413×10^8	1.002	1.920×10^8	0.687	-0.0946
3	κ_o	0.0	-0.051	-0.0003	-0.035	0.0048
4	τ_o	0.0	-0.081	-0.347	-0.055	0.0076
5	ε	0.159	-0.019	0.156	-0.013	0.0018
6	T_n	1.00	0.162	1.011	0.111	-0.0153
7	ξ	0.05	-0.226	0.033	-0.155	0.0213
8	S	0.50	0.352	1.078	0.241	-0.0332

variability of $\Delta\sigma$, it is noted that the probability of failure at the barrier level of 0.616 m reduces significantly from 0.196 to 0.0297 if the c.o.v. of $\Delta\sigma$ is reduced by 50% while maintaining the same modal value, and using the statistics of the remaining variables in Table 1.

7. Conclusions

It is shown in this paper that the variabilities of the input model parameters should be accounted in the estimation of the reliability of a linear oscillator. The variabilities can be incorporated using either the AFOSM method or the method of moments based on conditional probability concepts. The method of moments is simple but the estimated probability is sensitive to the assumed distribution of the maximum peak of the response. The AFOSM method, which overcomes the tail sensitivity problem, should therefore be used when the probability distributions of the basic variables are known or can be reasonably estimated. It is found through a numerical example that the variabilities of the structural parameters have a much less significant effect on the reliability than that of the geophysical parameters. Among the geophysical model parameters considered, it is shown that changes in $\Delta\sigma$ and γ_M affect the oscillator reliability most indicating the need to obtain better estimates for these parameters.

References

1. Yang, J.N.: First-excursion probability in nonstationary random vibration. J. Sound & Vib. 27(2) (1973) 165-182.

2. Quek, S.T.; Teo, Y.P.; Balendra, T.: Nonstationary structural response with evolutionary spectra using seismological input model. Earthq. Eng. & Struc. Dyn. 19 (1990) 275-288.

3. Ang, A.H-S.; Tang, W.H.: Probability concepts in engineering planning and design - volume 1. John Wiley and Sons 1975.

4. Hohenbichler, M.; Rackwitz, R.: Non-normal dependent vectors in structural safety. J. Eng. Mech. Div. ASCE 107(EM6) (1981) 1227-1238.

5. Vanmarcke, E.H.; Lai, S-S.P.: Strong-motion duration and rms amplitude of earthquake records. Bull. Seism. Soc. Amer. 70 (1980) 1293-1307.

6. Boore, D.M.: Stochastic simulation of high-frequency ground motions based on seismological models of the radiated spectra. Bull. Seism. Soc. Amer. 73 (1983) 1865-1894.

7. Anderson, J.G.; Hough, S.E.: A model for the shape of the Fourier amplitude spectrum of acceleration at high frequency. Bull. Seism. Soc. Amer. 74 (1984) 1969-1994.

8. Shinozuka, M.; Sato, Y.: Simulation of nonstationary random processes. J. Eng. Mech. Div. ASCE 93(EM1) (1967) 11-40.

9. Trifunac, M.D.; Brady, A.G.: A study on the duration of strong earthquake ground motion. Bull. Seism. Soc. Amer. 65 (1975) 581-626.

10. McGuire, R.K.: A simple model for estimating Fourier amplitude spectra of horizontal ground acceleration. Bull. Seism. Soc. Amer. 68 (1978) 803-822.

11. IMSL User's Manual, STAT/LIBRARY **1**, Houston, USA.

12. O'Connor, J.M.; Ellingwood, B.: Reliability of nonlinear structures with seismic loading. J. Struc. Eng. ASCE 113(5) (1987) 1011-1028.

13. Sues, R.H.; Wen, Y.K.; Ang A.H-S: Stochastic seismic performance evaluation of buildings. Struc. Res. Ser. 506, University of Illinois, Urbana 1983.

14. Wen, Y.K.; Chen, H.C.: On fast integration for time variant structural reliability. Prob. Eng. Mech. 2(3) (1987) 156-162.

A Stochastic Finite Element Analysis for the Seismic Response of Soil Sites

M.S. Rahman, C.W. Hwang

Department of Civil Engineering
North Carolina State University
Box 7908, Raleigh, NC 27695-7908

Summary

In this study, a stochastic finite element method is developed for the analysis of seismic response of soil sites. The earthquake-induced ground motion is treated as a random process. A finite element discretization is used to model the soil mass. The dynamic response of the system to the random ground excitation is formulated in the frequency domain with transfer functions for various parameters, viz. displacement, acceleration, strain, and stress. The spectral density functions of these response variables are then readily obtained, from which the calculation of root-mean-squared and the most probable extreme values of the response are made. The nonlinearities of the shear modulus and damping ratio are incorporated by using an equivalent linear model for stress-strain behavior of soils and an iterative solution of the response. The spatial variability of the shear modulus is described through a random field model. The dynamic response of soil sites (uncertain system) to earthquake-induced ground motion (stochastic excitation) is evaluated through a first-order perturbation method and a direct Monte-Carlo simulation technique. The results from these two approaches of stochastic finite element analysis are compared with respect to accuracy and computational efficiency. The results from a case study and some parametric studies are also presented.

Introduction

The evaluation of seismic response of soil sites constitutes an important problem with respect to both ground motion amplification and soil instability due to liquefaction. The ground motion generated by an earthquake is a random process. Details of input motions have significant effects on the response of soil sites, which has been reported by different researchers [1,2]. Also, there always exist uncertainties in defining the properties of a soil site. These result from: (i) natural heterogeneity or the in-situ variability of the soil, (ii) limited availability of information about

I. Elishakoff · Y. K. Lin (Eds.)
Stochastic Structural Dynamics 2
New Practical Applications
2nd International Conference on Stochastic Structural Dynamics
May 9-11, 1990, Boca Raton, Florida

subsurface conditions, and (iii) measurement errors. These uncertainties associated with system parameters are also likely to have some influence on the response [3,4]. Therefore the seismic response of soil sites should be studied in a general framework of the response of uncertain systems to stochastic excitation.

The recent development of stochastic finite element analysis provides a rational framework for the analysis of complex uncertain systems subjected to stochastic excitation. Many researchers [5,6,7,8,9,10,11,12,13] have contributed to the development of this field. Stochastic finite element analysis have also been applied to some geotechnical problems [14,15].

In this study, an application of stochastic finite element method is developed for the analysis of seismic response of soil sites. The seismically induced ground acceleration is treated as a random process. A finite element formulation is used to discretize the soil mass, and its dynamic response to the random ground excitation is formulated in the frequency domain with transfer functions for various parameters, viz. displacement, acceleration strain, and stress. The spectral density functions of these response variables are then readily obtained, from which the calculations for the root-mean-squared and the most probable extreme values are made. The nonlinearities of the shear modulus and damping ratio are incorporated by using an equivalent linear model for stress-strain behavior of soils and an iterative solution of the response. The shear modulus, considered as a random field, is also discretized through the finite element mesh. Thus the random input (for shear modulus) consists of a vector of random variables (i.e. the local averages across each element) whose covariance matrix depends on the finite element mesh. The random dynamic response is finally evaluated through a first order perturbation method and a direct Monte-Carlo simulation technique. The results from these two approaches of stochastic finite element analysis are compared with respect to accuracy and computational efficiency. The results from a case-study and some parametric studies are also presented.

Random Ground Motion

In this study, for simplicity and ease of analysis, the ground motion is considered to be a stationary random process (with constant frequency content) truncated for a finite duration. An approach suggested by Vanmarcke and Lai [16] is used, in which a chosen time history of recorded ground motion is considered as one sample of an entire ensemble representing the random process. The following power spectral density function suggested by Kanai and Tajimi is used to characterize a strong earthquake ground motion,

$$S_{\ddot{u}_g} = \frac{S_0(1+4\zeta_g^2(\omega/\omega_g)^2)}{[1-(\omega/\omega_g)^2]^2+4\zeta_g^2(\omega/\omega_g)^2} \quad . \tag{1}$$

In the above, ω is component frequency, $\mathbf{S_o}$ is the scale factor, ω_g and ζ_g are the equivalent natural frequency and damping ratio of the ground characterized by a single degree of freedom system. The three parameters **So**, ω_g and ζ_g defining the power spectrum are estimated from a chosen sample time history [17].

Two statistical parameters of the given spectral density function are of special interest, which are defined in terms of the following spectral moments [18],

$$\lambda_i = \int \omega^i S_{\ddot{u}_g}(\omega) d\omega \quad i = 0,1,2,\ldots \tag{2}$$

The RMS of the ground acceleration $\sigma_{\ddot{u}_g}$ and the central frequency Ω are given as:

$$\sigma_{\ddot{u}_g} = \sqrt{\lambda_0}$$
$$\Omega = \sqrt{\lambda_2/\lambda_1} \; . \tag{3}$$

The ground motion, thus characterized, represents a stationary random motion obtained by filtering a white noise. In order to approximate the nonstationarity of the actual ground motion, only a segment of the stationary motion, defined as equivalent stationary duration is used. The equivalent stationary duration is obtained by the following considerations: (i) the intensity of the original sample motion is preserved in the equivalent stationary motion, (ii) the peak ground acceleration $\ddot{u}_{g\ max}$ of the sample motion is expected to be exceeded once in the equivalent stationary transformed motion and (iii) the equivalent motion has the same central frequency Ω as the sample motion. These considerations lead to the following expression for the equivalent stationary duration, **T**

$$T = \frac{\pi}{\Omega} \exp\{\ddot{u}_{g\,max}^2 / (2\sigma_{\ddot{u}_g}^2)\}. \qquad (4)$$

In summary, the model for random ground motion is defined by (a) Kanai-Tajimi Spectrum with parameters S_o, ω_g and ζ_g and (b) the equivalent stationary duration, **T**.

Random Field Model For Soil Properties

In order to describe the variability of soil properties a random field model [19] is used. In this approach a homogeneous random field **u(z)** is characterized (Fig. 1) by these parameters: (1) the average value $\overline{u}$, (ii) the standard deviation or root-mean-square (rms) fluctuation of u(z), which measures the intensity of fluctuation or degree to which actual values of **u(z)** may deviate from $\overline{u}$ and (iii) the scale of fluctuation, δ_u, which measures the distance within which the soil property **u(z)**, shows relatively strong correlation or persistence from point to point. Small values of δ_u imply rapid fluctuation about the average, while large values δ_u suggest that a slowly varying component is superimposed on the average value of **u**.

The variation of soil property **u(z)** can be represented by a constant mean trend $\overline{u}$**(z)** and a zero-mean random field α**(z)** as:

$$u(z) = \overline{u}(z) + \alpha(z), \qquad (5)$$

Using the first- and second second-order statistical moments the variation of **u(z)** is described by

$$E[u(z)] = \bar{u}(z),$$

$$Var[u(z)] = Var[\alpha(z)] = \sigma_\alpha^2 = constant \qquad (6)$$

$$Cov[u(z), u(z+\xi)] = R_{\alpha\alpha}(\xi) = \sigma_\alpha^2 \rho_\alpha(\xi)$$

In the above **E[.]**, **Var[.]** and **Cov[.]** are the expected value, variance and covariance respectively; $\rho_\alpha(\xi)$ is the coefficient of correlation of α. For very small interval ξ, the coefficient of correlation will be close to 1 but will decay as the interval increases. Many specific analytical expressions have been proposed [19] for the autocorrelation function, $R_{\alpha\alpha}(\xi)$.

In the stochastic finite element analysis, an average soil property is assumed over an element. A one dimensional random field u(z) with mean $\bar{u}(z)$ and variance σ_α, its local average process $u_Z(z)$ of u(z) over the interval Z centered at z (Fig. 1) is defined as:

$$u_Z(z) = \frac{1}{Z}\int_{z-Z/2}^{z+Z/2} u(z')dz', \qquad (7)$$

The above local average will vary depending upon the specific location of the interval z within the "statistically homogeneous" soil layer. Simplifying the notation $u_Z(z)$ as u_Z, mean and variance of this average process are give by:

$$E[u_Z] = E[u(z)] = \bar{u}(z),$$

$$\qquad (8)$$

$$Var[u_Z] = Var[\alpha_Z] = \sigma_\alpha^2\gamma(Z).$$

where $\gamma(Z)$ is the variance reduction function of $\alpha(z)$, which measures the reduction of the point variance σ_α^2 under local averaging. The larger the spatial interval Z over which the $\alpha(z)$ is averaged, the more of fluctuation of $\alpha(z)$ tend to cancel in the process of spatial averaging. This tends to cause a reduction in variance as the size of the averaging domain increases; $\gamma(Z)$ is dimensionless function with the following properties:

$$\gamma(Z) \geq 0, \qquad \gamma(0) = 1, \qquad \gamma(-Z) = \gamma(|Z|) = \gamma(Z). \tag{9}$$

The variance function $\gamma(Z)$ is related to the correlation function, ρ_α as:

$$\gamma(Z) = \frac{1}{Z^2}\int_0^Z\int_0^Z \rho_\alpha(z_1 - z_2)dz_1dz_2 = \frac{1}{Z}\int_0^Z (1 - \frac{\zeta}{Z})\rho_\alpha(\zeta)d\zeta. \tag{10}$$

For simplification, Vanmarcke [19] proposed the following relations:

$$\left.\begin{aligned} \gamma(Z) &= 1 && \text{for} \quad Z \leq \delta \\ \gamma(Z) &= \frac{\delta}{Z} && \text{for} \quad Z \geq \delta \end{aligned}\right\}, \tag{11}$$

The covariance of α_{Z1} and α_{Z2} of the process of αZ, corresponding to two segments ΔZ_1, and ΔZ_2 (Fig. 1) is expressed as:

$$\text{Cov}[\alpha_{Z1}, \alpha_{Z2}] = \sigma_\alpha^2 B(Z_1, Z_2), \tag{12}$$

where,

$$B(Z_1, Z_2) = \frac{[Z_0^2\gamma(Z_0) - Z_{01}^2\gamma(Z_{01}) - Z_{02}^2\gamma(Z_{02}) + Z_{012}^2\gamma(Z_{012})]}{2 Z_1 Z_2}, \tag{13}$$

in which, $z_1 = \Delta z_1$, $z_2 = \Delta z_2$, $z_0 = \Delta z_0$, $z_{01} = \Delta z_1 + \Delta z_o$, $z_{02} = \Delta z_2 + \Delta z_0$, $z_{012} = \Delta z_1 + \Delta z_0 + \Delta z_2$. From the above equations the matrix of covariance of pairs of spatial average of the random field $\mathbf{u(z)}$ can be generated.

Digital Generation of Random Field

In order to perform the stochastic finite element analysis using simulation, the random field $\alpha(z)$ must be generated. For the zero-mean, one-dimensional, and univariate homogeneous random field with autocorrelation function $R_{\alpha_0\alpha_0}(\xi)$ the two sided spectral density function is

$$S_{\alpha_0\alpha_0}(\kappa) = \frac{1}{2\pi}\int_{-\infty}^{\infty} R_{\alpha_0\alpha_0}(\xi)\, e^{-i\kappa\xi} d\xi,$$

$$R_{\alpha_0\alpha_0}(\xi) = \int_{-\infty}^{\infty} S_{\alpha_0\alpha_0}(\kappa) e^{i\kappa\xi} d\kappa, \tag{14}$$

In the above both $R_{\alpha_0\alpha_0}(\xi)$ and $S_{\alpha_0\alpha_0}(\kappa)$ are even functions. According to Rice [20] the simulation $\alpha(z)$ of the stationary Gaussian field $\alpha_0(z)$ can be written as:

$$\alpha(z) = \sqrt{2}\sum_{j=1}^{L} c_j \cos(\kappa_j z + \phi_j), \tag{15}$$

where

$$c_j = \sqrt{2S_{\alpha_0\alpha_0}(\kappa_j)\Delta\kappa} = \sqrt{G_{\alpha_0\alpha_0}(\kappa_j)\Delta\kappa}, \tag{16}$$

and

$$\kappa_j = j\Delta k, \qquad j = 1, 2, ..., L.$$

The upper bound wave number is

$$\kappa_u = L\Delta\kappa, \tag{17}$$

which represents an upper cut-off frequency which $S_{\alpha_0\alpha_0}(\kappa)$, may be assumed to be zero. In above, $G_{\alpha_0\alpha_0}(\kappa) = 2S_{\alpha_0\alpha_0}(\kappa)$ represents the one-sided power spectral density function. Shinozuka [21] showed that if L is very large $E[\alpha(z)] = 0$ and when a upper cut-off frequency exists, $R_{\alpha\alpha}(\xi)$ and $S_{\alpha\alpha}(\kappa)$ are considered identical to $R_{\alpha_0\alpha_0}(\xi)$ and $S_{\alpha_0\alpha_0}(\kappa)$, respectively.

Seismic Response of Soil Sites: Governing Equations

The seismic response of a horizontally layered soil site is assumed to be caused primarily by the vertical propagation of shear waves from the underlying formation. Considering the soil to be a linear coulomb-damped elastic system, the response will be governed by the following wave equation:

$$\rho\frac{\partial^2 u}{\partial t^2} = G^*\frac{\partial^2 u}{\partial z^2} \tag{18}$$

with, $G^* = G(1+2i\beta)$.

In the above, **u[=u(z,t)]** is horizontal displacement, ρ is, the density of soil mass, $\mathbf{G}^*$ is the complex shear modulus, **G** is the shear modulus and β is damping ratio resulting from the frictional loss of energy. Here, **G** and β and hence $\mathbf{G}^*$ are assumed to be independent of frequency in accordance with the experimentally observed behavior of soils.

Stochastic Finite Element Formulation: Perturbation

Assuming the soil site to be consisting of various elements Ω^e for the discretized soil system subjected to a horizontal ground excitation, $\ddot{\mathbf{u}}$, the equation of motion can be written as:

$$\mathbf{M}\ddot{\mathbf{u}} + \mathbf{K}\mathbf{u} = -\mathbf{m}\ddot{\mathbf{u}}_g \tag{19}$$

where, $\ddot{\mathbf{u}}$ and **u** are acceleration and displacement vectors respectively; $\ddot{\mathbf{u}}_g$ is the ground acceleration and $\mathbf{m}(=M\delta)$ is mass vector; δ is a unit vector, and **M** and **K** are global mass and stiffness matrices, which are obtained by assembling the mass and stiffness matrices of **N** elements:

$$M = \sum_{e=1}^{N} m_e, \qquad K = \sum_{e=1}^{N} k_e, \tag{20}$$

Using only a linear approximation for the displacement, the element matrices can be obtained as:

$$k_e = \frac{1}{h_e}\begin{bmatrix} 1 & -1 \\ -1 & 1 \end{bmatrix} G_e^*$$
$$m_e = \frac{\rho_e h_e}{6}\begin{bmatrix} 2 & 1 \\ 1 & 2 \end{bmatrix}. \tag{21}$$

In the above $\mathbf{G_e^*}$ and ρ_e are the spatial averages of the complex modulus and density of the soil in element **e** with a length $\mathbf{h_e}$ (Fig. 2).

Element Stiffness Matrix

In this study, the spatial variation of the shear modulus is treated as a random field:

$$G(z) = G(z) + \alpha(x), \tag{22}$$

$\overline{G}$ is mean value and $\alpha(z)$ is zero mean real homogeneous random field, representing the deviation of the shear modulus from its mean. Accordingly, the complex shear modulus can be written as:

$$G(z) = G(z)(1 + 2i\beta) = \{G(z) + \alpha(x)\}(1 + 2i\beta) \tag{23}$$

Now, the spatial average G_e^* in eqn. (21) can be written as:

$$G_e^* = (\overline{G}_e + \alpha_e)(1 + 2i\beta), \tag{24}$$

where

$$\overline{G}_e = \frac{1}{h_e}\int_{Z_e}^{Z_{e+1}} \overline{G}(z)dz, \tag{25}$$

and α_e indicated the local average of $\alpha(z)$ for the element,

$$\alpha_e = \frac{1}{h_e}\int_{Z_e}^{Z_{e+1}} \alpha(z)\, dz. \tag{26}$$

Using Taylor's series expansion, element stiffness matrix $\mathbf{k_e}$ can be written as:

$$\mathbf{k_e} = \mathbf{k_e^o} + \mathbf{k_{ei}^I}\,\alpha_e + \frac{1}{2}\mathbf{k_{eee}^{II}}\,\alpha_e^2 + \ldots, \tag{27}$$

where

$$\mathbf{k_e^o} = \mathbf{k_e}\Big|_{\alpha_e=0} = \frac{\overline{G}_e}{h_e}\begin{bmatrix} 1 & -1 \\ -1 & 1 \end{bmatrix}(1 + 2i\beta),$$

$$\mathbf{k_{ei}^I} = \frac{\partial \mathbf{k_e}}{\partial \alpha_e}\Big|_{\alpha_e=0} = \frac{1}{h_e}\begin{bmatrix} 1 & -1 \\ -1 & 1 \end{bmatrix}(1 + 2i\beta), \tag{28}$$

Global Stiffness Matrix

When the first order perturbation is used, the global stiffness matrix can also be expressed in the following form with the assumption that each α_e is small ($\alpha_e << G_e$) and has zero mean.

$$K = K^o + \sum_{i=1}^{N} K_i^I \alpha_i + \dots . \tag{29}$$

The coefficient matrices K^O and K_i^I of the global stiffness are obtained by assembling k_e^O and k_{ei}^I over the entire region, with N being total number of elements,

$$K^o = \sum_{e=1}^{N} k_e^o = \sum_{e=1}^{N} k_e \Big|_{\alpha=0} ,$$
$$K_i^I = \sum_{e=1}^{N} k_{ei}^I = \sum_{e=1}^{N} \frac{\partial k_e}{\partial \alpha_i} \Big|_{\alpha=0} , \tag{30}$$

Since we consider the spatial variability of only one material parameter, the above summation have meaning only when e = i, because of the following relationships:

$$\frac{\partial k_e}{\partial \alpha_i} = 0 \qquad \text{if} \qquad e \neq i. \tag{31}$$

Transfer Function For Response

To solve the equations of motion, in the frequency domain, the modal displacements of the finite element model subjected to a unit amplitude ground acceleration $\ddot{u}_g$ (= exp(iωt) can be assumed as

$$u = H_d(i\omega)\, e^{i\omega t} . \tag{32}$$

in which $H_d(.)$ is the displacement transfer function vector. The acceleration response can therefore be written as:

$$\ddot{u} = -\omega^2 H_d(i\omega)\, e^{i\omega t} \tag{33}$$

Substituting eqns. (32) and (33) into eqn. (19), the equations of motion then can be written as:

$$(\mathbf{K} - \omega^2\mathbf{M})\mathbf{H}_d(i\omega) = -\mathbf{m} \tag{34}$$

This equation can also presented in a simple form as:

$$\mathbf{D}\mathbf{H}_d = -\mathbf{m}, \tag{35}$$

with

$$\mathbf{D} = \mathbf{K} - \omega^2\mathbf{M}; \tag{36}$$

where **D** is a dynamic system operator.
The **D** matrix in eqn. (35) also involves the non-dimensional random variables α_i ($i = 1,2,...,N$), which represent the spatial variation of the shear modulus. **D** can be expanded in the series form with the assumption each α_i is small ($\alpha_i \ll \overline{G}_i$) and has zero mean.

$$\mathbf{D} = \mathbf{D}^o + \sum_{i=1}^{N} \mathbf{D}_i^I \alpha_i + \dots \tag{37}$$

where $\mathbf{D}^o$ and $\mathbf{D}_i^I$ are defined in a way similar to eqn. (30).
Similarly, the unknown transfer function vector can also be expanded as:

$$\mathbf{H}_d = \mathbf{H}_d^o + \sum_{i=1}^{N} \mathbf{H}_{di}^I \alpha_i + \dots \tag{38}$$

Using the first order perturbation approximation,

$$\mathbf{H}_d^o = (\mathbf{D}^o)^{-1}(-\mathbf{m}),$$

$$\mathbf{H}_{di}^I = (\mathbf{D}^o)^{-1}(-\mathbf{D}_i^I \mathbf{H}_d^o), \tag{39}$$

The above is obtained by differentiating eqn. (35) with respect to α_i as:

$$\frac{\partial \mathbf{D}}{\partial \alpha_i}\mathbf{H}_d + \mathbf{D}\frac{\partial \mathbf{H}_d}{\partial \alpha_i} = 0, \tag{40}$$

and setting $\alpha = \mathbf{0}$.
The transfer function for absolute acceleration is

$$\mathbf{H}_a = -\omega^2 \mathbf{H}_d + \boldsymbol{\delta} = -\omega^2 \mathbf{H}_d^o + \boldsymbol{\delta} - \omega^2 \sum_{i=1}^{N} \mathbf{H}_{di}^I \alpha_i + \dots$$

$$= \mathbf{H}_a^o + \sum_{i=1}^{N} \mathbf{H}_{ai}^I \alpha_i + \dots . \tag{41}$$

where,

$$\mathbf{H}_a^o = -\omega^2 \mathbf{H}_d^o + \boldsymbol{\delta}; \qquad \mathbf{H}_{ai}^I = -\omega^2 \mathbf{H}_{di}^I . \tag{42}$$

In the case of the one-dimensional finite element analysis, the shear strain and shear stress of the *e* -th element are calculated in terms of $\mathbf{H}_{1e}$ and $\mathbf{H}_{2e}$, parts of the solution vector $\mathbf{H}_d$, related to the *e* -th element. $\mathbf{H}_{1e}$ is also represented in the following series form by taking the appropriate part of eqn. (38)

$$h_{1e} = h_{1e}^o + \sum_{i=1}^{N} h_{1ei}^I + \dots . \tag{43}$$

in which the coefficient vector $\mathbf{H}_{1e}^0$ and $\mathbf{H}_{1ei}^I$ are appropriate sub-vectors of $\mathbf{H}_d^O$ and $\mathbf{H}_{di}^I$, respectively. Similar expression can be written for $\mathbf{H}_{2e}$. The strain transfer function of *e*-th element is then described as:

$$h_{\varepsilon e} = \frac{h_{2e} - h_{1e}}{Z_e} = h_{\varepsilon e}^o + \sum_{i=1}^{N} h_{\varepsilon e i}^I \alpha_i + \dots , \tag{44}$$

with

$$h_{\varepsilon e}^o = \frac{h_{2e}^o - h_{1e}^o}{Z_e}, \qquad h_{\varepsilon e i}^I = \frac{h_{2ei}^I - h_{1ei}^I}{Z_e} . \tag{45}$$

Utilizing the stress-strain relationship, the stress transfer function of *e* -th element, $\mathbf{H}_{\tau e}$, is expressed as:

$$H_{\tau e} = G_e^* H_{\gamma e} = (\overline{G}_e + \alpha_e)(1 + 2i\beta)(H_{\gamma e}^o + \sum_{i=1}^{N} H_{\gamma e i}^I \alpha_i + \dots)$$

$$= H_{\tau e}^o + \sum_{i=1}^{N} H_{\tau e i}^I \alpha_i + \dots , \tag{46}$$

in which

$$H^{o}_{\tau e} = \overline{G}_e H^{o}_{\gamma e}(1+2i\beta), \tag{47}$$

and

$$H^{I}_{\tau e} = (\overline{G}_e H^{I}_{\gamma ei} + H^{o}_{\gamma e})(1+2i\beta), \qquad \text{if } i = e$$

$$H^{I}_{\tau e} = \overline{G}_e H^{I}_{\gamma ei}(1+2i\beta), \qquad \text{if } i \neq e \tag{48}$$

Spectral Density Function Of Response

For the linear system, with the transfer function obtained from preceding calculations, the spectral density function $\mathbf{S_x}$ for any response variable $\mathbf{x}$ can readily be obtained:

$$S_x(\omega) = H(i\omega)\, H^{*T}(i\omega)\, S_{\ddot{x}_g}(\omega)\,, \tag{49}$$

which for simplicity may be written as,

$$S_x = H\, H^{*T}\, S_{\ddot{x}_g}\,. \tag{50}$$

where $\mathbf{H}$ represents the transfer function vector for $\mathbf{x}$ and $\mathbf{H}^{*T}$ its conjugate transpose. From preceeding calculation, the variation of the transfer function can be represented in a perturbation form as:

$$H = H^{o} + \sum_{i=1}^{N} H^{I}_{i}\,\alpha_i + \ldots,$$

$$H^{*} = H^{*o} + \sum_{i=1}^{N} H^{*I}_{i}\,\alpha_i + \ldots\,. \tag{51}$$

Now the variation of the spectral density function, $\mathbf{S_x}$, of the response variable $\mathbf{x}$, expressed in series perturbation form:

$$S_x = (H^{o} + \sum_{i=1}^{N} H^{I}_{i}\,\alpha_i + \ldots)(H^{*oT} + \sum_{i=1}^{N} H^{*IT}_{i}\,\alpha_i + \ldots)S_{\ddot{x}_g}$$

$$= S^{o}_{X} + \sum_{i=1}^{N} S_{Xi}^{I}\,\alpha_i + \ldots\,, \tag{52}$$

with the relationships:

$$S_x^o = H^o H^{*oT} S_{\ddot{x}_g}$$
$$S_{xi}^I = (H^o H^{*IT}_i + H_i^I H^{*oT}_o) S_{\ddot{x}_g} . \tag{53}$$

When the first-order approximation is taken, the expectation and variance of spectral density function of the response can be evaluated. That is, the expected value is

$$E[S_x] = S_x^o , \tag{54}$$

and the covariance matrix

$$Cov[S_x , S_x] = E[(S_x - E[S_x])(S_x - E[S_x])^T]$$
$$= \sum_{i=1}^{N} \sum_{j=1}^{N} S_{xi}^I (S_{xj}^I)^T E[\alpha_i \alpha_j] , \tag{55}$$

or the variance vector which is the diagonal components of COV[$\mathbf{S_x}$,$\mathbf{S_x}$];

$$Var[S_x , S_x] = \sum_{i=1}^{N} \sum_{j=1}^{N} diag[S_{xi}^I (S_{xj}^I)^T] \; E[\alpha_i \alpha_j] . \tag{56}$$

In the above $\mathbf{E[\alpha_i \alpha_j]}$ is determined from the autocorrelation function of the underlying stochastic field of α.
The moments spectral density function $\mathbf{S_x(\omega)}$ are:

$$\lambda_i = \int_0^\infty \omega^i S_x(\omega) d\omega = \int_0^\infty \omega^i S_x^0 \, d\omega + \sum_{i=1}^{N} \{ \int_0^\infty \omega^i S_{xi}^I \, d\omega \} \alpha_i + \dots . \tag{57}$$

Extreme Values of the Response

Considering the response, **x(t)** as zero-mean Gaussian process, the probability density function for the extreme value of the normalized random value, $\eta(= x/\lambda_0^{1/2})$ over a period **T** can be written [22,23] as:

$$p_e(\eta) = \nu T \exp(-\frac{\eta^2}{2}) \cdot \exp[-\nu T \exp(-\frac{\eta^2}{2})]. \tag{58}$$

where

$$\nu = \frac{1}{2\pi}(\frac{\lambda_2}{\lambda_0})^{1/2}. \tag{59}$$

For the above, an approximation for the most probable extreme value of **x** was shown to be:

$$\hat{x} = \sqrt{2\ln(K\sqrt{\lambda_2/\lambda_0}}\ \sqrt{\lambda_0}, \tag{60}$$

with

$$K = \frac{T}{2\pi}$$

The perturbation formulation for **x** can be written as:

$$\hat{\gamma}_n = \hat{\gamma}_n^0 + \sum_{i=1}^{N} \hat{\gamma}_{ni}^I \alpha_i + \dots. \tag{61}$$

where

$$\hat{\gamma}_n^0 = \hat{\gamma}_n|_{\alpha_i=0} = \sqrt{2\ln(K\sqrt{\lambda_2^0/\lambda_0^0})} \bullet \sqrt{\lambda_0^0}, \tag{62}$$

$$\hat{\gamma}_{ni}^I = \frac{\partial \hat{\gamma}_n}{\partial \alpha_i}|_{\alpha_i=0} = \frac{\lambda_0^0\lambda_{2i}^I - \lambda_2^0\lambda_{0i}^I}{2\sqrt{2\ln(K\sqrt{\lambda_2^0/\lambda_0^0})} \bullet \sqrt{\lambda_0^0} \bullet \lambda_2^0} + \sqrt{2\ln(K\sqrt{\lambda_2^0/\lambda_0^0})} \bullet \frac{\lambda_{0i}^I}{2\sqrt{\lambda_0^0}} \tag{63}$$

with

$$\lambda_0^0 = \lambda_0|_{\alpha_i=0} \qquad \lambda_{0i}^I = \frac{\partial\lambda_0}{\partial\alpha_i}|_{\alpha_i=0}$$

$$\lambda_2^0 = \lambda_2|_{\alpha_i=C} \qquad \lambda_{2i}^I = \frac{\partial\lambda_2}{\partial\alpha_i}|_{\alpha_i=0} \tag{64}$$

Once the perturbation formulation of $\hat{\mathbf{x}}$ has been obtained, the expected value, covariance matrix, and variance vector of $\hat{\mathbf{x}}$ can also be calculated in a manner similar to eqns. (54), (55), and (56).

The Equivalent Linear Method

In the range of shear strains induced by strong earthquake motion, stress-strain behavior of soil is significantly nonlinear with **G** and β being dependent on the level of strain. Idriss and Seed [24] introduced an approximate equivalent linear method of evaluating the seismic response, in which an approximate solution is obtained by solving the linear system iteratively by using **G** and β compatible with the developed strain.

A set of initial shear modulus and damping ratio values are estimated for each soil element of the model. The dynamic responses of the system are then evaluated, and the most probable extreme values of the shear strains $\hat{\gamma}$ for each element are evaluated by using the formulations for the statistics of extreme presented in the previous section. The effective strain of each element is taken as:

$$\hat{\gamma}_{eff} = 0.65 \bullet E[\hat{\gamma}_n] = 0.65 \bullet \overset{0}{\hat{\gamma}_n}. \tag{65}$$

New values of **G** and β compatible with the effective shear strain, $\hat{\gamma}_{eff}$ are then read from experimentally generated curves. These values of **G** and β are then used in the next iteration of all the above calculation steps, and the process is repeated until convergence is achieved (usually within 3 to 5 iterations).

Stochastic Finite Element: Monte-Carlo Simulation

A direct Monte-Carlo simulation is also used to analyze the random response associated with the uncertainties in shear modulus of the soil. Using the formulation mentioned in the earlier section, a set of soil sites with randomly generated shear moduli is produced. Each one of these sites is then subjected to the same input random ground motion and their dynamic response are evaluated according to the scheme presented in the above sections. Finally the statistics of these response are evaluated.

Numerical Results

The stochastic finite element formulation presented in the earlier sections were used to analyze the seismic response of a horizontally layered soil site. The computational steps include: specification of random ground motion, calculation of covariance matrix of the shear modulus considered as a random field, calculation for transfer functions, the spectral density functions, response spectra, and extreme values of various response variables. The results from perturbation method are compared with those from the direct Monte-Carlo simulation with respect to accuracy and computational efficiency. Numerical results are presented in the following through a case-study. The results of some parametric studies are also presented to illustrate the influence of some variables on the response.

Problem Description

The shaking of a Niigata site during the earthquake of 1964 is used as case-study. The soil profile, finite element discretization, and the spectral density function of the input ground motion are presented in Fig. 2. The equivalent stationary duration of the motion recorded at the site was estimated to be 8 seconds. The variability of the shear modulus of sand is described by using a standard deviation $\sigma_G = 0.1G$ and a scale of fluctuation δ_G = 20 ft. The stochastic field **G(z)** of shear modulus is assumed to have the autocorrelation function and spectral density function defined as:

$$R_{\alpha\alpha}(\zeta) = R_{GG}(\zeta) = \sigma_G^2 \exp[-(\frac{|\zeta|}{b})],$$
$$S_{\alpha\alpha}(\kappa) = S_{GG}(\kappa) = \sigma_G^2 \frac{1}{\pi} \frac{b}{1+b^2\kappa^2} \qquad (66)$$

In above b = $\delta_{G/2}$ and ζ is the distance of separation between two points z and z' (=z + ζ). In stochastic finite element analysis, the statistical properties of local average are used, which can be derived from the cross statistical point property given by the procedure described in earlier section. The covariance matrix for local average is computed for the purpose of perturbation method and in the direct simulation, the random shear modulus of each element is generated.
Fig. 3 presents the CPU time and the standard deviation σ_γ(RMS value of the computed strain as a function of sample size used in the Monte-Carlo simulation. Those obtained for the perturbation method are also indicated in the figure. The CPU time appeared to be increasing linearly with sample size and the computed response does not change beyond the sample size of 200. It is clear that the perturbation method is computationally efficient and the computed strain response in this case is about the same as those obtained from the simulation.

Case-Study: Niigata Site Response

The soil profile shown is divided into 12 elements (Fig. 2) and the autocorrelation function and spectral density function for the random field model are developed. Using the stochastic finite element model, the random power spectra for various response parameters, viz. displacement, acceleration, shear strain and shear stress are computed. The random RMS (root mean square) the most probable extremes values of the parameters are then calculated. The results obtained from both first order perturbation and Monte-Carlo simulation are described in the following.
Fig. 4 shows the response spectra of the acceleration at the ground surface. The expected acceleration response spectra obtained from perturbation method and Monte Carlo simulation are the same. However, the standard deviation evaluated by the perturbation method is higher and with the increase in the expected value the standard deviation also increases.
Figs. 5 to 7 show the comparison of the expected and the RMS values of acceleration, shear strain, and shear stress as

evaluated by perturbation and simulation methods. It can be seen that the expected values of RMS of acceleration, shear strain and shear stress, are almost identical for both methods. Figs. 6 and 7 also indicate the results obtained by Facciolo [25] in which only the expected values are calculated.

The standard deviation of $\sigma_{\ddot{u}}$, σ_{γ}, and σ_{τ} as computed from the perturbation are in general larger than those obtained by simulation. The spatial variation from both methods are however very similar. The coefficient of variation of these response parameters are in the range of .01 to 0.1.

Parametric Study

A parametric evaluation was made to study the sensitivities of the response. Figs. 8 to 10 show the expected values and standard deviation of RMS of acceleration, shear strain and shear stress as a function plotted against the coefficient of variation of shear modulus, **G**. These should be the same for both perturbation and simulation method. The observed discrepancies for higher values of V_G are associated with the limitation of perturbation method.

The standard duration of RMS values of acceleration, shear strain and shear stress increase with the increase in the coefficient of variation of shear modulus. The results for the acceleration response are quite different for perturbation and simulation methods. However, the strain and stress response, are similar for both methods. The influence of V_G are most significant on the acceleration response.

Figs. 11 to 13 show the dependence of expected values and standard deviations of RMS values of acceleration, shear strain and shear stress on the scale of fluctuation of random field, G. Again, the expected values are not influenced by the scale of fluctuation and both simulation and perturbation methods yield similar results.

The standard deviations of the RMS values of the response as evaluated by simulation indicate some influence of the scale of fluctuation. The perturbation method indicated a small increase (in the standard deviations of the RMS values of the response variables) with increasing scales of fluctuation δ_G. The influence of δ_G is more on acceleration than on strain or stress.

Closure

An application of stochastic finite element method is developed for the analysis of seismic response of horizontally layered soil sites. The uncertainties associated with both the seismically induced ground acceleration and the stiffness characteristics of soil site are considered in a stochastic framework. The spatial variability of the shear modulus is described through a random field model. The ground acceleration is treated as random process.
A finite element formulation is used to discretize the soil The results from the perturbation method are compared with those from the direct Monte-Carlo simulation method with respect to accuracy and computational efficiency.
The numerical results indicate that perturbation method yields reasonable results for the strain and stress response. However, the acceleration response obtained by the perturbation method is significantly greater than those obtained by the simulation method with the discrepancy becoming larger with larger coefficient of variation of the shear modulus. The scale of fluctuation of the shear modulus also affects the response, and for the range of values of fluctuation investigated, the perturbation method yields higher response than the simulation. In spite of well known limitations of the perturbation method, a perturbation based stochastic finite element method may perhaps still serve as a useful tool for the evaluation of liquefaction potential since the stress ad strain response evaluated from this method compare favorably with those from the simulation method.

References

1. Romo-Organista, M.P.: Soil Structure interaction in at random seismic environment, Ph.D. Dissertation, Department of Civil Engineering, University of California Berkeley 1977.

2. Franklin, A.G. and Chang, F.K.: Earthquake resistance of earth and rock-fill dams: Permanent displacement of earth embankments by Newmark sliding block analysis; Misc. Paper 5-71-17, Dept. 5, U.S. Army Engineer Waterways Experiment Station, Viesburg 1977.

3. Deodatis, G. and Shinozuka, M.: Stochastic FEM analysis of a wave propagation problem; Technical Report, Columbia University 1987.

4. Igusa, T. and Kiureghian, A.D.: Response of uncertain systems to stochastic excitation J. Eng. Mech., ASCE (1988) 114, 5, 812-832.

5. Cambou, B.: Application of first-order uncertainty analysis in the finite element method in linear elasticity, Proceedings of The 2nd International Conference on Applications of Statistics and Probability in Soil and Structural Engineering 1971.

6. Astill, C.J.; Nossier, S.B.; and Shinozuka, M.: Impact Loading on Structures With Random Properties, J. of Struct. Mech., ASCE (1972) 1, 63-77.

7. Hisada, T. and Nakagiri, S.: A note of stochastic finite element method (part 2) variation of stress and strain caused by fluctuations of material properties and geometrical boundary conditions, Journal of the Institute of Industrial Science, University of Tokyo (1980) 32, 5, 262-265.

8. Handa, K. and Anderson K.: Application of finite element methods in the statistical analysis of structures, Proc. of The 3rd ICOSSAR, Trondheim, Norway, (1981) 409-417.

9. Vanmarcke, E. and Grigoriu, M.: Stochastic finite element analysis of simple beams, J. of Engg. Mech., ASCE, 109, 5, 1203-1214.

10. Der Kiureghian, A.: Finite element methods in structural safety studies, structural safety studies, edited by J.T.P. Yao et al., an ASCE publication, New York (1985) 40-52.

11. Liu, W.K.; Belytschko, T. and Mani, A.: A computational method for the determination of the probabilistic distribution of the dynamic response of structure, Proceedings of the 1985 Pressure Vessels and Piping Conference, New Orleans, LA (1985) 5, 243-248.

12. Shinozuka, M. and Dasgupta, G.: Stochastic finite element methods in dynamics, Proceedings of The 3rd Conference on Dynamic Response of Structures, Los Angeles, CA (1986) 44-54.

13. Der Kiureghian, A.D. and Ke J.B.: The stochastic finite element method in structural reliability, Proceedings of

The U.S. Austria Joint Seminar on Stochastic Structural Mechanics, Florida Atlantic University, Boca Raton, Florida 1987.

14. Baecher, G.B. and Ingra, T.S.: Stochastic FEM in settlement predictions, Journal of The Geotech. Engg. Div., ASCE, 107 (1981) 449-463.

15. Righetti, G.; Harrop-Williams, K.: Finite element analysis of random soil media, J. of Geoteh.. Eng., ASCE 114, 1 (1988) 59-75.

16. Vanmarcke, E.H. and Lai S.P.: Strong motion duration of earthquakes, Res. Rep. No. 1277-16, Department of Civil Engineering, M.I.T., Cambridge, MA 1977.

17. Binder, R.: Strong motion duration and spectral density functions for a set of 39 earthquake ground motions, seismic safety of buildings, internal study No. 16, Dept. of Civil Eng., MIT, Cambridge, MA 1978.

18. Vanmarcke, E.H.: Structural response to earthquakes, In Seismic Risk and Engineering Decisions, Developments in Geotechnical Engg. 15, Elsevier Scientific Publishing Co.; Amsterdam 1976.

19. Vanmarcke, E.H.: Probabilistic modelling of soil profiles, J. Geotech. Engg. ASCE (1977) 103, 11, 1227-1246.

20. Rice, S.O.: Mathematical analysis of random noise, Bell System Technical Journal, 1944-23 & 24. 46-156. Reprinted in selected papers in noise and stochastic processes, ed. N. Wax, Dover, New York 1954.

21. Shinozuka, M.: Stochastic fields and their digital simulation, Lecture notes, CISM Course on stochastic methods in structural mechanics, Udine, Italy 1985.

22. Cartwright, D.E. and Longuet-Higgins: The statistical distribution of the maxima of a random function, Proc. Royal Soc. of London (1956) A237, 212-232.

23. Davenport, A.G.: Note of the distribution of largest value of a random function with application in gust loading, Proc. Inst. Civil Engrs. London (1964) 28, 187-196.

24. Idriss, I.M. and Seed, H.B.: Seismic response of horizontal soil layers, J. of Soil Mech. and Found. Eng. ASCE (1968) 94, 10003-100031.

25. Faccioli, E.: A stochastic model for predicting seismic failure in a soil deposit, Earthquake Eng. Struct. Dyn. 1, 293.

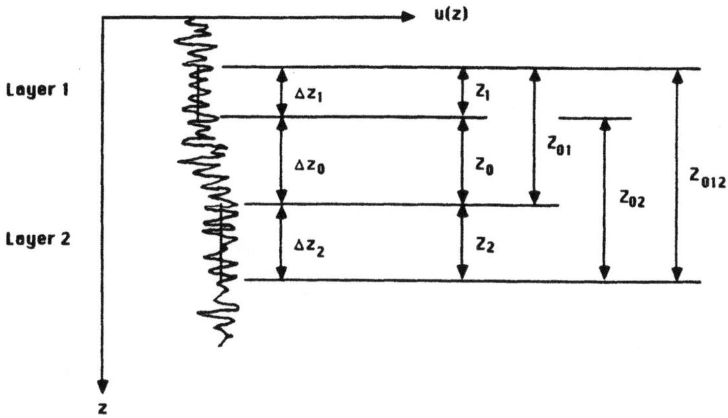

Figure 1. Parameters of randomly varying soil property

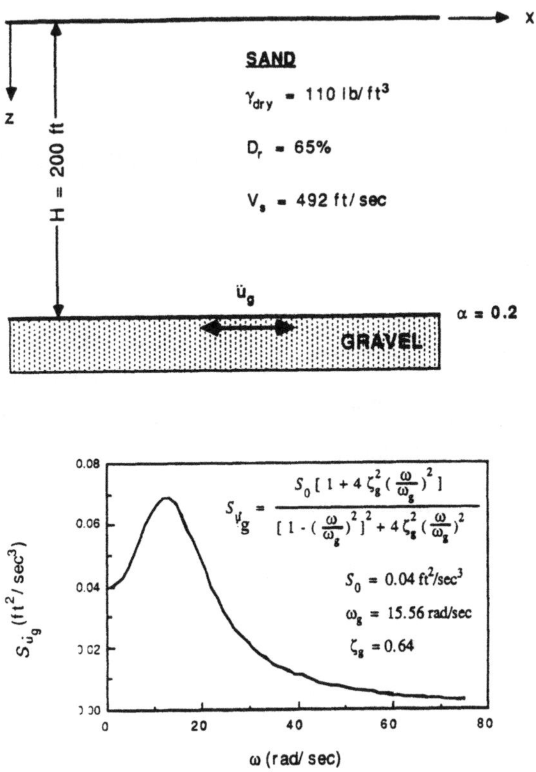

Figure 2. Problem description: soil profile and input motion

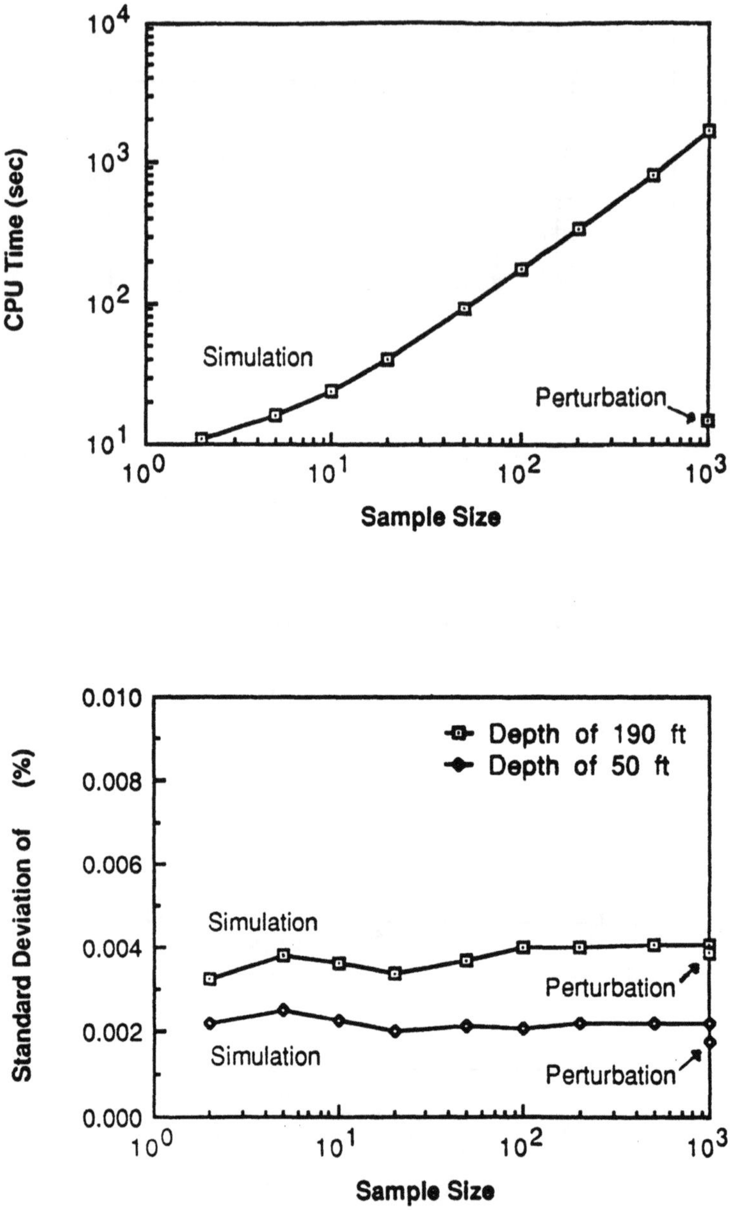

Figure 3. Effect of sample size on computing time and standard deviation of σ_γ (RMS value of shear strain) in simulation method

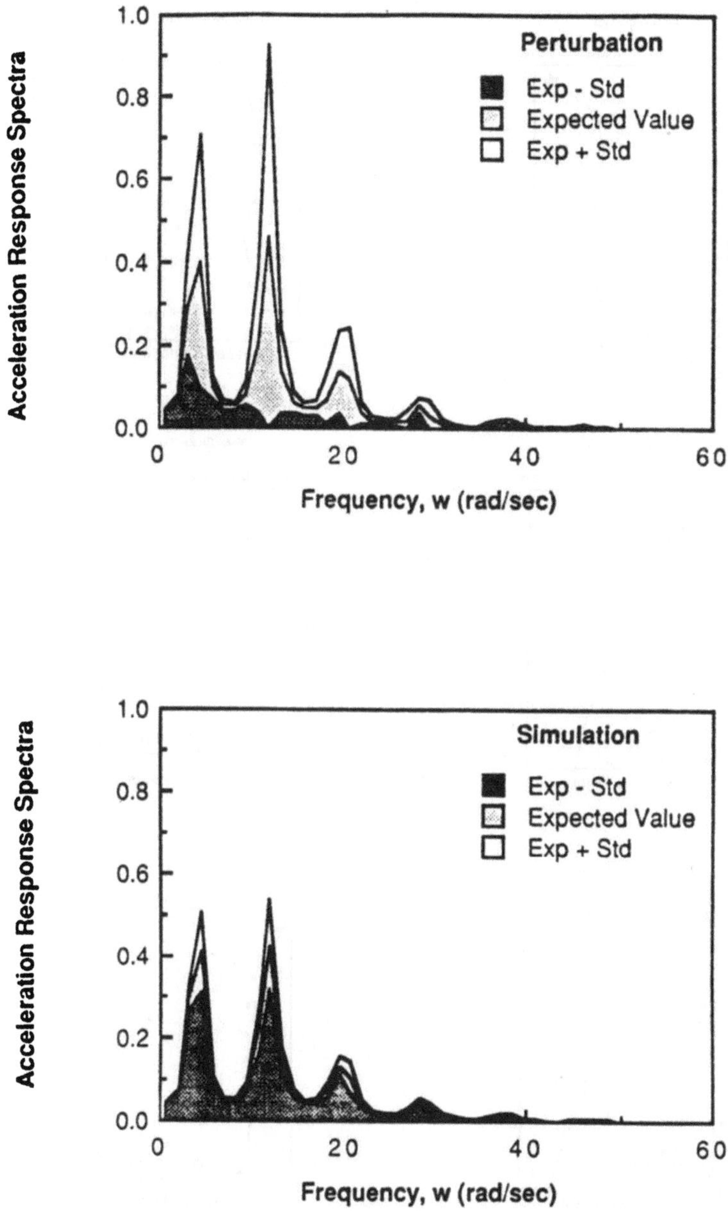

Figure 4. Comparison of ground surface acceleration response spectra as evaluated by perturbation and simulation methods

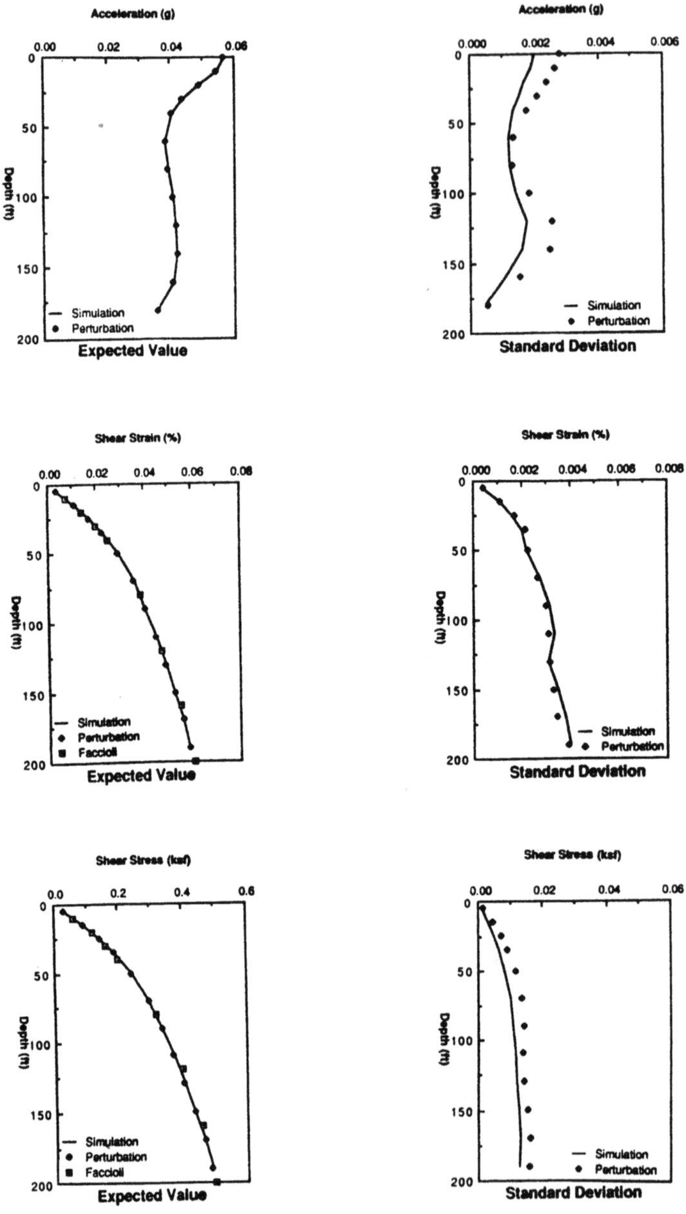

Figures 5,6,7. Comparison of expected values and standard deviations of rms values of accelaration, strain and stress as evaluated by perturbation and simulation methods

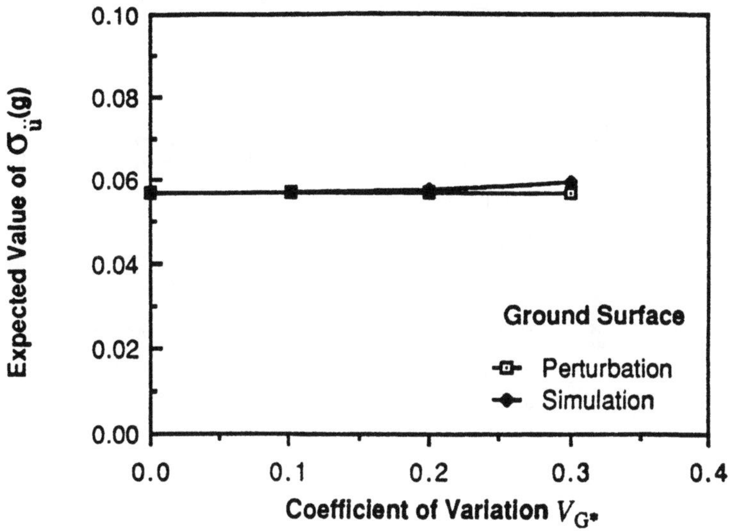

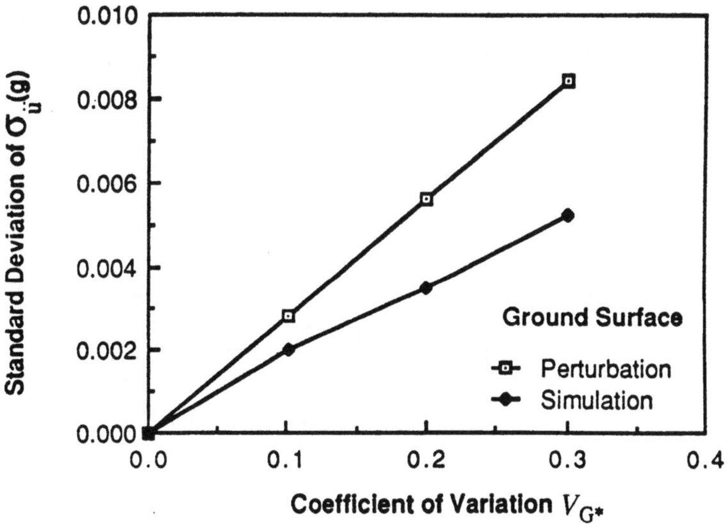

Figure 8. Comparison of expected values and standard deviations of $\sigma_{\ddot{u}}$ (RMS value of acceleration) at different coefficient of variation of shear modulus V_G.

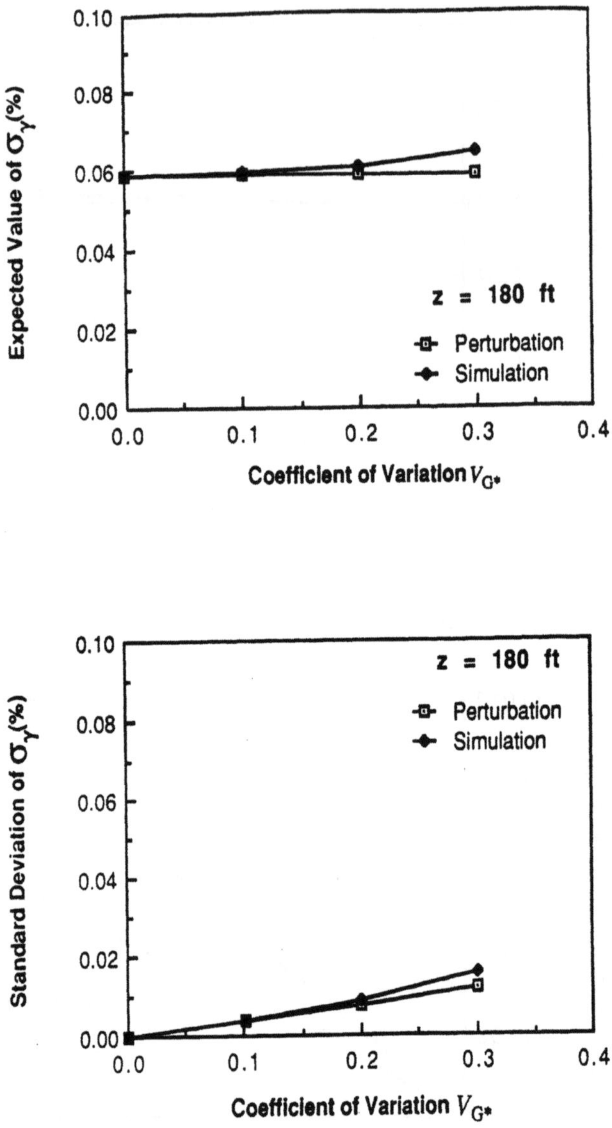

Figure 9. Comparison of expected values and standard deviations of σ_γ (RMS value of shear strain) at different coefficient of variation of shear modulus V_G.

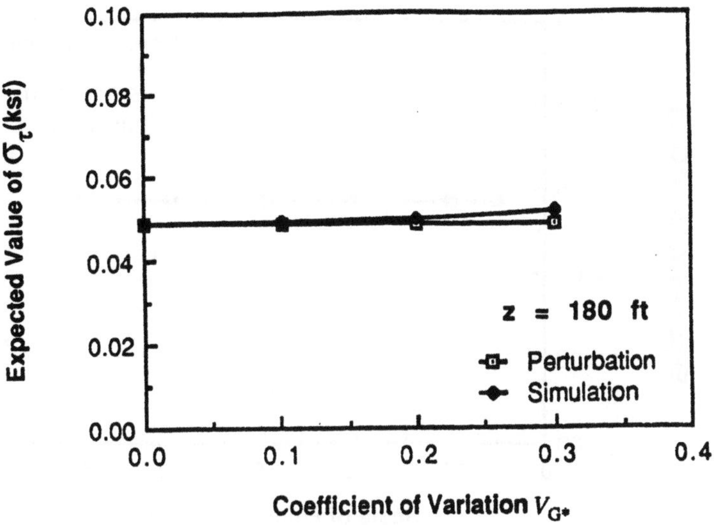

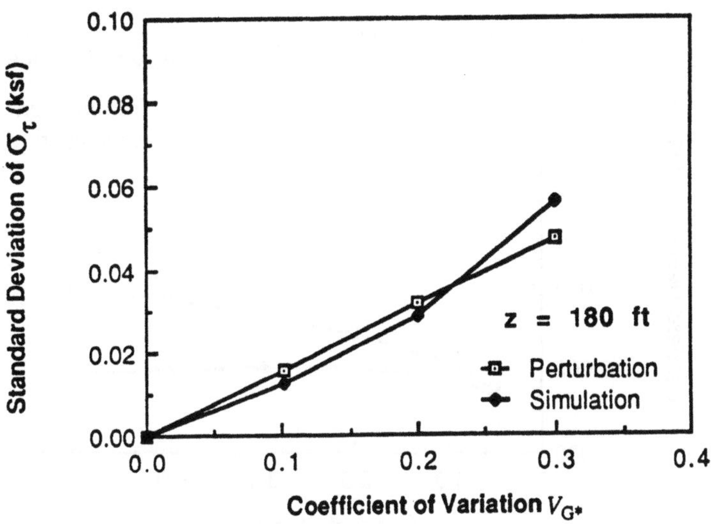

Figure 10. Comparison of expected values and standard deviations of σ_τ (RMS value of shear stress) at different coefficient of variation of shear modulus V_G.

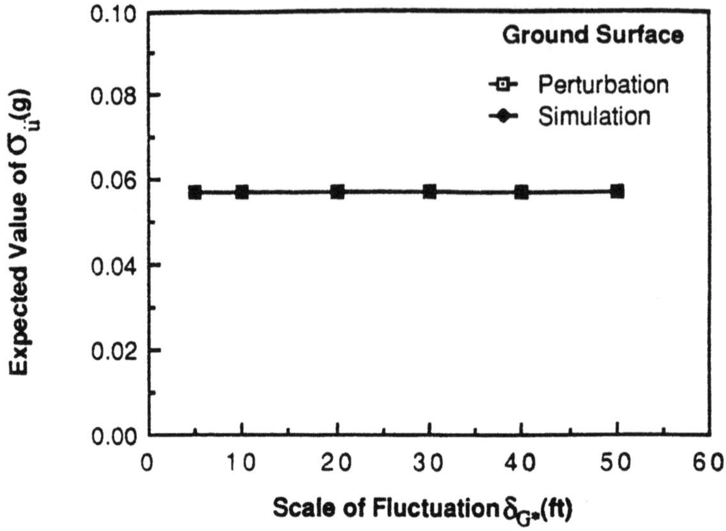

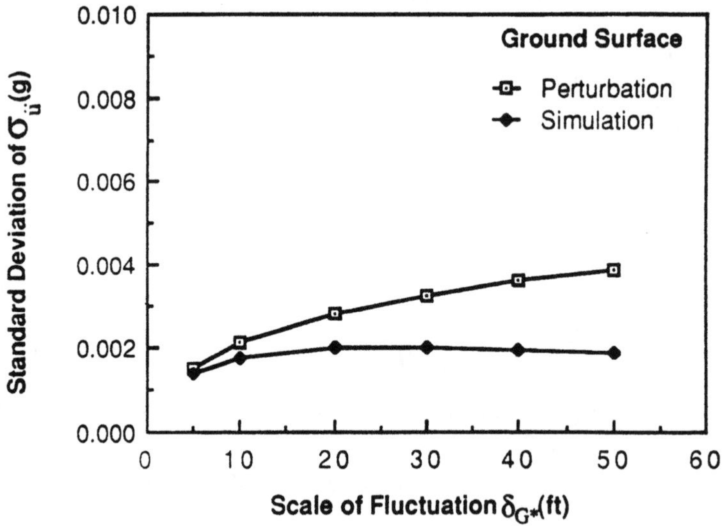

Figure 11. Comparison of expected values and standard deviations of $\sigma_{\ddot{u}}$ (RMS value of acceleration) at different scale of fluctuation of shear modulus δ_G

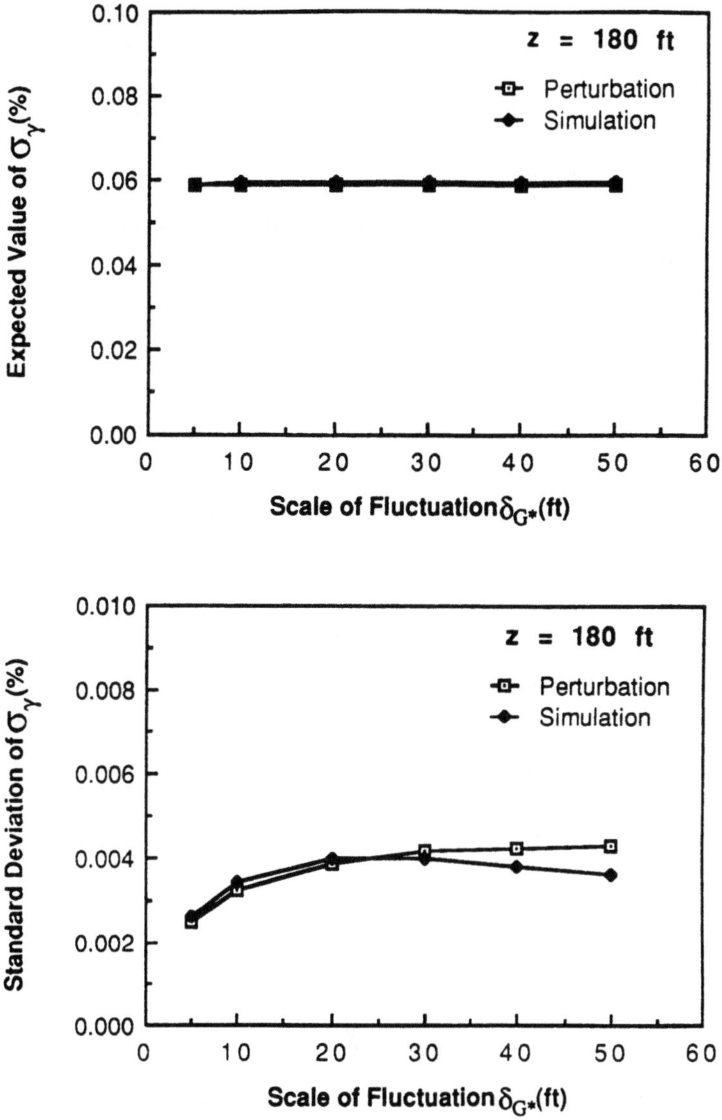

Figure 12. Comparison of expected values and standard deviations of σ_g (RMS value of shear strain) at different scale of fluctuation of shear modulus δ_G

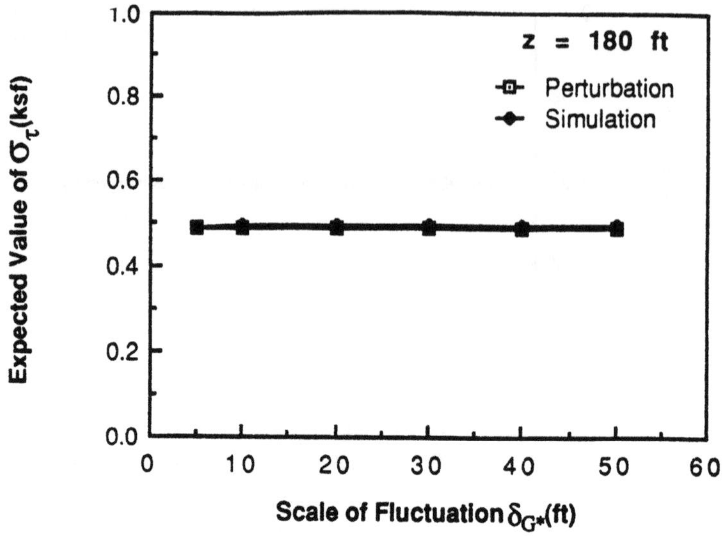

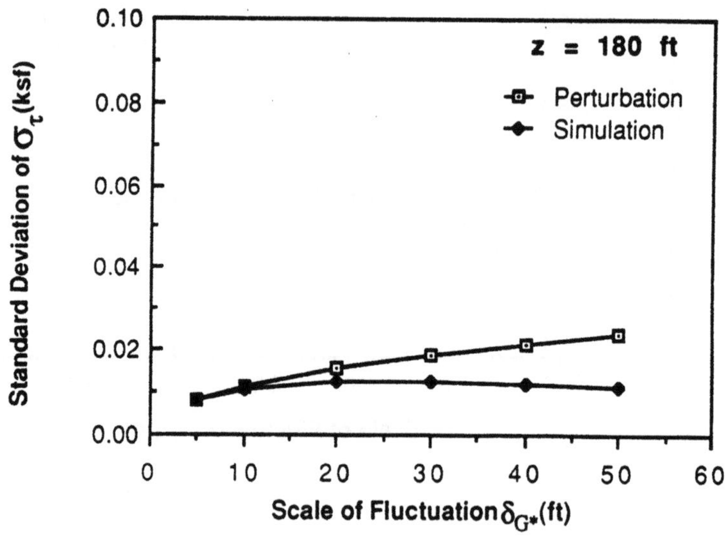

Figure 13. Comparison of expected values and standard deviations of σ_τ (RMS value of shear stress) at different scale of fluctuation of shear modulus δ_G

Stochastic Aspects of Bridge Deck Aeroelasticity under Turbulent Flow

R.H. SCANLAN and N.P. JONES

Department of Civil Engineering, The Johns Hopkins University, Baltimore, MD 21218

Summary

Returning to an earlier frequency-domain analysis of the problem of long-span bridge response to ambient winds, the paper emphasizes the indispensable experimental information necessary to bring results into conformity with full-scale prototype action. Focus is placed upon the use of a sectional wind tunnel model under simulated turbulent flow as the prime source of this information, which consists principally of the flutter derivatives and the spectra of the wind forces, both obtained under turbulent flow. The effect of three-dimensional turbulence upon the flutter derivatives is observed to influence structural stability in a manner consistent with that observed in full-bridge wind tunnel models.

Introduction

The problems of the aerodynamics of the decks of long-span bridges are complex for a number of reasons. First, the approach flow of the naturally occurring wind can be *turbulent* (with definable velocity spectra) and have varying mean velocity across the span of the bridge. Second, the local flow field effects due to the bluff-body geometry of the bridge deck itself will in general initiate *signature turbulence* not initially present in the approach flow. Finally, there will be *aeroelastic*, or interactive fluid-structure, effects attributable to any vibratory motion that the deck undergoes.

Clearly, none of the above effects has to date been satisfactorily calculated from first physical principles alone, nor particularly well characterized analytically; and recourse to *physical testing* has therefore been required. Such testing has most commonly been carried out on reduced-scale wind tunnel models in air at atmospheric pressure.

Faced with the complexities listed, some investigators have opted for a wholly empirical approach, testing scaled full-bridge aeroelastic models under simulated boundary layer winds. Basic problems with such models, which of necessity are constructed to small geometric scale (1/100 to 1/600), include high cost, poor geometric detail, inaccurate vibration mode and damping simulation, and Reynolds number and turbulence length scale mismatches. Under such conditions it has not been possible to pursue un-

I. Elishakoff · Y. K. Lin (Eds.)
Stochastic Structural Dynamics 2
New Practical Applications
2nd International Conference on Stochastic Structural Dynamics
May 9-11, 1990, Boca Raton, Florida

derstanding of the mechanisims affecting response, particularly when overall empirical results suggest stability without identifying its causes.

An alternative has been to pose the problem analytically and obtain the minimum essential experimental inputs for it from tests on larger-scale models, particularly the deck section model, at scales of 1/100 to 1/25. The present paper, following the method of an earlier one [1] employs a frequency-domain approach to the problem, emphasizing the nature and roles of its experimental inputs. In the previous paper the problem was posed as one in which the buffeting and aeroelastic forces were independent of each other. However, in the most general case, both occur together, the aeroelastic forces being modified by the effects of both approaching and signature turbulence. Since all forces on the system must in any case be ascertained by experimental measurement, one simple conception to employ is that the needed aerodynamic experiments are to be carried out under correctly simulated turbulent approach flow. Under this basic assumption the analytical format for the problem may remain essentially the same as previously given. However, the necessary supporting experiments may be different or at least differently interpreted. In particular, the buffeting forces will be directly measured rather than inferred from the wind velocity components, and the necessary aeroelastic forces will be measured in the presence of turbulence, which will in turn influence their effect upon stability.

Since a complete time-domain format for the equations of motion would of necessity require the use of indicial response functions derived in their turn from experimentally obtained frequency domain functions (such as the flutter derivatives), it does not appear useful to make the full conversion to the time domain from experimental frequency domain information. The problem formulation is therefore brought directly to the frequency domain, where it is allowed to remain, the necessary supporting experiments being designed to be appropriate to this end.

Aerodynamic Forces and Equations of Motion

The approach adopted will assume that sectional lift, moment, and drag, referred to the center of gravity of a spanwise section of the bridge deck, form the basis upon which the generalized forces on the structure can be calculated. Let the typical deck section motion be defined by deflection components h (vertical downward), α (twist with the windward edge upward), and p (lateral sway in the horizontal wind direction).

The sectional wind forces per unit span at spanwise station x are then

$$\text{Lift}: \quad L(x,t) = L_{ae} + L_b \tag{1}$$

$$\text{Moment}: \quad M(x,t) = M_{ae} + M_b \tag{2}$$

$$\text{Drag}: \quad D(x,t) = D_{ae} + D_b \tag{3}$$

where the subscripts *ae* refer to *aeroelastic*, or fluid-structure interactive, effects; and b to *buffeting* effects. Then, according to Ref. [5] (with slight modification), the linearized forms of the aeroelastic effects, for sinusoidal motions h, α, p, are:

$$L_{ae} = \frac{1}{2}\rho U^2 B \left[K H_1^*(K)\frac{\dot{h}}{U} + K H_2^*(K)\frac{B\dot{\alpha}}{U} + K^2 H_3^*(K)\alpha + K^2 H_4^*(K)\frac{h}{B} \right] \tag{4}$$

$$M_{ae} = \frac{1}{2}\rho U^2 B^2 \left[K A_1^*(K)\frac{\dot{h}}{U} + K A_2^*(K)\frac{B\dot{\alpha}}{U} + K^2 A_3^*(K)\alpha + K^2 A_4^*(K)\frac{h}{B} \right] \tag{5}$$

$$D_{ae} = \frac{1}{2}\rho U^2 B \left[K P_1^*(K)\frac{\dot{p}}{U} + K P_2^*(K)\frac{B\dot{\alpha}}{U} + K^2 P_3^*(K)\alpha + K^2 P_4^*(K)\frac{p}{B} \right] \tag{6}$$

where

ρ = air density
U = mean cross-wind velocity (horizontal)
B = deck width
K = $B\omega/U$
ω = oscillation frequency

and using experimentally obtained "flutter derivatives" ($i = 1,\ 2,\ 3,\ 4$)

$H_i^* = H_i^*(K)$
$A_i^* = A_i^*(K)$
$P_i^* = P_i^*(K)$

[It is specifically noted in Ref.[3] that the extension of eqs.(4–6) to other than sinusoidal motion can be obtained via use of the *indicial force functions* associated specifically with H_i^* , A_i^* , P_i^*. However, the explicit use of these functions will be avoided here by moving at a later point to the Fourier-transformed domain with the independent parameter K.]

Let the buffeting forces be written in the form:

$$L_b = \frac{1}{2}\rho U^2 B \mathcal{L}_b(x,t) \tag{7}$$

$$M_b = \frac{1}{2}\rho U^2 B^2 \left[C_M(\frac{2u}{U}) + C'_M \frac{w}{U} \right] = \frac{1}{2}\rho U^2 B^2 \mathcal{M}_b(x,t) \tag{8}$$

$$D_b = \frac{1}{2}\rho U^2 B \left[C_D(2\frac{u}{U}) \right] = \frac{1}{2}\rho U^2 B \mathcal{D}_b(x,t) \tag{9}$$

where $\mathcal{L}_b$, $\mathcal{M}_b$, and $\mathcal{D}_b$ are dimensionless time-varying forms of these forces. (It may be assumed that they are determinable by experiment, in which case the forces of both oncoming and signature turbulence will be included.)

It will be assumed that a preliminary vibration analysis of the complete bridge structure has been made, yielding definitions of a series of natural vibration modes i, together with their associated circular frequencies ω_i, and their generalized masses I_i . Let the dimensionless generalized coordinate associated with mode i be ξ_i , so that the respective components of displacement h, α, p over the bridge deck section at x are expressible as:

$$\text{vertical}: \quad h(x,t) = \sum_i h_i(x) B \xi_i(t) \tag{10}$$

$$\text{twist}: \quad \alpha(x,t) = \sum_i \alpha_i(x) \xi_i(t) \tag{11}$$

$$\text{sway}: \quad p(x,t) = \sum_i p_i(x) B \xi_i(t) \tag{12}$$

where $h_i(x)$, $\alpha_i(x)$, $p_i(x)$ are respectively the components of mode i that refer to the deck only at its local c.g. The equation of motion of the ith mode of the system is then

$$I_i \left[\ddot{\xi}_i + 2\zeta_i \omega_i \dot{\xi}_i + \omega_i^2 \xi_i \right] = Q_i(t) \tag{13}$$

where ζ_i is the mechanical damping ratio of mode i, and $Q_i(t)$ is the generalized force

$$Q_i(t) = \int_{deck} [L(x,\ t) h_i(x) B + M(x,\ t) \alpha_i(x) + D(x,\ t) p_i B] dx \tag{14}$$

It is useful to convert to the dimensionless time variable s defined by

$$s = \frac{Ut}{B} \tag{15}$$

whence eq.(13) becomes

$$I_i [\xi_i'' + 2\zeta_i K_i \xi_i' + K_i^2 \xi_i] = \left(\frac{B}{U}\right)^2 Q_i(s) \tag{16}$$

since

$$\dot{\xi}_i(t) = \frac{d}{ds} \xi_i(s) \frac{ds}{dt} = \frac{U}{B} \xi_i'(s) \tag{17}$$

etc., with

$$K_i = \frac{B\omega_i}{U} \tag{18}$$

Equations of Motion in the Fourier-Transformed Domain

Let the Fourier transform of $\xi_i(s)$ be defined by

$$\bar{\xi}_i(K) = \int_0^\infty \xi_i(s) e^{-iKs} ds \tag{19}$$

so that the transformed form of (16) becomes

$$\begin{aligned}
[-K^2 &+ 2i\zeta_i K_i K + K_i^2]\bar{\xi}_i(K) \\
&= \frac{\rho B^4 \ell}{2I_i} \int_{deck} \{ [\, iK^2 H_1^*(K) \frac{\bar{h}}{B} + iK^2 H_2^*(K)\bar{\alpha} \\
&+ K^2 H_3^*(K)\bar{\alpha} + K^2 H_4^*(K)\frac{\bar{h}}{B}\,] h_i(x) + [\, iK^2 A_1^*(K)\frac{\bar{h}}{B} + iK^2 A_2^*(K)\bar{\alpha} \\
&+ K^2 A_3^*(K)\bar{\alpha} + K^2 A_4^*(K)\frac{\bar{h}}{B}\,] \alpha_i(x) + [\, iK^2 P_1^*(K)\frac{\bar{p}}{B} + iK^2 P_2^*(K)\bar{\alpha} \\
&+ K^2 P_3^*(K)\bar{\alpha} + K^2 P_4^*(K)\frac{\bar{p}}{B}\,] p_i(x) + [\, \bar{\mathcal{L}}(x,\ K) h_i(x) + \bar{\mathcal{M}}(x,\ K)\alpha_i(x) \\
&+ \bar{\mathcal{D}}(x,\ K) p_i(x)\,]\,\} \frac{dx}{\ell}
\end{aligned} \tag{20}$$

where ℓ is the deck span and, according to eqs.(10)–(12):

$$\frac{\bar{h}(x,\ K)}{B} = \sum_i \bar{\xi}_i(K) h_i(x) \tag{21}$$

$$\bar{\alpha}(x,\ K) = \sum_i \bar{\xi}_i(K)\alpha_i(x) \tag{22}$$

$$\frac{\bar{p}(x,\ K)}{B} = \sum_i \bar{\xi}_i(K) p_i(x) \tag{23}$$

In eq.(20) there appear integrals of the form

$$\int_{deck} F_i^*(K) r_m(x) s_n(x) \frac{dx}{\ell} \tag{24}$$

where $F_i^*(K) = H_i^*(K),\ A_i^*(K),\ P_i^*(K)$ $(i = 1,\ 2,\ 3,\ 4)$, and $r,\ s$ represent h, α, or p, with $m,\ n = i$ or j. The flutter derivatives F_i^* may be influenced, in particular, by mode shape, wind turbulence, and by displacement amplitude. Since the primary interest here is on small *incipient* amplitude, eventual dependence of the flutter derivatives on higher amplitude will be ignored, thus keeping the formulation linear.

We are, in particular, interested in the effects of turbulence upon the flutter derivatives. The implications of eq.(20) upon the nature of the response may be examined by considering as an example the simplified case wherein only a single mode i is involved and all flutter derivatives except A_2^* and A_3^* vanish or are considered negligible. This corresponds to the case of a pure, uncoupled torsion mode. Examples of the trends of typical A_2^* flutter derivatives under laminar flow conditions are depicted in Fig.1. Eq.(20) becomes

$$[-K^2 + 2i\zeta_i K_i K + K_i^2]\bar{\xi}_i = \frac{\rho B^4 \ell}{2I_i} \int_{deck} \left\{ (iK^2 A_2^* + K^2 A_3^*) \alpha_i^2(x) \bar{\xi}_i + \bar{\mathcal{M}}(x, K) \alpha_i(x) \right\} \frac{dx}{\ell} \tag{25}$$

Under this description the net damping and stiffness terms, respectively, are

$$i \left[2\zeta_i K_i K - \frac{\rho B^4 K^2}{2I_i} \int_{deck} A_2^* \alpha_i^2(x) dx \right] \tag{26}$$

and

$$K_i^2 - K^2 \left[1 + \frac{\rho B^4}{2I_i} \int_{deck} A_3^* \alpha_i^2(x) dx \right] \tag{27}$$

from which it is seen that the stability and the frequency of response are influenced by the flutter derivatives A_2^* , A_3^* . In the absence of turbulence effects the values of these functions are unaffected by correlation loss along the span, and the laminar flow values may be retained and taken outside the integrals. Under turbulence, however, this correlation loss may be accounted for, based on experiment. In this circumstance A_2^* and A_3^* are replaced by their net effective values A_{2e}^* , A_{3e}^* (see below).

The corresponding stability criterion (eqs.(55) and (56) of Ref.1) in the present case is

$$\frac{K \rho B^4 \ell}{4 K_i \zeta_i I_i} [A_{2e}^*(K) G(\alpha_i , \alpha_i)] > 1] \tag{28}$$

with

$$K^2 = \frac{K_i^2}{1 + \frac{\rho B^4}{2I_i} [A_{3e}^*(K) G(\alpha_i , \alpha_i)]} \tag{29}$$

where

$$G(\alpha_i , \alpha_i) = \int_{deck} \alpha_i^2(x) dx \tag{30}$$

These conditions merely reflect the fact that the net effective damping of the system becomes negative at a modified dimensionless frequency K.

Experimental Sources of Information

The section model has the two degrees of freedom h and α (p may be included if desired) and responds in general according to the same theoretical equations of motion as the prototype but is restricted to the two modes $h_i(x) = \text{constant}$, $\alpha_i(x) = \text{constant}$. The flutter derivatives obtained from it under laminar approach flow by one of the usual identification methods [4] may be considered as effectively free of the influence of turbulence (although some signature turbulence effects are always present.) When approach flow turbulence is present (as created by an upstream grid) the section model is buffeted and its flutter derivatives are influenced by the presence of the turbulence.

The Fourier and power spectra of the lift, pitching moment and drag on the section model are obtained on the model as fixed in a high-frequency force balance. These are obtained independently of model motion and are assumed here to be independent of aeroelastic effects.

Aeroelastic effects, however, are not assumed (or found) to be completely independent of turbulence effects, which influence their coherence along the span. For purposes of discussion the case of single-degree torsion will be treated below. Eq.(20) in this case reduces to eq.(25) where now $\bar{\xi}_i$, $\bar{\mathcal{M}}$ are known from long time-history records as (complex) functions of K, ℓ_m is model span and $\alpha_i(x) \equiv 1$. Hence the following integrals are identifiable as functions of K:

$$\int_{model\ deck} A_2^* \frac{dx}{\ell_m} \quad ; \quad \int_{model\ deck} A_3^* \frac{dx}{\ell_m} \tag{31}$$

For use in prototype analysis, an effective value A_{2e}^*, for example, is developed as follows [2]: Assume that in fact the sectional value of A_2^* is a quantity whose coherence with adjacent values falls off in an exponential manner, as is the case with the spectra of the time-dependent gust forces. In this manner, assume that the coherence of the sectional value of A_2^* with distance $|x_a - x_b|$ along the deck span is defined by the form

$$A_2^* = (A_2^*)_{lam} e^{-C|x_a - x_b|/\ell} \tag{32}$$

where $(A_2^*)_{lam}$ is the sectional value of A_2^* obtained under laminar flow, and with constant C to be evaluated. Set the condition

$$\begin{aligned} \int_{model\ deck} A_2^* \frac{dx}{\ell_m} &= \int\int_{model\ deck} (A_2^*)_{lam}\ e^{-C|x_a - x_b|/\ell_m} \frac{dx_a}{\ell_m} \frac{dx_b}{\ell_m} \\ &= \left[\frac{2}{C}(C - 1 + e^{-C})\right] (A_2^*)_{lam} \end{aligned} \tag{33}$$

This establishes a value for C. Then define for the prototype:

$$A^*_{2e} = \int\int_{deck} (A^*_2)_{lam} e^{-C|x_a - x_b|/\lambda \ell_m} \frac{dx_a}{\ell} \frac{dx_b}{\ell} \tag{34}$$

where λ is the geometrical scale ratio: prototype to model. This is directly evaluated as

$$A^*_{2e} = \left[\frac{2}{C'} (C' - 1 + e^{-C'}) \right] (A^*_2)_{lam} \tag{35}$$

where

$$C' = \frac{C\ell}{\lambda \ell_m} \tag{36}$$

Similar definitions hold for A^*_{3e} , etc.

Summary Remarks

The bridge deck wind problem is stochastic in nature. In seeking to understand its mechanisms, physical experiments are indispensable. However, global model experiments, encompassing all parameters at once, have many sources of error and are not revealing as to the physical mechanisms at work.

Therefore in the present paper the minimum essential experimental tool, the geometrically accurate deck section model, is retained. This model is required typically to be subjected to four types of test:

1. a fixed high-frequency force balance test under laminar flow;
2. the same under turbulent approach flow (proper turbulence being developed by an upstream grid);
3. spring-mounted test for flutter derivatives under laminar flow;
4. the same under turbulent flow.

From this indispensable minimum of testing, enough data are forthcoming to permit an analysis of bridge response in the frequency domain.

Time domain analysis can of course be an alternate choice. However "fluid memory" (wake vorticity) effects must be accounted for in time domain analyses, and such effects are not as conveniently measured as frequency-domain (flutter derivative) effects. In fact, time domain indicial response functions must usually be derived by the equivalent of Fourier transformation from experimentally acquired frequency domain results.

The approach chosen here is in fact heavily dependent on— and remains close to— the physical nature of the problem. For example, the detailed character of the oncoming turbulent wind spectral components are incorporated *a priori* into the flow itself. Also, the detailed manner in which "the wind seizes the structure" is incorporated, via experiment, into the analytical model and is not dependent upon any assumptions, such as a quasi-steady force format.

It has been demonstrated, in the analysis given, how the coherence-loss effects of turbulence upon the flutter derivatives can lead to results of analysis that are consistent with the experience of full-model wind tunnel tests, wherein response to buffeting is greater at subcritical speed, but critical flutter speed is postponed to higher velocity. (See Fig.2.)

References

1. Scanlan, R.H.: "On Flutter and Buffeting Mechanisms in Long-Span Bridges" *Probabilistic Engrg. Mechanics*, Vol.3, No.1, (Mar.1988) pp.22–27.

2. Scanlan, R. H. and Jones, N.P.: "Aeroelastic Analysis of Cable-Stayed Bridges." *Jnl. Struct. Engrg. ASCE*, Vol.116, No.2, (Feb.1990) pp.279–297.

3. Scanlan, R. H.: "Role of Indicial Functions in Buffeting Analysis of Bridges", *Jnl. Struct. Engrg., ASCE*, July 1984, Vol.110, No.7, pp.1433–1446.

4. Imai, H., Yun, C.-B., Maruyama, O. and Shinozuka, M.: "Fundamentals of System Identification in Structural Dynamics", *Probabilistic Engrg. Mechanics*, Vol.4, No.4, (Dec.1989) pp.162–173.

5. Scanlan, R. H.: "The Action of Flexible Bridges under Wind", *Jnl. of Sound and Vibration*, 1978, Vol.60, No.2, I: Flutter Theory, pp.187–199; II: Buffeting Theory, pp.201–211.

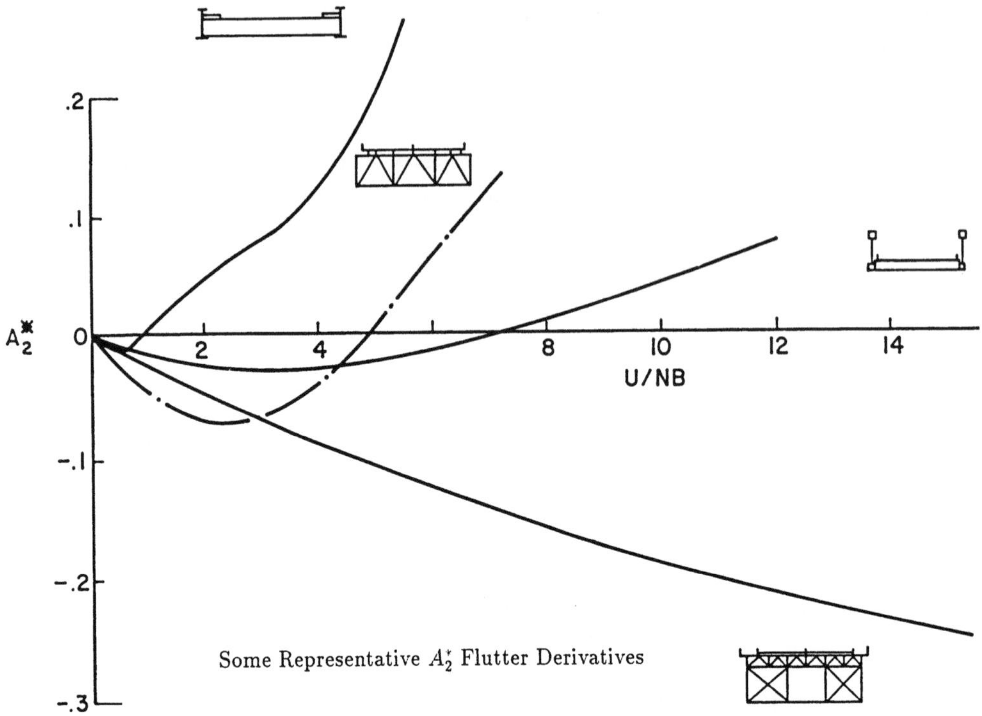

FIG. 1

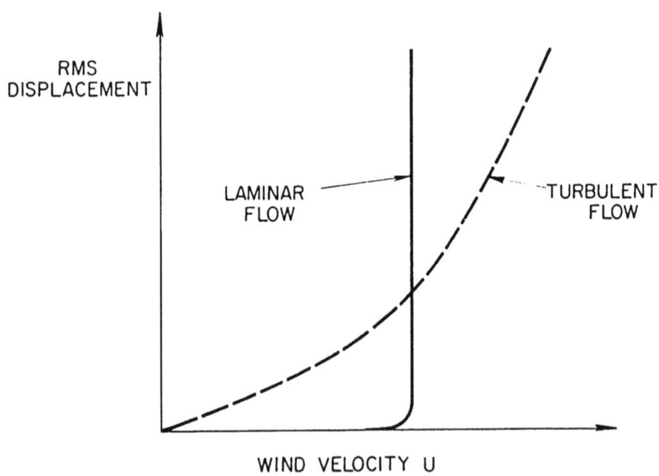

FIG. 2
Bridge Deck Displacement Trends under Smooth and Turbulent Flows

Computational Stochastic Structural Analysis - A Cotribution to the Software Development for the Reliability Assessment of Structures under Dynamic Loading

G.I. SCHUËLLER and C.G. BUCHER
Institute of Engineering Mechanics
University of Innsbruck
Austria, Europe

Summary
Mainly due to the lack of respective available procedures generally traditional methods of analysis and design of engineering structures neglect the statistical, i.e. intrinsic, uncertainties of the parameters involved. As shown in the present paper, however, the treatment of complex systems becomes feasable by interrelating mechanical modeling procedures, e.g. by finite elements, with efficient and accurate stochastic analysis, e.g. Monte Carlo type simulation. Hence the suggested method, which is in fact based on the response surface procedures, allows the utilization of the respective most modern structural as well as stochastic analysis procedures. This, of course implies the total collapse failure criterion of structures under dynamic loading.

1. INTRODUCTION

It is a well known fact that traditional analysis and design procedures of engineering structures neglect the intrinsic, i.e. physical, uncertainties of the parameters and variables, e.g. loading, material, geometrical properties, etc., involved. These procedures are generally labeled as deterministic procedures. Stochastic approaches by which, in general, these uncertainties may be treated more realistically are - particularly for dynamic problems - open to criticism with respect to the degree of simplification of the mechanical modeling. In other words, while the load models show a considerable degree of sophistication, in most cases the structural mechanical models are far from reflecting the state of the art. Procedures are generally limited to SDOF or two DOF systems. PΔ effects with large deflections, which are so important, particularly when investigating collapse failure mechanisms, are generally not included in the analysis. Moreover, in many cases only second moment properties of the response are provided, which are certainly of limited practical use only for the analyst or designer. It is mainly due to this lack of sophistication of the mechanical modeling, i.e. due to the fact that the mechanical models of the structures for stochastic analysis do not match those which are used in so called deterministic design, that stochastic procedures still lack the recognition by practicing engineers they deserve. Basically stochastic procedures represent a particular way of information processing. While deterministic procedures select "representative" (mean, maximum, minimum, etc.) values to describe the parameters, stochastic methods process the

I. Elishakoff · Y. K. Lin (Eds.)
Stochastic Structural Dynamics 2
New Practical Applications
2nd International Conference on Stochastic Structural Dynamics
May 9–11, 1990, Boca Raton, Florida

entire information which is available, i.e. instead of arbitrarily selecting one particular value, the entire spectrum of possible values - in terms of probability distributions - is utilized.

It is quite obvious that the practical use of this information processing requires not only efficient computational procedures, but also the respective software. The lack of availability of such a - possibly easy to use - software appears to be an additional obstacle for the general acceptance and utilization of stochastic structural analysis and design procedures. In this context it should be noted that the general acceptance of e.g. finite element procedures increased considerably when the respective software became available. The developments as shown in this paper are intended to be a step in this direction with respect to stochastic analysis.

For the sake of completeness, several computational concepts and procedures available are discussed shortly. First, consider the wide area of so called "Stochastic Finite Elements" (SFE). This field has been investigated extensively by research groups in both USA and Japan. As representative references the publications of Shinozuka [1], Liu and Belytschko [2], Nakagiri [3], Der Kiureghian [4] should be mentioned. In principle, all SFE procedures rely on the assumption of small random disturbances of system parameters, so that a kind of First-order second-moment analysis can be performed. While this information in many cases gives an excellent impression of the possible variation (variability) of the response of such a system it may, however, be inadequate for structural reliability analysis, especially when dealing with severely nonlinear problems such as collapse.

The software CALREL-FEAP [5] very much relies on analytical expressions for system sensitivity (i.e. partial derivatives) which inhibits (or at least significantly complicates) a general extension of its underlying concept to e.g. nonlinear hysteretic dynamics. While the software PROBAN [6] can certainly be considered one of the most advanced packages for reliability analysis it does not provide specific efficient tools for complex systems under dynamic load.

Finally, concepts requiring analytical solutions of the Fokker-Planck Equation [e.g. 7] at the present appear not to be extendable to complex structures which have to be modeled by Finite Elements.

2. SOFTWARE CONCEPT

Before discussing the theoretical developments and computational procedures, a reflection of the respective requirements by the engineering practice is in order:

(i) Since in modern structural analysis procedures all structural system parameters are discretized, it is quite natural that the load parameters should be discretized as well. This implies that complex finite element models should also be suitable for stochastic structural analysis.

(ii) The design criteria, such as serviceability, collapse, fatigue, brittle fracture, etc., as used in deterministic analysis and design procedures should also apply when using probabilistic methods.

(iii) Stochastic procedures have to meet standard requirements for practical application with respect to efficiency and accuracy. In other words the computational effort must not exceed acceptable limits while providing the required accuracy.

In conclusion, the software for computational stochastic structural analysis (COSSAN) has to consist of structural mechanical, e.g. finite elements, as well as probabilistic, e.g. simulation, procedures. A target reliability can be reached by changing the parameters and carrying out a certain number of computational cycles. This, of course underlines the importance of the efficiency requirement.

3. THEORETICAL CONCEPT

Structural Analysis and Design Criteria

It is well known that the design criterion of structures with random systems parameters which may be represented by a random vector $\underline{x}$ can be described by a so called limit state function $g(\underline{x})$ of the random variables, where the following definition applies:

$$g(\underline{x}) > 0 \quad \text{no failure}$$
$$g(\underline{x}) \leq 0 \quad \text{failure} \tag{1}$$

If time variance of the structural resistance, e.g. due to crack propagation, embrittlement, corrosion, etc., enters the problem the above equation writes:

$$g(\underline{x}, T) > 0 \quad \text{no failure}$$
$$g(\underline{x}, T) \leq 0 \quad \text{failure} \tag{2}$$

where T is the time range of observation, e.g. the design life of a particular structure.

To determine the effect of the various randomly distributed input parameters - e.g. load and systems parameters respectively - on the structural response, a sensitivity analysis is in order. For this purpose, a so called response surface method is most instrumental. In particular the procedure which is used here allows the simultaneous determination of the response surface. In other words the limit state function as defined by eq.(1) or (2) will be approximated by interpolation and hence by a function $\bar{g}(\underline{x})$.

In view of the large number of variables involved, particularly when treating realistic practical problems, the procedure as such may be considered somewhat impractical. However, as shown in Fig. 1, in most cases the number of random variables may be reduced considerably, i.e. by comparing the size of critical values y_m^*.

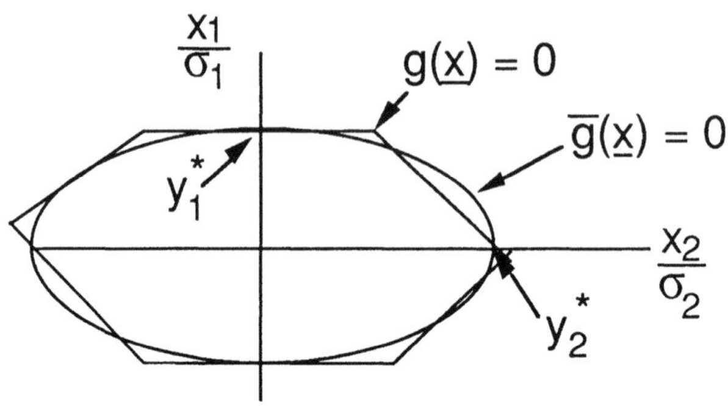

Fig. 1: Limit State Function and Response Surface

The effect of the remaining variables on the system behavior is investigated by a pairwise combination. It should be noted that for this purpose the combinations of the most significant variables, i.e. those with minimum values of y_m^*, are considered. The response surface may be characterized most conveniently by the following expression:

$$\bar{g}(\underline{x}) = a + \sum_{i=1}^{r} b_i x_i + \sum_{i=1}^{r} \sum_{j=1}^{r} c_{ij} x_i x_j \qquad (3)$$

where the coefficients a, b and c are to be determined by structural analysis and r represents the number of variables considered in the analysis. As already mentioned above, this analysis may include all important mechanical effects, i.e. the most modern structural analysis procedures can be utilized. Within the further analysis the actual limit state function $g(\underline{x})$ is represented by the response surface $\bar{g}(\underline{x})$. At this stage it should be emphasized that the function $\bar{g}(\underline{x})$ represents just an approximation of the limit state, i.e. the region of transition from the survival to the failure state respectively and hence does not represent the actual system response. Naturally the approximations as described above require an error estimation which will be treated in detail in the subsequent developments.

Estimation of Failure Probabilities

Depending on the respective design criteria the failure probabilities have to be determined in a final step. Following the discretization of the load and system variables respectively it can be calculated by evaluating the following integral:

$$p_f(t) = \int_{g(\underline{x},T)\leq 0} f_{\underline{X}}(\underline{x})\, d\underline{x} \qquad (4)$$

where $f_{\underline{X}}(\underline{x})$ represents the joint probability density of the random variables $\underline{x}$ and $g(\underline{x},T)$ defines the time variant failure domain (eq.(2)). For evaluating eq.(4) various procedures are available (see e.g. [8] and [9]). Most advantageously both efficient and accurate methods, such as advanced simulation procedures (see e.g. [10] and [11], etc.) are utilized. Needless to say that the replacement of the actual limit state surface $g(\underline{x})$ by the response surface $\bar{g}(\underline{x})$ results in a significant reduction of computational effort.

Error Estimation

Since the response surface as defined by eq.(3) is based on interpolation procedures there are regions in the space of basic variables where errors, i.e. deviations, may occur when compared to the exact limit state surface. In other words, there is a difference p_f^c between the exact value p_f and $\bar{p}_f$, where the latter value, i.e. $\bar{p}_f$, is calculated by utilizing the response surface approximation of the limit state function. As shown in Fig. 2 these errors may contribute both positive and negative values to the failure probability.

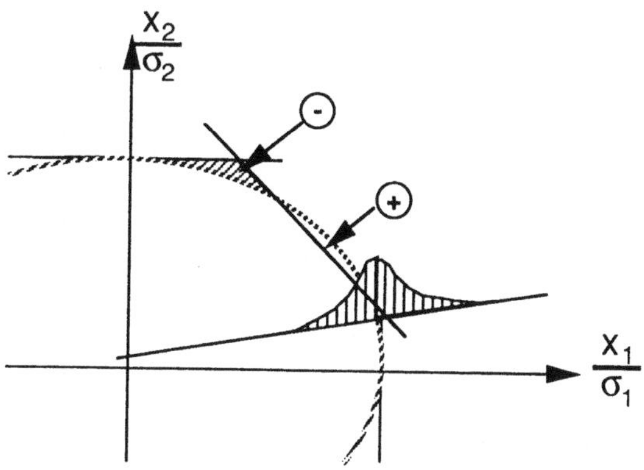

Fig. 2: Error Estimation for Response Surface Method

Quite clearly, the major contributions to the error result from regions close to the limit state surface. Consequently, in order to check the accuracy of the response surface, a sampling scheme which produces samples close to $g(\underline{x}) = 0$ is most appropriate. Such a scheme, called conditional sampling, is described in [12]. For the present problem the conditional sampling procedure is carried out as follows:

(1) In order to obtain the optimum sampling densities $h_{\underline{Y}}(\underline{x})$ for the vector $\underline{x}$ and for the safety margin Z perform adaptive sampling [13] using the response surface.

(2) Identify the most important variable y_m^* in terms of sensitivity factors.

(3) Generate samples for $\underline{x}$ from the sampling density $h_{\underline{Y}}(\underline{x})$ and discard the m-th component y_m^*.

(4) Generate samples for the safety margin $Z = \bar{g}(\underline{x})$ near $Z = 0$ and calculate y_m^* from $\underline{x}$ and Z (cf. eq.(3)). Check whether $g(\underline{x})$ and $\bar{g}(\underline{x})$ have identical sign. If not, add to the error estimate $p_f{}^c$ (cf. Fig.2).

This procedure ensures that the generated samples are close to the limit state surface thus yielding a reliable error estimate.

4. NUMERICAL EXAMPLES

In the following the procedure is applied to two cases where in the first case the error estimation procedure is emphasized by e.g. treating a larger frame structure. The second example shows the versatility of the procedure with respect to the respective constitutive material laws, the properties of the load process as well as geometric nonlinearities.

Example 1

A two-bay two-storey frame under both vertical and horizontal loading is considered.

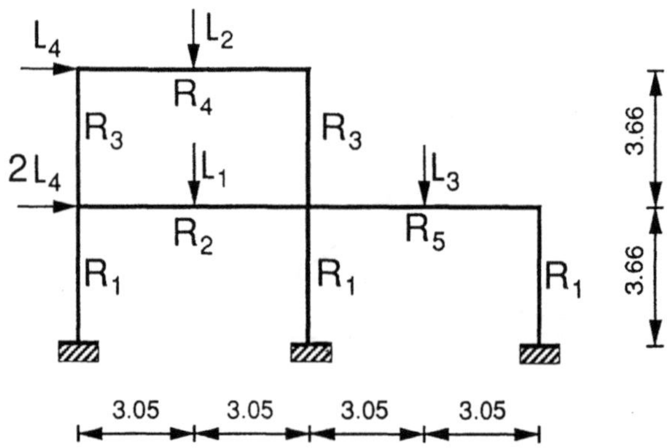

Fig. 3: Two-Bay Two-Storey Frame (units in [m])

The structure is modeled according to perfectly rigid-plastic behavior with possible formation of plastic hinges [14]. Seven important failure modes are taken into account. The loading L_i and the plastic moments R_i are assumed to be Gaussian distributed with statistical properties as listed in Tab. 1.

	Mean Value	Standard Deviation
L_1	124.5	37.4
L_2 [kN]	71.2	19.9
L_3	93.4	26.2
L_4	31.1	7.8
R_1	94.7	14.2
R_2	203.3	30.7
R_3 [kNm]	94.7	14.2
R_4	121.9	18.2
R_5	162.8	24.5

Tab. 1: Statistical Data for Example 1

The response surface includes all mixed terms which results in a highly accurate estimate of the failure probability $\bar{p}_f = 4.3 \cdot 10^{-2}$ as compared to the exact value of $p_f = 3.9 \cdot 10^{-2}$ obtained from adaptive sampling using 8000 samples. The error estimates, from 128 samples only, are given in Tab. 2 for five different runs.

No. of run	$\bar{p}_f - p_f$
1	$- 0.6 \cdot 10^{-2}$
2	$- 0.2 \cdot 10^{-2}$
3	$- 0.3 \cdot 10^{-2}$
4	$- 0.4 \cdot 10^{-2}$
5	$- 0.8 \cdot 10^{-2}$

Tab. 2: Error Estimates for Response Surface

Although there is still some fluctuation in the results in all cases the true error of $0.4 \cdot 10^{-2}$ is approximated very well with a very low number of samples.

Example 2

It has already been pointed out above that the procedure suggested in this paper is generally applicable, i.e. the assumptions as introduced in the following do not represent a basic limitation. In other words, any constitutive material law as well as evolutionary load process can be utilized. Although the procedure is also applicable to large structural systems, the present example, as shown in Fig. 4a is confined to a SDOF-system in order to be able to check the results with a direct simulation procedure. The structure is subjected to earthquake loading. Furthermore it is assumed that plastic hinges may form in the columns, where the moment deflection relation develops as shown in Fig. 4b.

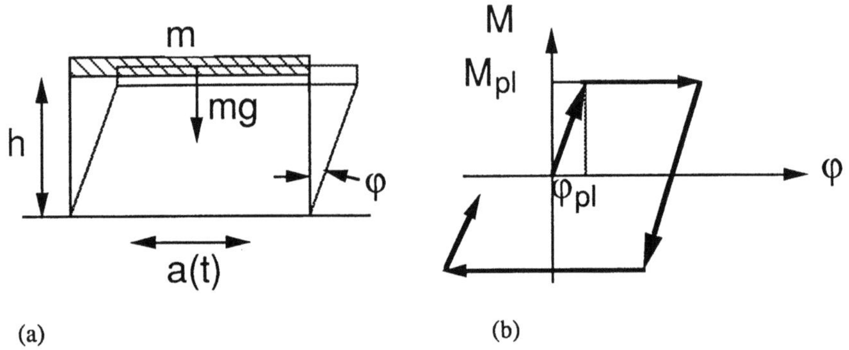

Fig. 4: Sketch of Structural System and Material Law for Example 2

The PΔ effect with large deflections is considered. The failure criterion is the total collapse of the frame structure. The structural data are as follows: m = 7.876t, M_{pl}/φ_{pl} = 1134kNm, h = 5.0 m; g = 9.81 m/s^2. The ground acceleration process is described by an amplitude modulated stationary process, i.e. by

$$a(t) = e(t)\, s(t) \tag{5}$$

where

$$e(t) = 1.778\,[\exp(-0.25t) - \exp(0.5t)] \tag{6}$$

$$E[s(t)\, s(t+\tau)] = e^{-\zeta\omega|\tau|}\,[\cos \omega'\tau - \frac{\zeta\omega}{\omega'} \sin \omega'|\tau|] \tag{7}$$

and $\zeta = 0.1$, $\omega = 10$ rad/s, $\omega' = \omega\sqrt{1-\zeta^2}$.

The natural period of the linearized system is 1.25s. Hence an interval for discretization of $\Delta\tau$ = 0.25s appears to be sufficiently small. The system is analysed for a load duration of 15s. Hence the ground acceleration is characterized by 60 random variables.
The equations of motion are integrated by the respective FE-modul (for the present example see [15]). For the reduction of the number of variables involved the determination of eigenvalues prove to be very useful (see APPENDIX).

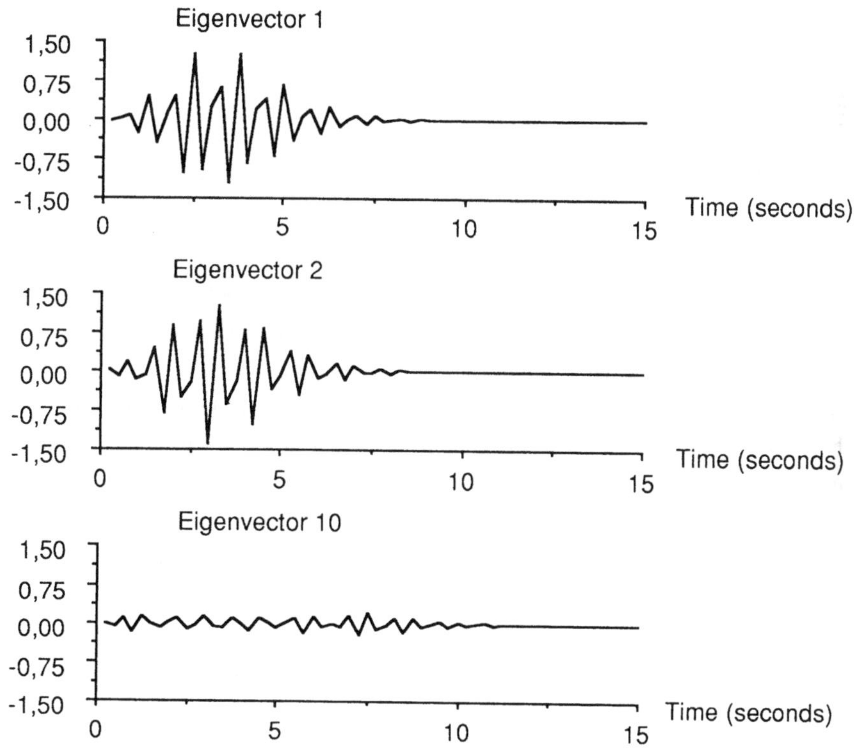

Fig. 5: Eigenvectors of the Ground Acceleration Process

The results of the discretization of the load process and the respective eigenvectors are shown in Fig. 5. Note that, for the purpose of comparing their magnitudes, the eigenvectors are already multiplied by the respective standard deviations $\sqrt{\lambda_m}$. The investigation of the sensitivity result in the critical values y_m^* as listed in Tab. 3.

Eigenvector No.	Critical Value y_m^*
1	± 5.19
2	± 5.30
3	± 6.29
4	± 5.94
.	.
.	.
.	.
10	± 15.20
> 10	> 20.00

Tab. 3: Critical Values y_m^* for the Eigenvectors of the Ground Acceleration

Since the eigenvectors of higher values show insignificant influence, only 10 variables with the lowest values, as shown in Tab. 3, are used in the further analysis. For the determination of the response surface $\bar{g}(\underline{x})$ a pairwise combination of the first four variables, i.e. those with the lowest values and hence of greatest significance, are investigated.

Utilizing the response surface $\bar{g}(\underline{x})$, Monte Carlo simulation procedures result in a failure probability of $\bar{p}_f = 1.2 \cdot 10^{-3}$. The difference p_f^c between $\bar{p}_f$ and the exact solution is determined - by using 256 simulation - to be $p_f^c = - 3.0 \cdot 10^{-4}$. From this a failure probability $p_f = 0.9 \cdot 10^{-3}$ may be calculated. A comparison with a full Monte Carlo simulation (using 8000 samples) results in a failure probability $p_f = 1.0 \cdot 10^{-3}$, which proves the goodness of the correction term as well as the good agreement with the approximate solution. It should be noted, that due to convergence problems generally used methods of approximation to determine the failure probability, such as first order methods, do not lead to satisfactory results.

5. CONCLUSIONS

The suggested procedure to determine failure probabilities of structures under static as well as dynamic loading proved to be both efficient and versatile. Its main advantage is the fact, that even for dynamic problems in context with ultimate load, i.e. collapse failure criteria, the state of the art with respect to structural mechanical and load modeling as well as numerical methods can be utilized. There are also no restrictions with respect to material laws, theories of systems analysis, probability distributions of the input parameters, etc..

The approximation of the actual limit state surfaces by response surfaces of second order polynomial type is shown to be satisfactory. For practical application it is of particular advantage that any desired design criterion can be included in the analysis. This allows the design for a particular target reliability and hence a probability based optimization of structural systems. Since the concept is formulated on a very general basis along with an easy to use software a direct application by the practicing engineer should be feasible without difficulties.

ACKNOWLEDGEMENT

This research is partially supported by the Austrian Industrial Research Promotion Fund (FFF) under contract No. 6/546 which is gratefully acknowledged by the authors.

REFERENCES

[1] Mochio, T., Samaras, E., Shinozuka, M. Stochastic Equivalent Linearization in a Finite Element-Based Reliability Analysis. Structural Safety and Reliability. Konishi, I., Ang, A.H-S., Shinozuka, M. (eds), Proc., 4th ICOSSAR'85, Kobe, May 27-29, 1985, pp. I 375-I 384.

[2] Der Kiureghian, A., Ke, J.-B. The stochastic finite element method in structural reliability. *Prob. Eng. Mech.*, 1988, Vol.3, No.2, pp. 83-91.

[3] Liu, W.K., Belytschko, T., Mani, A. Probabilistic Finite Elements for Nonlinear Structural Dynamics. *Computer Methods in Applied Mechanics and Engineering*, Vol 56, 1986, pp. 61-88.

[4] Hisada, T., Nakagiri, S. Role of the Stochastic Finite Element Method in Structural Safety and Reliability. Structural Safety and Reliability. Konishi, I., Ang, A.H-S., Shinozuka, M. (eds), Proc., 4th ICOSSAR'85, Kobe, May 27-29, 1985, pp. I 385-I 394.

[5] Liu, P.-L., Der Kiureghian, A. Finite-Element Reliability Methods for Geometrically Nonlinear Stochastic Structures. Report No. UCB/SEMM-89/05, Dept. of Civil Engineering, University of California, Berkeley, USA, January 1989.

[6] Madsen, H.O. PROBAN: Theoretical Manual for External Release. Technical Report No.: 88-2005, AS Veritas, Oslo, Norway, January 1988.

[7] Lin, Y.K. and Cai, G.Q. On Randomly Excited Hysteretic Structures. *J. Applied Mechanics*, Vol. 57, No. 2, June 1990, pp. 442-448.

[8] Shinozuka, M. Basic Analysis of Structural Safety. *J. Structural Division* , ASCE, Vol.109, No. 3, March 1983, pp. 721-740.

[9] Schuëller G.I., Bucher, C.G., Bourgund, U. and Ouypornprasert, W. On Efficient Computational Schemes to Calculate Structural Failure Probabilities. *Probabilistic Engineering Mechanics* , Vol. 4, No. 1, 1989, pp. 10-18.

[10] Schuëller G.I. and Stix, R. A Critical Appraisal of Methods to Determine Failure Probabilities. *Structural Safety* , Vol. 4, 1987, pp. 293-309.

[11] Bucher C.G., Nienstedt, J., Ouypornprasert, W. Adaptive Strategies in ISPUD V3.0 - A User's Manual. Report No.25-89, Institute of Engineering Mechanics, University of Innsbruck, Austria, 1989.

[12] Ouypornprasert W., Bucher, C.G., Schuëller, G.I. On the Application of Conditional Integration in Structural Raliability. Structural Safety and Reliability (Eds. A.H-S. Ang, M. Shinozuka and G.I. Schuëller), Proc. ICOSSAR '89, 1990,Vol. III, pp. 1683-1689.

[13] Bucher C.G. Adaptive Sampling - An Iterative Fast Monte-Carlo Procedure. *Structural Safety* , Vol. 5, No. 2, 1988, pp. 119-126.

[14] Ang, A.H-S. and Ma, H-F. On the Reliability of Structural Systems. in: Structural Safety and Reliability, T.Moan and M. Shinozuka (Eds.), Proc. ICOSSAR '81, Elsevier Scientific Publ. Co., 1981, pp. 295-314.

[15] Pradlwarter, H.J. MOFEP - A User's Manual. Institute of Engineering Mechanics, University of Innsbruck, Austria, 1990 (draft).

APPENDIX

Load Modeling

Numerical analyses require a discrete representation of the load. While this is apriori the case for static loading, the loading history has to be discretized for random dynamic loading, e.g. for earthquake loading. In other words, a continuous process a(t) is described by discrete values a_k, k=1 ..,N (see Fig A1).

This corresponds to the generally applied procedure in structural dynamics.

For a given covariance function,

$$R_a(t,s) = E[a(t)a(s)] \tag{A1}$$

(assume zero mean for a(t)) of a continuous process the covariance matrix of a discrete process is defined by

$$C_{km} = E[a_k a_m] = E[a(t_k)a(t_m)] \tag{A2}$$

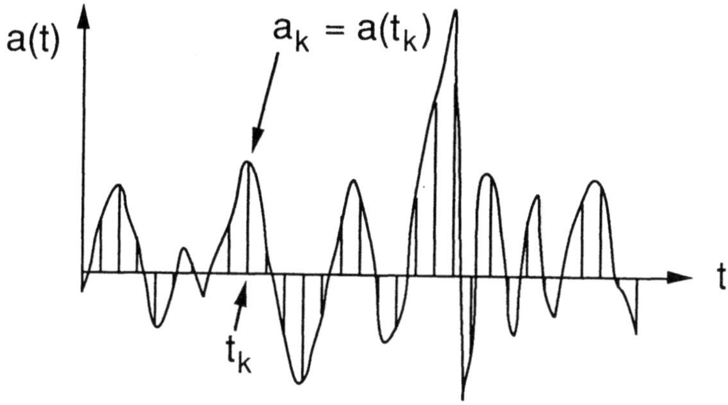

Fig. A1 Discretization of a Continuous Process

Hence, the loading process is now defined by N correlated random variables a_k. With respect to simulation it is advantageous to diagonalize the matrix $\underline{C}$, i.e.

$$\underline{\Phi}^T \ \underline{C} \ \underline{\Phi} = \text{diag} \ (\lambda_m) \tag{A3}$$

where Φ represents the orthogonal transformation matrix (eigenvectors).
By denoting

$$\underline{a} = \underline{\Phi} \ \underline{y} \tag{A4}$$

the load process a(t) is represented by a linear combination of uncorrelated variables y_m with a variance of λ_m. For realistic cases the values of λ_m show a strongly decreasing tendency, i.e. in general significantly less than N random variables are required to describe a(t) sufficiently well.

First Passage Problem: A Probabilistic Dynamic Analysis for Hot Aerospace Components

Michael C. Shiao

Sverdrup Technology, Inc., Lewis Research Center Group
Brookpark, Ohio 44142

ABSTRACT

Structures subjected to non-white random excitations with uncertain system parameters affected by surrounding environments are studied. Methods are developed to determine the statistics of dynamic responses such as time-varying mean, standard deviation and autocorrelation functions. Moreover, the first-passage problems with deterministic and stationary/nonstationary random barriers are evaluated. Time-varying (joint) mean crossing rate and the probability density function of the first-passage time for various random barriers are derived.

INTRODUCTION

Many structures which require high performance/reliability and durability often operate under complex environments including random excitations and random temperatures. These excitations and temperature variation not only degrade the material but also cause an additional randomness in the uncertain material behaviors. In addition, a small variation in the structural shape may have a significant effect on the structural responses. For this reason, a methodology is developed for a probabilistic structural dynamic analysis of stochastic structures degraded by surrounding random environments. This methodology evolves from the concept of NESSUS (Numerical Evaluation of Stochastic Structures Under Stress) which is a probabilistic structural analysis code developed under NASA's PSAM (Probabilistic Structural Analysis Methods) project (ref. 1) The methodology consists of five parts: (1) random process decomposition, (2) probabilistic material behavior model, (3) perturbed dynamic analysis of uncertain systems, (4) first order reliability method (FORM), (5) a new set of probability equations. Random process decomposition procedure expresses the random phenomenon by a set of independent random variables and their respective characteristic functions. Probabilistic material behavior model defines the relationship between material properties (Young's modulus, damping ratio, thermal expansion coefficient, strength etc.) and the random phenomena (temperature, fatigue etc.).

I. Elishakoff · Y. K. Lin (Eds.)
Stochastic Structural Dynamics 2
New Practical Applications
2nd International Conference on Stochastic Structural Dynamics
May 9–11, 1990, Boca Raton, Florida

Perturbed dynamic analysis of uncertain systems produces perturbed time histories (dynamic responses) which are used to numerically determine the response functions in terms of independent random variables at each time. The first order reliability method is used to redefine the limit state function in the standard normal probability space. A new set of probability equations are derived to determine the important response statistics and to solve the first passage problems.

The versatility of the newly developed method is demonstrated by performing a transient analysis of a stochastic structure subjected to random excitations under severe random temperature conditions. The structure is modelled by a single degree-of-freedom oscillator. The natural frequency of this oscillator, which is a function of the uncertain system parameters (geometry, material properties etc.), is also randomly perturbed by the temperature process. The non-white excitations can be stationary or nonstationary random processes. Second order statistics of any response such as time-varying mean, standard deviation and autocorrelation functions are determined. Also, the first passage problems with constant or stationary/nonstationary barriers are evaluated. A random barrier, such as the uncertain material strength or clearance for maximum displacement, is composed of its original randomness and by a variation due to the effect of random temperature and fatigue. The time-varying mean crossing rate and joint mean crossing rate are derived. The probability density functions of the first-passage time for various random barriers are determined by both crossing rate based method and equivalent system based method. In the following, each part of the methodology is described and an example is shown to verify the derivation.

RANDOM PROCESS DECOMPOSITION

For the random process in time such as the loads and temperature, a Karhunen-Loeve expansion is used (ref. 2). This expansion is a representation of the random process in terms of independent random variables U_n and their associated time-varying characteristic functions $f_n(t)$ as given by equation (1).

$$S(t) = \overline{S}(t) + \sum_{n=1}^{r} U_n \, f_n(t) \tag{1}$$

where $\overline{S}(t)$ is the mean value of the random process and r is the number of terms included. The smoother the random process is, the less the number of terms, r, is needed. Karhunen-Loeve expansion can also be used to decompose the random processes in sapce if spacially correlated random variables such as geometry, material properties are considered for a multi-degrees-of-freedom system.

PROBABILISTIC MATERIAL BEHAVIOR MODEL

A generic material behavior model (multi-factor interaction model) (ref. 3) is used to synthesize the varitions of material properties for structures operating under hostile environments such as high/cryogenic random temperature conditions. The fundamental assumption for this model is that the material behavior can be synthesized by independent random variables V. The general form of this model is shown in equation (2).

$$M_P = M_{PO} \prod_{j=1}^{N} \left[\frac{V_{F_j} - V_j}{V_{F_j} - V_{O_j}} \right]^{a_j} \tag{2}$$

where N is the number of environmental effects (temperature, fatigue, etc.); M_{PO}, V_{F_j}, V_{O_j} and V_j, independent random variables, represent the reference material property, j^{th} final, reference and current environmental variables respectively; M_P is the synthesized current material property. Each environmental effect on the material property is described by a monotonic function from an initial value to a terminal or ultimate state. The shape of the specific function depends on the exponent a_j which are determined from available experiment data or can be estimated from anticipated material behavior due to a particular environmental effect. Therefore, the form of equation (2) makes it possible to select a variety of parameters so that function dependence is consistent with the physical consideration.

PERTURBED DYNAMIC ANALYSIS OF UNCERTAIN SYSTEMS

The uncertain structural parameters such as geometry or material properties are random processes in space. The uncertain service environments such as random excitations and temperature are random processes in time. Further more, the material properties are affected and degraded by temperature. Since a probabilistic structural analysis by first order reliability method requires the input to be independent random variables, a decomposition procedure is applied to expand the random process in terms of independent random variables and their respective time-varying characteristic functions. Some other variables such as initial boundary conditions, damping ratio, random amplitude for harmonic loads etc. can also be included in the analysis.

A dynamic analysis is then performed repeatedly to generate a set of parallel time histories (dynamic responses) corresponding to a small change in each of the independent random variables. These perturbed time histories provide information on how the uncertainties in the problem parameters propagate through the time and affect the transient responses. The structural responses R(t), in terms of independent random variables U's, can be expressed as

$$R(t) = a_o(t) + \sum_{i=1}^{N} a_i(t)\, U_i \, + \, \sum_{i=1}^{N} b_i(t)\, U_i^2 \tag{3}$$

where a_i and b_i are the coefficients computed using perturbed time histories by a least-squares procedure. The explicit response function as shown in equation (3) is required for a first order reliability analysis.

THE FIRST ORDER RELIABILITY METHOD

For a given limit state function $g(\overline{Z})$, where $\overline{Z}$ is a random vector, the probability that $g(\overline{Z}) \leq 0$ is defined by

$$P(g(\overline{Z}) \leq 0) = \int_{g(\overline{Z}) \leq 0} f_{\overline{Z}}(\overline{z})\, d\overline{z} \tag{4}$$

where $f_{\overline{Z}}$ denotes the joint probability density function (jpdf). In general, $\overline{Z}$ are not normally distributed and equation (4) is difficult to evaluate. However, an approximation can be obtained by the first order reliability method (ref. 4). The limit state surface $g(\overline{Z}) = 0$ is first mapped into a standard normal space by a probability transformation. It is then assumed that the transformed limit state surface $G(\overline{U}) = 0$ in the normal space is sensibly approximated by a tangent hyperplane at the point on the surface closest to the origin. The hyperplane is the linear limit state surface $M = 0$. M is a linear safety margin in the standard normal space which is defined by

$$M = \beta - \sum_{i=1}^{N} \alpha_i\, U_i \tag{5}$$

where β is the reliaility index; α_i is the sensitivity factor of ith independent random variable; U is a standardized normal variable; N is the number of independent random variables. It should be pointed out that this safety margin is corresponding to a given limit state at a given time, not the one for an overall reliability consideration. The probability in equation (4) is then approximated by

$$P(\, g(\overline{Z}) \leq 0\,) \simeq \Phi(\, -\beta) \tag{6}$$

In the following, a first order reliability method is used, together with the explicit response function determined by equation (3), to derive the probabilistic informations of structural responses. Also derived are the time-varying (joint) crossing rate and the probability density function of the first-passage time.

THE STATISTICS OF DYNAMIC RESPONSES

The dynamic response of uncertain structures subjected to random excitation is also a random process and its second order statistics are important for reliability assessment. In the following, the mean, standard deviation and autocorrelation function of any random response process are derived. Also derived are the joint probability density function of any response and its derivative.

In order to calculate the response statistics, a limit state function g is first defined by :

$$g^t = R(t) - z \tag{7}$$

where R(t) is the structural response ($X(t)$, $\dot{X}(t)$, $\ddot{X}(t)$) at time t which is a function of independent random variables as shown in equation (3) and z is any realization. This limit state is subsequently transformed into a standardized normally distributed probability space and is replaced by a linear safety margin M using a first order reliability method. M is defined by

$$M^t = \beta^t - \sum_{r=1}^{N} \alpha_r^t U_r \tag{8}$$

where β^t and α_r^t are the reliability index and sensitivity factor respectively; U is a standardized normally distributed random variable; N is the number of independent random variables. The probability that $g^t \leq 0$ or $R(t) \leq z$ is determined by

$$\begin{aligned} Prob[g^t \le 0] &= Prob[R(t) \le z] \\ &= Prob[M^t \le 0] \\ &= \Phi(-\beta^t) \end{aligned} \tag{9}$$

where Φ is the standard normal cumulative distribution function. The cumulative distribution function (cdf) F_R for any structural response R(t) at any time can be generated from equation (9) by selecting a number of realization z from an appropriate response range as shown in equation (10).

$$F_{R(t)}(z_i) = \Phi(-\beta_i^t) \qquad i = 1, \ldots, k \tag{10}$$

Denoting $f_{R(t)}$ as the probability density function (pdf) of R(t), the mean and standard deviation of R(t) can be calculated by the equations

$$E[R(t)] = \overline{R}(t) = \sum_{i=1}^{k} z_i\, f_{R(t)}(z_i)\, \Delta z \tag{11}$$

and

$$\sigma_{R(t)} = \sqrt{\sum_{i=1}^{k} (z_i - \overline{R}(t))^2\, f_R(z_i)\, \Delta z} \tag{12}$$

The joint cdf $F_{R_1R_2}$ is derived as follows. Let R_1 and R_2 be any two of $X(t)$, $X(\tau)$, $\dot{X}(t)$, $\dot{X}(\tau)$, $\ddot{X}(t)$, $\ddot{X}(\tau)$, etc., where X(t) is a random response process. Defining the limit state function g_i and g_j by

$$g_i = R_1 - z_i \qquad g_j = R_2 - y_j \tag{13}$$

where z_i and y_j are different realizations. Also defining their corresponding linear safety margins M_i and M_j by

$$M_i = \beta_i - \sum_{r=1}^{N} \alpha_{ri}\, U_r \qquad M_j = \beta_j - \sum_{r=1}^{N} \alpha_{rj}\, U_r \tag{14}$$

The joint cdf $F_{R_1R_2}$ will be determined by the equation

$$
\begin{aligned}
F_{R_1R_2}(z_i, y_j) &= Prob[R_1 \le z_i \text{ and } R_2 \le y_j] \\
&= Prob[M_i \le 0 \text{ and } M_j \le 0] \\
&= \Phi(-\beta_i)\,\Phi(-\beta_j) + \int_0^{\rho_{ij}} \phi(-\beta_i, -\beta_j; u)\, du
\end{aligned}
\tag{15}
$$

$$
i = 1, \dots, k \qquad j = 1, \dots, l
$$

where

$$
\rho_{ij} = \sum_{r=1}^{N} \alpha_{ri}\,\alpha_{rj} \tag{16}
$$

Letting $R_1 = X(t)$ and $R_2 = X(\tau)$, the autocorrelation function of X(t) can be calculated by the equation

$$
\rho(t, \tau) = \frac{\sum\sum [z_i - \overline{X}(t)][y_j - \overline{X}(\tau)]\, f_{X(t)X(\tau)}(z_i, y_j)\, \Delta z\, \Delta y}{\sigma_{X(t)}\, \sigma_{X(\tau)}} \tag{17}
$$

where $f_{X(t)X(\tau)}$ is the joint pdf of $X(t)$ and $X(\tau)$. Letting $R_1 = X(t)$ and $R_2 = \dot{X}(t)$, equation (15) becomes the joint cdf of X(t) and $\dot{X}(t)$.

MEAN CROSSING RATE $v_\xi(t)$

Mean crossing rate of a random process X(t) crosses a random barrier $\xi(t)$ is determined by the equation

$$
\begin{aligned}
&(\textit{The average number} \text{ of } \textit{positive crossing} \text{ in time } [t, t + \Delta t]) \\
&= Prob[\textit{positive slope crossing} \text{ in time } [t, t + \Delta t]] \\
&= Prob[X(t) < \xi(t) \text{ and } X(t + \Delta t) \ge \xi(t + \Delta t)] \\
&= P'_c
\end{aligned}
\tag{18}
$$

where $\xi(t)$ is the barrier. Therefore,

$$
v_\xi(t)\ \Delta t = P'_c \tag{19}
$$

Let $t = n\Delta t$, the crossing rate becomes

$$
v_\xi(n\Delta t) = \frac{P'_c}{\Delta t} \tag{20}
$$

P'_c is determined as follows. The limit state functions g^t and $g^{t+\Delta t}$ are defined by the equations

$$g^t = \xi(t) - X(t) \qquad g^{t+\Delta t} = \xi(t+\Delta t) - X(t+\Delta t) \tag{21}$$

The corresponding linear safety margins M^t and $M^{t+\Delta t}$ in the standardized normally distributed probability space are

$$M^t = \beta^t - \sum_{r=1}^{N} \alpha_r^t U_r \qquad M^{t+\Delta t} = \beta^{t+\Delta t} - \sum_{r=1}^{N} \alpha_r^{t+\Delta t} U_r \tag{22}$$

Substituting equations (21) and (22) into equation (18), one obtains

$$\begin{aligned} P'_c &= Prob[g^t > 0 \text{ and } g^{t+\Delta t} \leq 0] \\ &= Prob[M^t > 0 \text{ and } M^{t+\Delta t} \leq 0] \end{aligned} \tag{23}$$

Define the probability events A, B, $\overline{A}$ and $\overline{B}$ by

$$\begin{aligned} A &= M^t \leq 0 & \overline{A} &= M^t > 0 \\ B &= M^{t+\Delta t} \leq 0 & \overline{B} &= M^{t+\Delta t} > 0 \end{aligned} \tag{24}$$

Substituting equation (24) into equation (23), P'_c is solved by

$$\begin{aligned} P'_c &= Prob[\overline{A}B] \\ &= Prob[B] - Prob[AB] \\ &= Prob[M^{t+\Delta t} \leq 0] - Prob[M^{t+\Delta t} \leq 0 \text{ and } M^t \leq 0] \\ &= \Phi(-\beta^{t+\Delta t}) - [\Phi(-\beta^{t+\Delta t})\,\Phi(-\beta^t) + \int_0^{\rho} \phi(-\beta^{t+\Delta t}, -\beta^t; u)\, du] \end{aligned} \tag{25}$$

where ϕ is the standard normal pdf and

$$\rho = \sum_{r=1}^{N} \alpha_r^t \alpha_r^{t+\Delta t} \tag{26}$$

JOINT MEAN CROSSING RATE $\nu_{\xi\xi}(t, \tau)$

The joint mean crossing rate that a random process X crosses a random barrier ξ at time t and τ is derived by the following equations.

$$
\begin{aligned}
&\left(\begin{array}{r}\textit{The average number of positive crossing in time } [t, t+\Delta t] \\ \text{and } \textit{positive crossing} \text{ in time } [\tau, \tau+\Delta t]\end{array}\right) \\
&= Prob \left[\begin{array}{l}\textit{positive crossing} \text{ in time } [t, t+\Delta t] \text{ and} \\ \textit{positive crossing} \text{ in time } [\tau, \tau+\Delta t]\end{array}\right] \\
&= Prob \left[\begin{array}{l} X(t) < \xi(t) \text{ and } X(t+\Delta t) \geq \xi(t+\Delta t) \text{ and} \\ X(\tau) < \xi(\tau) \text{ and } X(\tau+\Delta t) \geq \xi(\tau+\Delta t)\end{array}\right] \\
&= P''_c
\end{aligned} \tag{27}
$$

where $t > \tau$. Therefore the joint mean crossing rate is determined by the following equation with $t = n\Delta t$ and $\tau = i\Delta t$.

$$
\nu_{\xi\xi}(n\Delta t, i\Delta t) = \frac{P''_c}{(\Delta t)^2} \tag{28}
$$

P''_c is determined by first defining the limit state functions g_1, g_2, g_3 and g_4 .

$$
\begin{aligned}
g_1 &= \xi(t) - X(t) \qquad & g_2 &= \xi(t+\Delta t) - X(t+\Delta t) \\
g_3 &= \xi(\tau) - X(\tau) \qquad & g_4 &= \xi(\tau+\Delta t) - X(\tau+\Delta t)
\end{aligned} \tag{29}
$$

The corresponding linear safety margins M_1, M_2, M_3 and M_4 in the standardized normally distributed probability space are

$$
M_i = \beta_i - \sum_{r=1}^{N} \alpha_{ri} U_r \qquad i = 1, 2, 3, 4 \tag{30}
$$

Then substituting equations (29) and (30) into equation (27), one obtains

$$
\begin{aligned}
P''_c &= Prob[(g_1 > 0 \text{ and } g_2 \leq 0) \text{ and } (g_3 > 0 \text{ and } g_4 \leq 0)] \\
&= Prob[(M_1 > 0 \text{ and } M_2 \leq 0) \text{ and } (M_3 > 0 \text{ and } M_4 \leq 0)]
\end{aligned} \tag{31}
$$

Also define the probability events

$$
\begin{aligned}
A &= \; M_1 \leq 0 \qquad & \overline{A} &= \; M_1 > 0 \\
B &= \; M_2 \leq 0 \qquad & \overline{B} &= \; M_2 > 0 \\
C &= \; M_3 \leq 0 \qquad & \overline{C} &= \; M_3 > 0 \\
D &= \; M_4 \leq 0 \qquad & \overline{D} &= \; M_4 > 0
\end{aligned} \tag{32}
$$

Substitute equation (32) into equation (31), one obtains

$$P''_c = Prob[\bar{A}B\bar{C}D] \\ = Prob[BD] + Prob[ABCD] - Prob[ABD] - Prob[BCD] \tag{33}$$

To alleviate the computational difficulty in equation (33), two equivalent probability events E_{e1} and E_{e2} are derived to replace probability events AB and CD respectively. Therefore equation (33) becomes

$$P''_c = Prob[BD] + Prob[E_{e1}E_{e2}] - Prob[E_{e1}D] - Prob[BE_{e2}] \tag{34}$$

Each term in the right hand side of the equation can be easily determined from equation (15). The equivalent probability events E_{e1} and E_{e2} are determined as follows. Since the safety margins M_1 and M_2 are highly correlated, the probability event AB can be replaced by an equivalent event E_{e1} (ref. 5) in such a way that

$$Prob[AB] = Prob[M_1 \le 0 \text{ and } M_2 \le 0] \\ = Prob[M_{e1} \le 0] \\ = Prob[E_{e1}] \tag{35}$$

M_{e1} is the equivalent linear safety margin defined by

$$M_{e1} = \beta_{e1} - \sum_{r=1}^{N} \alpha_r^{e1} U_r \tag{36}$$

where

$$\beta_{e1} = -\Phi^{-1}[Prob(AB)] \tag{37}$$

and

$$\alpha_r^{e1} = \frac{\dfrac{\partial \beta_{e1}}{\partial U_r}}{\sqrt{\sum_{j=1}^{N} (\dfrac{\partial \beta_{e1}}{\partial U_j})^2}} \qquad r = 1, \ldots ,N \tag{38}$$

The probability event CD is replaced by equivalent event E_{e2} in a similiar way.

THE PROBABILITY DENSITY FUNCTION OF THE FIRST-PASSAGE TIME

The objective of a probabilistic dynamic analysis is to assess the reliability of structures subjected to random excitations. One of the failure modes concerning the structural reliability is so-called first-excursion failure. It states that the structure fails at the first excursion of a response to a given threshold. The response can be displacement, stress, strain or eigenvalue. The threshold can be thought of as maximun clearance for displacement at any time, or the material strength to resist crack initiation caused by high stress, or upper and lower bounds for eigenvalue within which responses can be minimized. The reliability assessment for such problems are often represented by the probability density function of the time when the response crosses the threshold at first time (pdf of the first passage time). In this analysis, the threshold is modeled as a deterministic time function and a nonstationary random process. The probability density function of the first-passage time is calculated by two methods: (1) crossing rate based method, (2) equivalent system based method as described next.

(a) crossing rate based method

The probability density function $f_1(t)$ of the first-passage time can be obtained by solving the integral equation

$$f_1(t) + \int_0^t K(t \mid \tau) f_1(\tau)\, d\tau = v_\xi(t) \tag{39}$$

where $\xi(t)$ is a random barrier process; $v_\xi(t)$ is the mean crossing rate; $K(t \mid \tau)$ is the crossing rate at t conditional on the first crossing having occurred at τ . It is an integral identity expressing the fact that a barrier crossing at time t must either be the first, or the first crossing must have occurred at some previous time τ. Thus, the crossing rate can be given as the sum of two contributions from the left hand side of equation (39). The simplest approximation of pdf of the first passage time is from the approximation $K(t \mid \tau) = v_\xi(t)$; i.e. the crossing rate at t is assumed to be independent of a prior first crossing at τ . In this case

$$f_1(t) = v_\xi(t) - v_\xi(t) \int_0^t f_1(\tau)\, d\tau \tag{40}$$

In the case where the Poisson hypothesis about independent barrier crossings is invalid, an improved approximation for the kernel $K(t \mid \tau)$ is necessary. If the requirement that a barrier crossing at τ should be first is dropped, then

$$K(t \mid \tau) = \frac{\nu_{\xi\xi}(t, \tau)}{\nu_{\xi}(\tau)} \tag{41}$$

The pdf of the first-passage time is then calculated by the equation

$$f_1(t) = \nu_{\xi}(t) - \int_0^t \frac{\nu_{\xi\xi}(t, \tau)}{\nu_{\xi}(\tau)} f_1(\tau)\, d\tau \tag{42}$$

where $\nu_{\xi\xi}(t, \tau)$ is the joint mean crossing rate. Letting $t = n\,\Delta t$ and $\tau = i\,\Delta t$, equation (42) can be discretized as

$$f_1(n\Delta t) = \nu_{\xi}(n\Delta t) - \sum_{i=0}^{n-1} \frac{\nu_{\xi\xi}(n\Delta t, i\Delta t)}{\nu_{\xi}(i\Delta t)} f_1(i\Delta t)\, \Delta t \tag{43}$$

(b) equivalent system based method

The probability density function $f_1(n\Delta t)$ of the first-passage time can also be defined by the equation

$$f_1(n\Delta t)\Delta t = Prob \left[\begin{array}{l} \text{positive crossing in time } [(n-1)\Delta t, n\Delta t] \\ \bigcap_{i=1}^{n-1} \text{no crossing in time } [(i-1)\Delta t, i\Delta t] \end{array} \right] \tag{44}$$

Assume Δt is small enough to allow at most one crossing occurrence in this time interval. Defining the limit state function g_i and its corresponding safety margin M_i by the equations

$$g_i = \xi(i\Delta t) - X(i\Delta t) \qquad M_i = \beta_i - \sum_{r=1}^{N} \alpha_{ir}\, U_r \qquad i = 1, \dots, n \tag{45}$$

where ξ is a random barrier process and X is a random response process; β and α are the reliability index and sensitivity factor respectively; U is a standardized normally distributed random variable; N is the number of independent random variables. Substituting equation (45) into equation (44), one obtains

$$f_1(n\Delta t)\Delta t = Prob[(g_n \leq 0 \text{ and } g_{n-1} > 0) \bigcap_{i=1}^{n-1} (g_i > 0 \text{ and } g_{i-1} > 0)\,]$$
$$= Prob[(g_n \leq 0) \bigcap_{i=1}^{n-1} g_i > 0] \tag{46}$$
$$= Prob[(M_n \leq 0) \bigcap_{i=1}^{n-1} M_i > 0]$$

The probability event $\bigcap_{i=1}^{n-1} M_i > 0$ is approximated by an equivalent event $M_e(n-1) > 0$ which is determined by a recursive equation as follows :

$$M_e(k-1) > 0 \; = \; M_{(k-1)} > 0 \bigcap M_e(k-2) > 0 \qquad k = 2, \ldots, n \tag{47}$$

Therefore,

$$f_1(n\Delta t)\Delta t = Prob[M_n \leq 0 \text{ and } M_e(n-1) > 0] \tag{48}$$

The probability at right hand side of equation (48) will be calculated by equation (25).

NUMERICAL EXAMPLES AND DISCUSSION

The algorithms derived previously are applied to a single degree-of-freedom oscillator shown in figure (1) . The spring constant is affected by a stationary random temperature process T(t) and characterized by the following material behavior model

$$K(t) = K_o\left[\frac{T_F - T(t)}{T_F - T_o}\right]^a \tag{49}$$

where K_o, T_F, T_o are independent random variables. T(t) is a stationary random process whose covariance kernel is defined by the equation

$$COV[T(t),\; T(\tau)] = \sigma_T^2 e^{-C_T|t-\tau|} \tag{50}$$

where σ_T is the standard deviation of T(t) and the constant C_T is the parameter which controls the smoothness of the random process. If $C_T = 0$, the random process reduces to a single random variable. When C_T approaches infinity, the random process becomes a white noise. Although the damping ratio is set to be constant, it can be random and treated in the same manner as for the spring constant. At this stage, only elastic analysis is performed to verify the concept of this methodology and the new set of probability equations. The extension to a viscoelastic analysis can be achieved by using an appropriate probabilistic material model to characterize this physical phenomenon. However, the remainder of the methodology should not be changed. The natural frequency of the oscillator is obtained by the equation

$$\omega_n(t) = \sqrt{\frac{K(t)}{M}} \tag{51}$$

where M is the lump mass. Substitute equation (49) into to equation (51) then

$$\omega_n(t) = \sqrt{\left[\frac{K_o}{M}\right]\left[\frac{T_F - T(t)}{T_F - T_o}\right]^a} \tag{52}$$

Therefore, $\omega_n(t)$ also becomes a random process in time. The equation of the motion is then defined by the equation

$$\ddot{x} + 2\beta\omega_n(t)\,\dot{x} + \omega_n^2(t)\,x = F(t) \tag{53}$$

where β is the damping ratio. In the following numerical examples, these values are assumed : M = 300000; $E[K_o]$ = 30000000; $\sigma(K_o)$ = 600000; $E[T_F]$ = 2100; $\sigma(T_F)$ = 42; T_o = 260; a = 0.25; $E[T(t)]$ = 1800; $\sigma(T(t))$ = 36; C_T = 0.25; β = 0.10.

The oscillator is subjected to a nonstationary random excitations F(t) which is defined by

$$F(t) = H(t)\,S(t) \tag{54}$$

where H(t) is a deterministic function defined by

$$\begin{aligned} H(t) &= \frac{t}{t_c} \qquad t \le t_c \\ &= 1 \qquad t > t_c \end{aligned} \tag{55}$$

as shown in figure (2). S(t) is a stationary random process with mean $E[S(t)] = \bar{S}(t) = 20.0$ and standard deviation $\sigma_S = 2.0$. The covariance kernel of S(t) is defined by the equation

$$COV[S(t),\ S(\tau)] = \sigma_S^2\, e^{-C_S\,|t-\tau|} \tag{56}$$

Again, the constant C_S will determine the smoothness of the random excitation.

A procedure is used next to decompose the random excitation S(t) and random temperature T(t) into a set of independent random variables and their respective time-varying characteristic functions. In the time interval $[\,-l_t, l_t]$, the process S(t) can be expanded by the equation

$$S(t) = \bar{S}(t) + \sum_{i=0}^{r} b_i f_i(t) + \sum_{j=0}^{r'} b_j f_j(t) \tag{57}$$

where b_i and b_j are the zero-mean independent random variables and

$$f_i(t) = \sqrt{\frac{2\, C_S\, \sigma_S^2}{\omega_i^2 + C_S^2}} \; \frac{\cos(\omega_i t)}{\sqrt{l_t + \dfrac{\sin(2\omega_i l_t)}{2\omega_i}}} \tag{58}$$

and

$$f_j(t) = \sqrt{\frac{2\, C_S\, \sigma_S^2}{\omega_j^2 + C_S^2}} \; \frac{\cos(\omega_j t)}{\sqrt{l_t + \dfrac{\sin(2\omega_j l_t)}{2\omega_j}}} \tag{59}$$

in which ω_i and ω_j are the solution to the transcendental equations

$$C_S - \omega_i \;\tan(\omega_i\, l_t) = 0 \tag{60}$$

and

$$\omega_j + C_S \;\tan(\omega_j\, l_t) = 0 \tag{61}$$

The random temperature process T(t) will be decomposed in a similar way.

<u>(a) the statistics of dynamic responses</u>

The covariance of the random excitations is defined by equation (56) with C_S is equal to 5 . This value will produce a non-smooth random process with correlation function similar to the earthquake records. The covariance of the random temperature is defined by equation (50) with C_T is equal to 0.25 because temperature oscillates slowly in time. The mean, standard deviation and autocorrelation function of the dynamic displacement are determined from equations (11), (12) and (17). The results are compared with Monte Carlo simulation solutions and shown in figure (3).

Good agreement is observed. It is also noticed that in the time interval (0,2), the displacement process X(t) is nonstationary with varying mean, standard deviation and autocorrelation function. However, the process becomes stationary after 2 second.

(b) the probability density function of the first-passage time

The first passage problems have been studied with different barriers and loading conditions. Two types of random barriers are used which are defined by equation (62) and (63).

Type A :

$$\xi(t) = \xi_o \left[1 - \frac{t}{t_F} \right] \tag{62}$$

Type B :

$$\xi(t) = \xi_o \left[1 - \frac{t}{t_F} \right] \left[\frac{T_F - T(t)}{T_F - T_o} \right]^a \tag{63}$$

where ξ_o is the reference barrier and t_F is the final time. $1/t_F$ is the degradation slope which reflects the environmental degradation on the barriers. Type B random barrier also includes the temperature effect. Two types of random loading are considered. The first type is the one with C_S in equation (56) equal to 5. This type of loading is similar to the earthquake records which are non-smooth random processes. The second type loading has C_S equal to 1, corresponding to a smooth random process. Two different degradation slopes are used, 0.02 and 0.15, to represent slow and fast environmental degradation. The mean reference barriers are chosen such that they are corresponding to one and three σ level of the response process at t $= 0$. The statistics of random loads and barriers for each problem are listed in Table (1).

The probability density function of the first-passage time are calculated by both crossing rate based method and equivalent system based method. The results are compared with Monte Carlo simulation as shown in figure (4) to (7). It is noticed that with (1) Type A deterministic barriers, (2) a non-smooth random excitation ($C_S = 5.0$) and (3) small barrier degradation slope (0.02), the crossing rate based method performs well as shown in figures (4a) and (4b), whereas the equivalent system based method does not. However, the equivalent system based method gives satisfactory results for problems with large barrier degradation slope (0.15) as shown in figures (4c) and (4d), while the crossing rate based method does not. With the same barrier conditions for problems in figures (4a), (4b) and using a smooth random excitation ($C_S = 1.0$) , it is observed in figure (5a) and (5b) that the results from equivalent system based method are improved compared with figure (4a) and (4b). It is also noticed in figure (5c) and (5d) that the pdf of the first-passage time predicted by crossing rate method for large degradation slope are improved compared with figure (4c) and (4d). Figure (6) are the pdf of first passage time with non-smooth random excitation ($C_S = 5.0$) and Type B random barriers. As expected, the equivalanet system based method performs well for large degradation slope as shown in figure (6c) and figure (6d). It also gives a satisfactory prediction for small degradation slope as shown in figure (6a) and (6b). For problems with Type B random barriers and smooth random excitations ($C_S = 1.0$) , some improvements in the results predicted by equivalent system based method are also observed for small degradation slope (0.02) as shown in figure (7a) and (7b). As usual, decent results are predicted by equivalent system based method with large degradation slope (0.15) as shown in figure (7c) and (7d).

In general, crossing rate based method is good for problems with small degradation slope (0.02) subjected to non-white ($C_S \leq 5$) excitations. Equivalent system based method is good for problems with large degradation slope (0.15) subjected to non-white ($C_S \leq 5.0$) cexcitations. However, when any of the following conditions is present, the results predicted by equivalent system based method will converge to the solutions by Monte Carlo simulation. The conditions are (1) smoother random excitations, (2) more uncertainties in the barrier and (3) larger degradation slope. The reason is stated as follows : In order to have equivalent system based method perform well, either the reliability index β's defined in equation (45) decrease rapidly or the correlations between the safety margins are large. Otherwise, methods other than FORM such as SORM (second order reliability method) or those suggested by Ono (ref. 6) and Wu (ref. 7) should be investigated to improve the accuracy. The degradation slope essentially controls the decreasing speed of β . The degree of uncertainties in the barriers and the smoothness of the random excitations, on the other hand, determine the correlations of the safety margins.

Although the proposed method is verified with a single degree-of-freedom oscillator, it can be applied directly to a multi-degree-of-freedom system with an application of the random process decomposition procedure to uncorrelate the spacially correlated random variables such as modulus, geometry etc. The computing time required by Monte Carlo simulation is highly dependent on the degree of freedom of the finite element model. However, the computing time required for the proposed methods is dependent on the number of random variables and is not very sensitive to the degree of freedom of the finite element model. Therefore, they are more attractive for problems with complex structures and for non-linear problems where analytical solutions do not exist. It is necessary to point out that the proposed method is not suitable for problems with white noise excitations. Because infinity number of independent random variables are required to represent this special random process which makes the proposed method inefficient.

CONCLUSION

Methods have been developed for the probabilistic dynamic analysis of uncertain structures subjected to non-white random excitations with random environmental effects on the uncertain structural parameters. The statistics of dynamic analysis such as time-varying mean, standard deviation and autocorrelation function are determined. The first-passage problems with barriers being deterministic functions and stationary/nonstationary random processes are studied. Mean crossing rate and joint mean crossing rate are derived. Two methods are proposed to determine the pdf of the first-passage time. The crossing rate based method is for problems of non-smooth random excitations with small uncertainties in the barrier and small environmental effect on the barrier. The equivalent system based method is for problems of either large barrier degradation rate, smooth random excitations, or large uncertainties in the barrier.

REFERENCES

1. Chamis C. C. "Probabilistic Structural Analysis Methods for Space Propulsion System components", presented in 3rd Space System Technology AIAA Conference, San Diego, California, June, 1986.

2. Spanos, P. D., Ghanem, R., "Stochastic Finite Element Expansion for Random Media", J. of Engineering Mechanics, Vol. 115, No. 5, May 1989, PP. 1035-1053

3. Chamis, C. C., Murthy, P. L. N., Hopkins, D. A., "Computational Simulation of High Temperature Metal Matrix Composites Cyclic Behavior", NASA TM 1022115, 1988.

4. Lind, N. C., Krenk, D., Madsen, H. O., "Methods of Structural Safety", Prentice-Hall, 1985.

5. Gollwitzer, S., Rackwitz, R., "Equivalent Components in First Order System Reliability", Reliability Engineering, Vol. 5, 1983, pp. 99-115

6. Ono, T., Idota, H., Totsuka, A., "System Reliability Using High-Order Moments", Proceeding of ICOSSAR '89, the 5th International Conference on Structural Safety and Reliability, pp. 959-966

7. Wu, Y.-T., Torng, Y. T., "A Fast Convolution Procedure for Probabilistic Engineering Analysis", Proceeding of The First International Symposium on Uncertainty Modeling and Analysis, 1990

FIGURE	RANDOM BARRIERS				LOADS
	BARRIER TYPE	REFERENCE BARRIER MEAN	REFERENCE BARRIER COV	DEGRADATION SLOPE	Cs in EQ (56)
4a	A	0.37	0.00	0.02	5
4b	A	0.47	0.00	0.02	5
4c	A	0.37	0.00	0.15	5
4d	A	0.47	0.00	0.15	5
5a	A	0.37	0.00	0.02	1
5b	A	0.47	0.00	0.02	1
5c	A	0.37	0.00	0.15	1
5d	A	0.47	0.00	0.15	1
6a	B	0.58	0.10	0.02	5
6b	B	0.74	0.10	0.02	5
6c	B	0.58	0.10	0.15	5
6d	B	0.74	0.10	0.15	5
7a	B	0.58	0.10	0.02	1
7b	B	0.74	0.10	0.02	1
7c	B	0.58	0.10	0.15	1
7d	B	0.74	0.10	0.15	1

Table (1). Statistics for Random Loads and Barriers

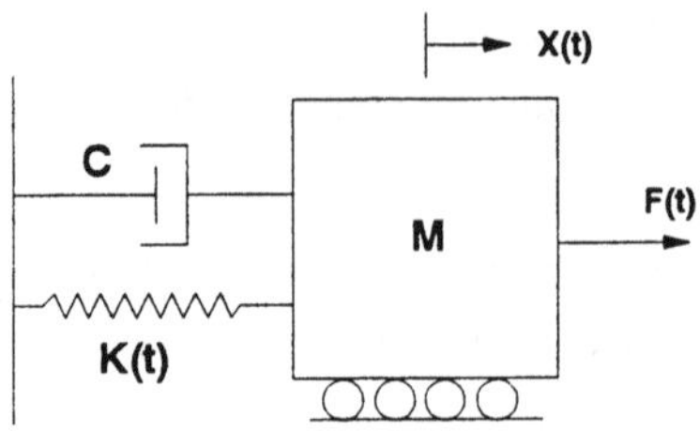

Figure (1). A Single Degree-of-Freedom Oscillator subjected to random excitation

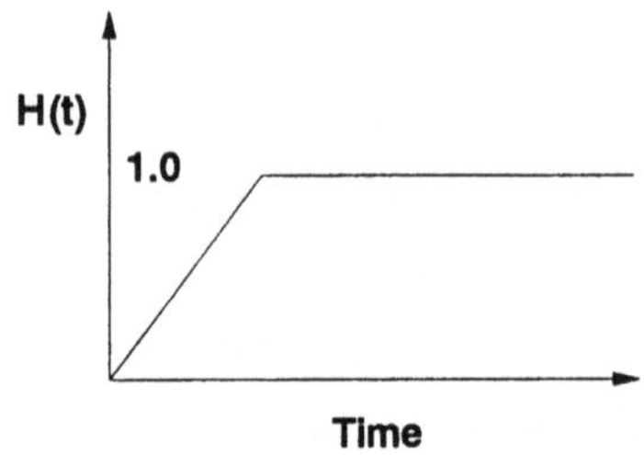

Figure (2). Modulating Function H(t)

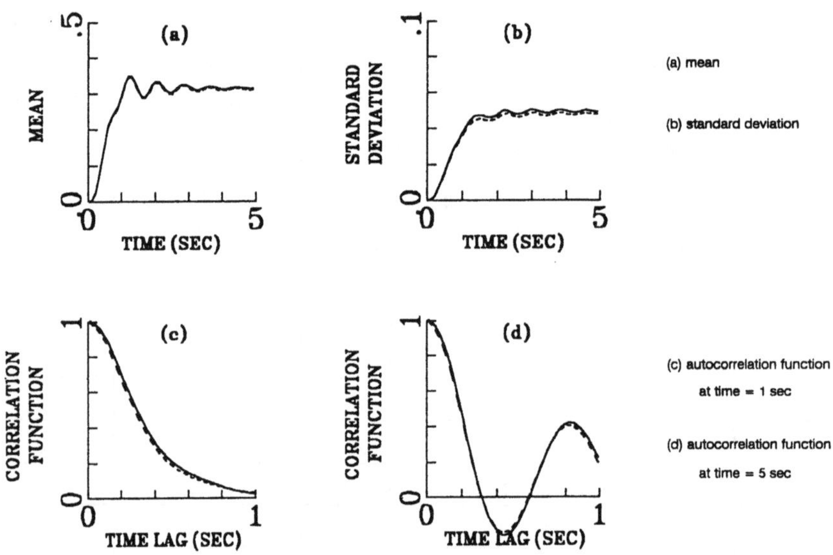

Figure (3). Statistics of The Displacement in eq.(53)
(Cs in eq.(56) = 5)

- - - - - MONTE CARLO SIMULATION ———— PROPOSED METHOD

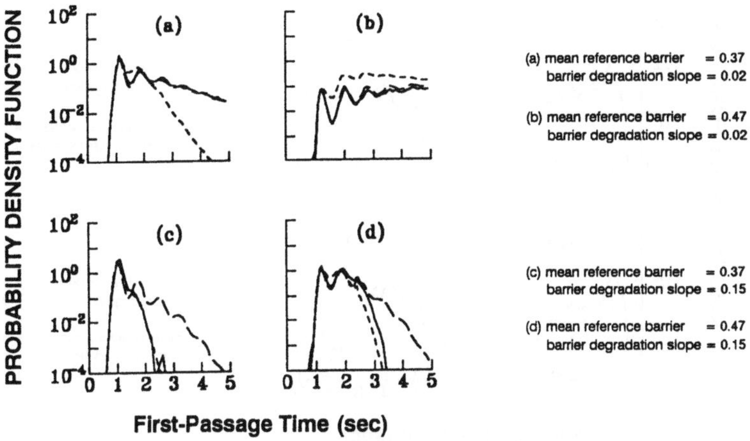

Figure (4). PDF of The First-Passage Time (Cs in eq.(56) = 5, COV of Reference Barrier = 0.00, Type A Deterministic Barrier Model)

- - - EQUIVALENT SYSTEM BASED METHOD — — CROSSING RATE BASED METHOD ——— MONTE CARLO SIMULATION

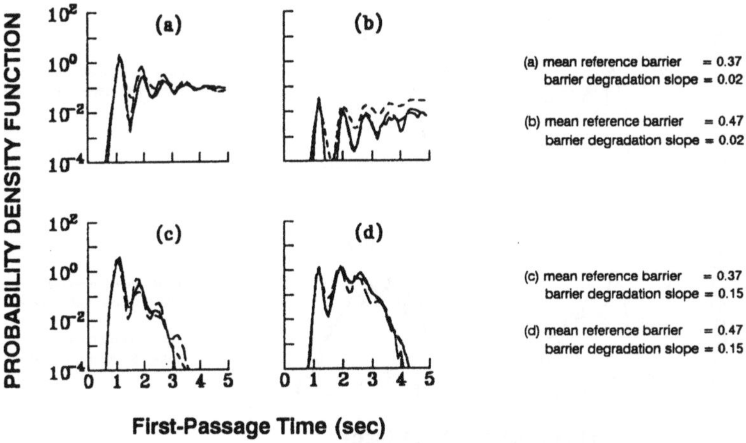

Figure (5). PDF of The First-Passage Time (Cs in eq.(56) = 1, COV of Reference Barrier = 0.00, Type A Deterministic Barrier Model)

- - - EQUIVALENT SYSTEM BASED METHOD — — CROSSING RATE BASED METHOD ——— MONTE CARLO SIMULATION

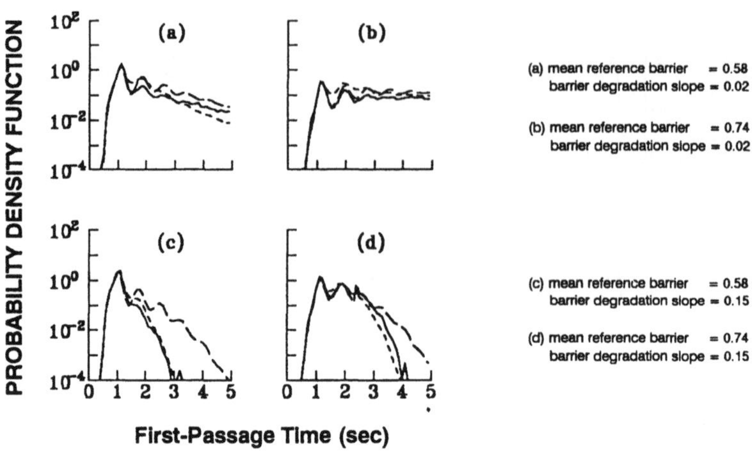

Figure (6). PDF of The First-Passage Time (Cs in eq.(56) = 5, COV of Reference Barrier = 0.10, Type B Random Barrier Model)

- - - EQUIVALENT SYSTEM BASED METHOD — — CROSSING RATE BASED METHOD ——— MONTE CARLO SIMULATION

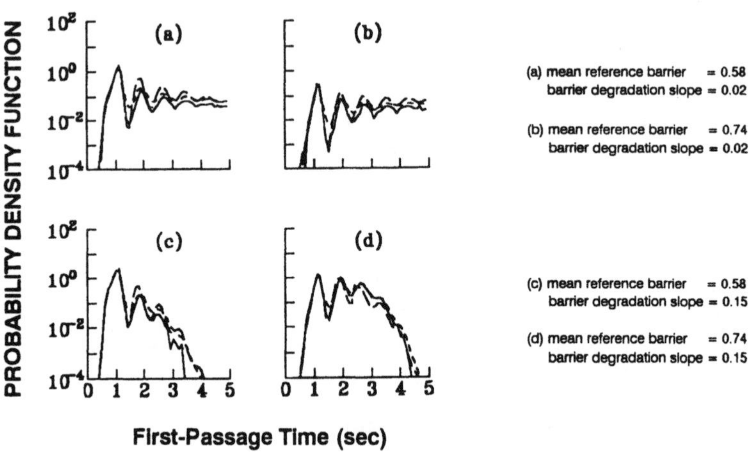

Figure (7). PDF of The First-Passage Time (Cs in eq.(56) = 1, COV of Reference Barrier = 0.10, Type B Random Barrier Model)

- - - EQUIVALENT SYSTEM BASED METHOD — — CROSSING RATE BASED METHOD ——— MONTE CARLO SIMULATION

Stability Investigation of Long-Span Bridges Using Indical Functions with Oscillatory Terms

A. STERNBERG

Raphael, Kiriat Mozkin, Haifa, 31021, Israel

1. Introduction

When wind flow passes around a structure, it generates in general, moment, lift and drag loads on the structure. If the shape of the structure is unstreamlined, which is the case with most bridge decks, then flow separation occurs, and organized vortices are shed. Sometimes the organized vortices are accompanied by smaller scale vortices. These vortices generate random pressure fluctuations in the wake, causing the structure, to move. In turn, the structure's motion generates motion-dependent loads, referred to as self-excited loads. On the other hand, turbulence that generally exists in the oncoming wind generates additional external excitations which are independent of the structural motion and are referred to as the buffeting loads.

Stability analysis of bridges has gained scientific attention since the failure of the Tacoma Narrows bridge in 1940. It is surmised that wind induced failure are mostly of the flutter type; thus, the concept of flutter derivatives for airfoils in the frequency domain [1,2] is applicable. This concept was adopted by Scanlan and his co-researchers [3,4,5] for the modeling of aerodynamic bridge loads in terms of measured flutter derivaties [4]. The corresponding time-domain analysis in terms of indicial functions has also been carried out by Scanlan, Beliveau and Budlong [6], and Scanlan [7]. In 1979, Lin [8] and later Lin & Ariaratnam [9] proposed a self-excited load model which is turbulence dependent. It was shown that turbulence had a destabilizing effect on a single-degree-of-freedom bridge model. More recently Bucher &

I. Elishakoff · Y. K. Lin (Eds.)
Stochastic Structural Dynamics 2
New Practical Applications
2nd International Conference on Stochastic Structural Dynamics
May 9-11, 1990, Boca Raton, Florida

Lin [10,11], extended the analysis to a multiple-degree-of-freedom system and found that turbulence played two opposite roles in affecting the stability of bridge motion: It would destabilize individual modes if these modes were to act alone; however, it could also serve as a conduit to transfer energy from the least stable mode to more stable modes, thus stabilizing the over-all system.

Incorporating turbulence in the self-excited loads results in equations of motion that contain randomly, time-varying coefficients, referred to as random parametric excitations. In this case, the frequency-domain analysis is no longer applicable. Instead, formulation of the self-excited loads in the time domain is appropriate. Using the stochastic averaging procedure and the Itô calculus [12] a set of first order linear differential equations for the first and second statistical moments can be obtained, and stabilities of these moments can be investigated quite simply. The time-domain analysis requires the use of aerodynamic indicial functions or equivalently, aerodynamic impulse response functions, which were assumed to be of the non-oscillatory first-order filter type in the previous investigations [6-11]. This paper focuses on the problem of proper modeling of the aerodynamic impulse response functions for the self excited loads. Oscillatory second-order-filter type impulse response functions are found to agree much better with the measured flutter derivatives for some bridge decks. This study compares the stability boundaries for the first and second statistical moments obtained separately from the oscillatory and non-oscillatory models for two bridges, thus providing some insight into the importance of the oscillatory self-excited loads.

2. Deterministic Stability Analysis

In the following discussion, a bridge model capable of two degrees of freedom, one torsion and one vertical bending, will be considered. The equations of motion are given by

$$\ddot{\alpha} + Z\zeta_\alpha \omega_\alpha \dot{\alpha} + \omega_\alpha^2 \alpha = M/I \tag{2.1}$$

$$\ddot{h} + 2\zeta_h \omega_h \dot{h} + \omega_h^2 h = L/m \qquad (2.2)$$

where α and h are the modal torsional angle and modal vertical displacements, respectively; an over-dot denotes one differentiation with respect to time; each ω_j is a natural frequency, and each ζ_j is a damping ratio with the subscript $j = \alpha$ or h, indicating either the torsional mode or the bending mode; M and L are the generalized modal aerodynamic pitching moment and lift, respectively; I is the modal mass moment of inertia and m the modal mass. It is implied in equations (2.1) and (2.2) that the two degrees of freedom are uncoupled in inertia, stiffness and damping; therefore, only aerodynamic coupling is considered in what follows. This restriction can be relaxed with some additional algebra.

The deterministic analysis refers to the ideal case in which no turbulence exists in the oncoming wind. Two types of instability are possible, the non-oscillatory divergence or oscillatory flutter. It has been found [5] that flutter is more likely to occur for long-span bridges. This is also confirmed in the present study in that divergence is associated with a much higher wind speed than flutter for both bridges investigated. Therefore, divergence will be omitted in the remaining discussion.

Flutter commences at a wind speed that renders the combined structural and aerodynamic damping of a coupled vibration mode, zero. Beyond that wind speed the total damping becomes negative and the amplitude of the oscillatory motion increases indefinitely.

The unsteady, self-excited aerodynamic loads may be expressed as the following convolution integrals

$$M(t) = M_\alpha(t) + M_h(t) = \int_0^t f_{M\alpha}(t-\tau)\alpha(\tau)d\tau$$

$$+ \int_0^t f_{Mh}(t-\tau)h(\tau)d\tau \qquad (2.3)$$

$$L(t) = L_\alpha(t) + L_h(t) = \int_0^t f_{L\alpha}(t-\tau)\alpha(\tau)d\tau$$

$$+ \int_0^t f_{Lh}(t-\tau)h(\tau)d\tau \quad (2.4)$$

in which M_α and M_h are the torsional moments generated by the torsional and vertical motions respectively, and L_α and L_h are the lifts generated by torsional and vertical motions, respectively. The physical meanings of the corresponding aerodynamic impulse response functions $f_{M\alpha}$, f_{Mh}, $f_{L\alpha}$, f_{Lh} are clear in these equations.

Expressing the aerodynamic loads as convolution integrals as in equation (2.3) and (2.4) and inserting them into equations (2.1) and (2.2), the latter becomes integro-differential equations. Application of the Fourier Transform to these equations yields

$$(\omega_\alpha^2 - \omega^2 + 2i\zeta_\alpha \omega_\alpha\omega)\ \bar{\alpha}(\omega) = \bar{f}_{M\alpha}(\omega)\ \bar{\alpha}(\omega) + \bar{f}(\omega)\ \bar{h}(\omega) \quad (2.5)$$

$$(\omega_h^2 - \omega^2 + 2i\zeta_h \omega_h\omega)\ \bar{h}(\omega) = \bar{f}_{L\alpha}(\omega)\ \bar{\alpha}(\omega) + \bar{f}_{Lh}(\omega)\ \bar{h}(\omega) \quad (2.6)$$

in which an overbar denotes a Fourier transformed quantity. The $\bar{f}(\omega)$ functions may be referred to as the aerodynamic transfer functions. A comparison of equations (2.5) and (2.6) with the flutter derivatives from Ref. 4, yields the following relations:

$$\bar{f}_{M\alpha}(\omega) = \rho B^4\ \omega^2[A_3^* + i\ A_2^*] \quad (2.7)$$

$$\bar{f}_{Mh}(\omega) = \rho B^3\ \omega^2[A_4^* + i\ A_1^*] \quad (2.8)$$

$$\bar{f}_{L\alpha}(\omega) = \rho B^3\ \omega^2[H_3^* + i\ H_2^*] \quad (2.9)$$

$$\bar{f}_{Lh}(\omega) = \rho B^2\ \omega^2[H_4^* + i\ H_1^*] \quad (2.10)$$

in which A_i^* and H_i^* are the flutter derivatives in nondimensional form, and they are expressed as functions of the reduced (namely, nondimensionalized) velocity $V = 2\pi u/\omega B$, where u is the physical flow velocity. For bluff bridge profiles these quantities have to be measured experimentally and were traditionally used in bridge stability and response analyses since they were easier to measure.

Substituting equations (2.7) through (2.10) into (2.5) and (2.6) and casting into the following matrix form

$$\begin{bmatrix} D_{11} & D_{12} \\ D_{21} & D_{22} \end{bmatrix} \begin{bmatrix} \bar{\alpha} \\ \bar{h} \end{bmatrix} = 0 \tag{2.11}$$

it can be shown that

$$D_{11} = - X^2 + 2i\zeta_\alpha X(\omega_\alpha/\omega_h) + (\omega_\alpha/\omega_h)^2 - (\rho B^4/I)X^2(A_3^* + iA_2^*)$$

$$D_{12} = - (\rho B^3/I)X^2(A_4^* + iA_1^*) \tag{2.12}$$

$$D_{21} = - (\rho B^3/m)X^2(H_3^* + iH_2^*)$$

$$D_{22} = -X^2 + 2i\zeta_h X + 1 - (\rho B^2/m)X^2(H_4^* + iH_1^*)$$

where $X = \omega/\omega_h$. A nontrivial solution is possible when the determinant of the coefficient matrix is zero. A complex quartic equation of the unknown frequency ratio X and the reduced velocity is obtained. By equating separately the real and imaginary parts to zero, X and V can be solved numerically. Results for the deterministic critical wind speeds and frequencies are listed in Table 2 for two typical bridge models. The structural data utilized in the calculation is given in Table 1.

3. Identification of Indicial Functions for Self Excited Loads

The self-excited loads must be represented in the time domain in order that the stability analysis can be extended to include random fluctuation in the flow velocity. Assuming that the flutter derivatives have been determined experimentially, the aerodynamic impulse response functions can be identified through Fourier transformation. In the aeronautical engineering literature, the indicial functions (or step response functions) are more commonly used. These are related to the impulse response functions by integration; for example,

$$g_{Mh}(t) = \int_0^t f_{Mh}(\tau)d\tau \tag{3.1}$$

The indicial function for the self-excited lift on an airfoil in uniform flow induced by a vertical motion of the airfoil is known as the Wagner function [14]. The following approximate form for the Wagner function has been proposed by Jones [15].

$$g(t) = C_0 + C_1\, e^{-a_1 t} + C_2\, e^{-a_2 t} \tag{3.2}$$

The parameters C_i and a_i were identified by fitting equation (3.2) to the known Wagner function. Scanlan and his associates adopted the same form for the indicial functions of bridge decks [6]. They used the measured flutter derivatives from Ref. 4 to identify the parameters C_k and a_k of the indicial functions, using a least square error algorithm. The same approach was used by Bucher & Lin [10]. Alternatively, Huston [16] adopted a more general form of a Prony series for the impulse response functions as follows:

$$f(t) = \sum_{n=0}^{N} C_n\, \delta^{(n)}(t) + \sum_{n=N+1}^{M} C_n \exp(-a_n t) \tag{3.3}$$

in which the $\delta^{(n)}(t)$ is the n-th derivative of the delta function. It is clear that the Jones approximation, equation (3.2) is a special case of equation (3.3).

In Refs. 6, 10 and 16, the parameters C_n and a_n were assumed to be purely real; thus, these models could not capture the oscillatory nature which was evident in some measured flutter derivatives [4]. In this paper, the Prony series representation (3.3) is extended to include complex conjugate exponential terms as follows

$$f(t) = \sum_{k=0}^{N} C_k \, \delta^{(k)}(t) + \sum_{k=N+1}^{r} C_k \, e^{-a_k t} + \sum_{m=-i}^{i} D_m \, e^{-b_m t} \tag{3.4}$$

in which D_m and b_m are two sets of complex conjugate parameters. Without loss of generality, we may let $D_0 = 0$. The complex terms account for oscillatory behavior found in some of the experimental results, for example M_h reported by Scanlan and Tomko [4]. The impulse response function for this load with oscillatory terms will have the following form

$$f_{Mh}(t) = \rho U^2 B^2 [C_{1,Mh}\, \delta(t) + C_{2,Mh} \frac{B}{U} \dot{\delta}(t) + \delta(t)(\sum_{k=3}^{4} C_{k,Mh}$$

$$+ \sum_{k=-1}^{1} D_{k,Mh}) - \sum_{k=3}^{4} C_{k,Mh}\, a_{k,Mh} \frac{U}{B}\, e^{-a_{k,Mh}(U/B)t}$$

$$- \sum_{m=-1}^{1} D_{m,Mh}\, b_{m,Mh} \frac{U}{B}\, e^{-b_{m,Mh}(U/B)t}] \tag{3.5}$$

in which B is the bridge deck width, U is the mean wind speed and ρ is the air density, $D_{-1,Mh}$ is the complex conjuguate of $D_{1,Mh}$. Oscillatory self-excited loads due to motion-induced vortices were also reported by Shiraishi & Matsumoto [17,18].

The indicial functions follow from integrating equation (3.5), resulting in

$$g_{Mh}(t) = \rho U^2 B^2 [C_{1,Mh} + C_{2,Mh} \frac{B}{U} \delta(t) + \sum_{k=3}^{n} C_{k,Mh} e^{-a_{k,Mh}(U/B)t}$$

$$+ \sum_{m=-1}^{1} D_{m,Mh}\, e^{-b_{m,Mh}(U/B)t}] \tag{3.6}$$

The first two terms within the brackets of equation (3.6) are recognized as the quasi-steady terms. The reason for not combining all the terms involving $\delta(t)$ in equation (3.5) in one group is now clear; it results in a simpler indicial function equation (3.6). Without the last term (the "oscillatory term"), the indicial function is similar to those in earlier works [6,8,9,10,11,19,20,21] for either deterministic or stochastic stability analysis.

Let

$$b_{Mh} = d_{Mh} + i\, k_a \tag{3.7}$$

in which k_a is a reduced circular frequency, defined as,

$$k_a = \frac{\omega_a B}{U} \tag{3.8}$$

where ω_a is the frequency of the oscillatory load due to motion-induced vortices. A note on the physical significance of C_1 in each one of the four impulse response functions is in order. For example, the coefficient $C_{1,M\alpha}$ is associated with the quasi-steady moment due to the torsional motion; therefore it should be identified with $dC_M/d\alpha$. Similarly, $C_{1,Mh}$, $C_{1,L\alpha}$ and $C_{1,Lh}$ should be identified, respectively, with dC_M/dh, $dC_L/d\alpha$ and dC_L/dh. It is argued that the lift and moment coefficients should not be dependent on the instantaneous vertical position of the bridge in the flow. Therefore, the coefficients $C_{1,Lh}$ and $C_{1,Mh}$ are expected to be zero. This argument will be used towards a proper identification of the parameters in the indicial functions. The following discussion will pertain to f_{Mh}, the oscillatory indicial function. The same procedure is applicable to other indicial functions.

Applying the Fourier transform to the second convolution integral in equation (2.3) and using the relation (2.8), it can be readily shown that the oscillatory flutter derivatives A_1^* and A_4^* are related to the parameters of the indicial function as follows:

$$A_1^*(V) = \frac{V^2}{4\pi^2}\left[C_{1,Mh} + \sum_{k=3}^{4} \frac{4\pi^2 C_{k,Mh}}{a_{k,Mh}^2 V^2 + 4\pi^2} + 2D_R \right.$$

$$- \frac{V^2(N_5 + N_6)}{[(d_{5,Mh}^2 V^2 + k_a^2 V^2 - 4\pi^2)^2 + 16\pi^2 d_{5,Mh}^2 V^2} \Bigg]$$

$$N_5 = C_{5,Mh} d_{5,Mh} (d_{5,Mh}^2 + k_a^2 V^2 + 4\pi^2)$$

$$N_6 = C_{6,Mh} k_a (d_{5,Mh}^2 V^2 + k_a^2 V^2 - 4\pi^2)\} \qquad (3.9)$$

$$A_4^*(V) = \frac{V}{2\pi} \Bigg[C_{2,Mh} + \sum_{k=3}^{4} C_{k,Mh} \frac{a_{k,Mh} V^2}{a_{k,Mh}^2 V^2 + 4\pi^2}$$

$$+ \frac{V^2\{C_{5,Mh}(d_{5,Mh}^2 V^2 - k_a^2 V^2 + 4\pi^2 + 2C_{6,Mh}\, d_{5,Mh}\, k_a\, V^2\}}{(d_{5,Mh}^2 V^2 + k_a^2 V^2 - 4\pi^2)^2 + 16\pi^2 d_{5,Mh}^2 V^2} \Bigg]$$

(3.10)

in which V = reduced velocity. The real parameters $C_{5,Mh}$ and $C_{6,Mh}$ are related to the real and imaginary parts of $D_{1,Mh} = D_R + iD_I$ as follows

$$D_R = \frac{k_a C_{6,Mh} + d_{5,Mh} C_{5,Mh}}{2\,(k_a^2 + d_{5,Mh}^2)}; \quad D_I = \frac{d_{5,Mh} C_{6,Mh} - k_a C_{5,Mh}}{2\,(k_a^2 + d_{5,Mh}^2)} \qquad (3.11)$$

Note that the term $2D_R$ in (3.9) is kept separate from the parameter $C_{1,Mh}$ because of their different physical meaning. Relations similar to (3.9) and (3.10) can be obtained for the other pairs of flutter derivatives by setting C_5 and C_6 to zero for the nonoscillatory flutter derivatives. Equations like (3.9) and (3.10) can then be used for the identification of the parameters $C_{k,Qq}$, $a_{k,Qq}$, $d_{5,Mh}$ and k_a from measured flutter derivatives taken from Scanlan & Tomko [4]. (The subscript Qq correspond to one of the load types M_α, M_h, L_α or L_h).

Two bridge decks for which the measured flutter derivative A_1^* displaying an oscillatory behavior were chosen from Ref. 4 as examples for parameter identification in this study. (Bridges

No. 2 and No. 8 in Ref. 4). Unfortunately, the flutter derivatives A_4^* and H_4^* (the real parts of the corresponding aerodynamic transfer functions) that relate the pitching moment and the lift, respectively, to the vertical bridge motion, were not measured [4] on the assumption that they are negligible. It was argued earlier that A_4^* and H_4^* need not be identically zero, but only certain parameters related to these flutter derivatives, namely, $C_{1,Lh}$ and $C_{1,Mh}$ are negligible.

Solutions for equations (3.9) and (3.10) and similar equations for the other self-excited loads, were carried out on a VAX 11/750 minicomputer with the aid of a numerical routine from the IMSL library. The routine performs optimal curve fitting using the least square principle as the parameters in the indicial functions are varied.

Figure 1 depicts the measured flutter derivatives [4] for the pitching moment of bridge No. 1 in Table 1, and the corresponding Fourier transforms of fitted indicial functions obtained in this work. Note particularly the excellent agreement between the measured and fitted results for the oscillatory flutter derivative A_1^*, as compared with the earlier results obtained in Ref. 10 for the same bridge deck where the oscillatory component is neglected. The variation of the flutter derivative A_4^* is seen to be within a very narrow range close to zero, compatible to Scanlan's assumption on the negligibility of this flutter derivative. Figure 2 depicts the fitted results of the flutter derivative A_1^* for bridge No. 8, as obtained in this work, applying once the non-oscillatory indicial function (Fig. 2a) and then the oscillatory model (Fig. 2b). Again it is clear that the non-oscillatory model fails to fit adequately to the measured data. The flutter derivative A_4^* which is presumably very small, is identified as such in Figure 2b. The other flutter derivatives corresponding to the induced lifts were satisfactorily fitted with non-oscillatory indicial functions (not shown), and the results were similar to those obtained in Ref. 10. Table 3 lists the identified parameters for both bridge models. Note that parameters $C_{1,Lh}$ and $C_{1,Mh}$ are either identically zero or very small, when the oscillatory indicial function model is used, in accord with the earlier assumption.

4. Stochastic Stability Analysis

Stability analysis incorporating flow fluctuations in the self-excited loads must be pursued in the time domain since the equations of motion contain randomly varying parameters. Hence, the system is time variant and the usual spectral analysis is not applicable.

Stochastic stability of a dynamic system must be interpreted in terms of convergence of stochastic variables. For engineering purposes, stability in terms of the first two statistical moments is generally adequate.

In what follows, only the pitching moment M_h induced by the vertical motion will be treated in detail since it is the only one that contains an oscillatory component in the present examples. The other non-oscillatory self-excited loads can be obtained in a similar manner.

Substitution of the aerodynamic impulse response function f_{Mh} of the form of (3.5) in the second convolution integral of equation (2.3) yields

$$M_h(t) = \rho U^2 B^2 [C_{1,Mh} h(t)/B + C_{2,Mh} \dot{h}(t)/U +$$

$$+ \sum_{k=3}^{4} C_{k,Mh} \int_0^t \exp[-a_{k,Mh}(t-\tau)U/B][\dot{h}(\tau)/B]d\tau +$$

$$+ \sum_{m=-1}^{1} D_{m,Mh} \int_0^t \exp[-b_{Mh}(t-\tau)U/B][\dot{h}(\tau)/B]d\tau] \quad (4.1)$$

Note the inclusion of the oscillatory terms in the above expression.

Equation (4.1) may now be modified to account for random wind velocity fluctuation. Let

$$U(t) = \bar{U}[1 + \xi(t)] \tag{4.2}$$

where $\bar{U}$ is the mean wind speed and $\xi(t)$ is the ratio of the fluctuating wind speed to $\bar{U}$. The ratio $\xi(t)$ rarely reaches 0.2, thereby permitting the approximation

$$U^2(t) \simeq \bar{U}^2[1 + 2\xi(t)] \tag{4.3}$$

The length scale of atmospheric turbulence is generally large compared with the lateral dimensions of a bridge. Thus, it is reasonable to assume that the generating mechanism of self-excited loads remains the same as that of smooth flows. Specifically, we assume that equation (4.1) can be replaced by

$$M_h(t) = \rho B^2 U^2\{[C_{1,Mh}h(t)/B + C_{2,Mh}\dot{h}(t)/\bar{U}][1+2\xi(t)] +$$

$$+ \sum_{k=3}^{4} C_{k,Mh} \int_0^t \exp[-a_{k,Mh}(t-\tau)\bar{U}/B][1+2\xi(\tau)]\dot{h}(\tau)/B \, d\tau$$

$$+ \sum_{m=-1}^{1} D_{m,Mh} \int_0^t \exp[-b_{m,Mh}(t-\tau)\bar{U}/B][1+2\xi(\tau)]\dot{h}(\tau)/B \, d\tau\}$$

$$\tag{4.4}$$

Similar expressions can be found for M_α, L_α and L_h. A set of coefficients P_i for these loads are summarized as follows:

$P_1 = A_M C_{1,M\alpha}$; $P_2 = A_M C_{2,M\alpha}$; $P_3 = A_M C_{3,M\alpha}$;

$P_4 = A_M C_{4,M\alpha}$;

$P_{11} = A_M C_{1,Mh}$; $P_{12} = A_M C_{2,Mh}$; $P_{13} = A_M C_{3,Mh}$;

$$P_{14} = A_M\, C_{4,Mh}; \quad P_{15} = A_M\, D_{Mh}; \quad P_{16} = A_M\, D_{Mh}^{*};$$

$$P_{21} = A_L\, C_{1,L\alpha}; \quad P_{22} = A_L\, C_{2,L\alpha}; \quad P_{23} = A_L\, C_{3,L\alpha};$$

$$P_{24} = A_L\, C_{4,L\alpha}; \tag{4.5}$$

$$P_{31} = A_L\, C_{1,Lh}; \quad P_{32} = A_L\, C_{2,Lh}; \quad P_{33} = A_L\, C_{3,Lh};$$

$$P_{34} = A_L\, C_{4,Lh};$$

where $A_M = \dfrac{\rho B^2 \overline{U}^2}{I}$ and $A_L = \dfrac{\rho B \overline{U}^2}{m}$. Note that oscillatory terms are associated with the coefficients $P_{k+10(n-1)}$, where k=5,6 and n=1,2,3 or 4, corresponding to the four convolution integrals in equations (2.3) and (2.4). In the present examples the oscillatory load is M_h, therefore only n=2 applies. Using the notation of (4.5) the following structural parameters can be defined:

$$\zeta_{o\alpha} = (2\zeta_\alpha \omega_\alpha - P_2\, B/\overline{U}) \,/\, (2\omega_{o\alpha})$$

$$\omega_{o\alpha} = (\omega_\alpha^2 - P_1)^{1/2}$$

$$\zeta_{oh} = (2\zeta_h \omega_h - P_{32}/\overline{U}) \,/\, (2\omega_{oh})$$

$$\omega_{oh} = (\omega_h^2 - P_{31}/B)^{1/2} \tag{4.6}$$

These are the effective damping ratios and natural frequencies in torsion and heaving motions. Dependence on the mean wind speed is noted. Note also that the effective damping ratios can become negative; however, this does not necessarily lead to instability in the stochastic case, since the horizontal turbulence also contributes to damping as will be seen in the ensuing analysis.

The equations of motion become now:

$$\ddot{\alpha} + 2\zeta_{o\alpha}\omega_{o\alpha}\dot{\alpha} + \omega_{o\alpha}^{2}\alpha = 2P_{1}\xi(t) + 2P_{2}\ (B/\bar{U})\xi(t)\ \dot{\alpha} +$$

$$+ \sum_{k=3}^{4} P_k \int_0^t \exp[-a_{k,M\alpha}(t-\tau)\bar{U}/B][1+2\xi(\tau)]\dot{\alpha}(\tau)d\tau$$

$$+ P_{11}h(t)/B + P_{12}\dot{h}(t)/\bar{U} + 2P_{11}\xi(t)h(t)/B + 2P_{12}\xi(t)\dot{h}(t)/\bar{U}$$

$$+ \sum_{\substack{k=3\\ j=k+10}}^{4} P_j \int_0^t \exp[-a_{k,Mh}(t-\tau)\bar{U}/B][1+2\xi(\tau)][\dot{h}(\tau)/B]d\tau +$$

$$+ \sum_{\substack{k=5\\ j=k+10}}^{6} P_j \int_0^t \exp[-a_{k,Mh}(t-\tau)\bar{U}/B][1+2\xi(\tau)][\dot{h}(\tau)/B]d\tau \qquad (4.7)$$

$$\ddot{h} + 2\zeta_{oh}\omega_{oh}\dot{h} + \omega_{oh}^{2}h = P_{21}\alpha + P_{22}\dot{\alpha}\ B/\bar{U} + 2P_{21}\xi(\tau) +$$

$$+ 2P_{22}\xi(t)\dot{\alpha}\ B/\bar{U} +$$

$$+ \sum_{\substack{k=3\\ n=k+20}}^{4} P_n \int_0^t \exp[-a_{k,L\alpha}(t-\tau)\bar{U}/B][1+2\xi(\tau)]\dot{\alpha}(\tau)d\tau$$

$$+ 2P_{31}\xi(t)h(t)/B + 2P_{32}\xi(t)\dot{h}(t)/\bar{U}$$

$$+ \sum_{\substack{k=3\\ m=k+30}}^{4} P_m \int_0^t \exp[-a_{k,Lh}(t-\tau)\bar{U}/B][1+2\xi(\tau)]\ [\dot{h}(\tau)/B]d\tau \qquad (4.8)$$

In equation (4.7) $a_{5,Mh}$, $a_{6,Mh}$, P_{15} and P_{16} are two complex conjugate pairs.

Equations (4.7) and (4.8) are differential integral equations. For the purpose of stability analysis it is convenient to convert these equations to a set of purely differential equations in the first order form. First, each convolution integral may be represented by a new state variable

$$Q_{qi} = P_j \int_0^t \exp[-a_{k,Qq}(t-\tau)\bar{U}/B]\ [1+2\xi(\tau)]\dot{q}(\tau)d\tau \qquad (4.9)$$

in which Q is either M or L and q can be either α or h/B; $a_{k,Qq}$ are the parameters for Qq load term; i = 1, 2, 3 or 4, corresponding to the first, second, third or fourth convolution integral in each load expression; k = i + 2 and j = k + 10(m-1). Furthermore, the indices m=1 and 2 are used for the state variables $M_{\alpha i}$ and M_{hi}, respectively, and m = 3 and 4 for $L_{\alpha i}$ and L_{hi}, respectively. In the expressions for M_{h3} and M_{h4}, the parameters $a_{5,Qq}$ and $a_{6,Qq}$ are a complex conjugate pair. Differentiating (4.9) with respect to time t yields

$$\dot{Q}_{qi} = -\, a_{k,Qq}(\overline{U}/B)Q_{qi} + P_j[1+2\xi(t)]\dot{q}(t) \tag{4.10}$$

The entire set of the first order differential equations may be cast in a matrix form

$$\dot{Y} = A\,Y + B\,Y\,\xi(t) \tag{4.11}$$

in which Y is a 14-dimensional state vector and A and B are 14x14 matrices. The list of nonzero elements of the matrices is given in Appendix I. The state vector Y contains the following 14 state variables $Y_1 = \alpha$, $Y_2 = \dot{\alpha}$, $Y_3 = h$, $Y_4 = \dot{h}$, $Y_5 = M_{\alpha 1}$, $Y_6 = M_{\alpha 2}$, $Y_7 = M_{h1}$, $Y_8 = M_{h2}$, $Y_9 = M_{h3}$, $Y_{10} = M_{h4}$, $Y_{11} = L_{\alpha 1}$, $Y_{12} = L_{\alpha 2}$, $Y_{13} = L_{h1}$, $Y_{14} = L_{h2}$. The first 4 rows of equation (4.11) are:

$$\dot{Y}_1 = Y_2$$

$$\begin{aligned}\dot{Y}_2 = &-\omega_{o\alpha}^2 Y_1 - 2\zeta_{o\alpha}\omega_{o\alpha}Y_2 + 2P_1\xi(t)Y_1 + 2P_2\xi(t)(B/\overline{U})Y_2 \\ &+ (P_{11}/B)Y_3 + (P_{12}/\overline{U})Y_4 + (2P_{11}/B)\xi(t)Y_3 + (2P_{12}/\overline{U})\xi(t)Y_4 \\ &+ Y_5 + Y_6 + Y_7 + Y_8 + Y_9 + Y_{10}\end{aligned}$$

$$\dot{Y}_3 = Y_4$$

$$\dot{Y}_4 = -\,\omega_{oh}^{\ 2}Y_3 - 2\zeta_{oh}\omega_{oh}Y_4 + P_{21}Y_1 + P_{22}(B/\overline{U})Y_2 + 2P_{21}\xi(t)Y_1$$

$$+ 2P_{22}(B/\bar{U})\xi(t)Y_2 + (2P_{31}/B)\xi(t)Y_3 + (2P_{32}/\bar{U})\xi(t)Y_4$$

$$+ Y_{11} + Y_{12} + Y_{13} + Y_{14} \qquad (4.12)$$

and the other differential equations have a common form

$$\dot{Y}_j = P_n[1+2\xi(t)]Y_i - a_{k,Qq}(\bar{U}/B)Y_j \qquad (4.13)$$

in which $n = k + 10(m-1)$. The indices k, m, j, i are related to the subscripts Qq in the following index table, such that each horizontal entry corresponds to one of the differential equations expressed in equation (4.13).

An Index Table

k	m	j	Q	q	i
3	1	5	M	α	2
4		6			
3	2	7	M	h	4
4		8			
5		9	M	h	4
6		10			
3	3	11	L	α	2
4		12			
3	4	13	L	h	4
4		14			

If no turbulence is present, namely $\xi(t)=0$, equation (4.11) is reduced to

$$\dot{Y} = A\,Y \qquad (4.14)$$

which describes a deterministic system. This system is stable if and only if all eigenvalues of matrix **A** have negative real parts. Increasing the wind speed U from zero will generally increase the real part of at least one pair of complex conjugate eigenvalues such that they become purely imaginary at a critical speed, U_{cr} when flutter commences. The imaginary part of the eigenvalue gives the flutter frequency. Results of the current analysis for the deterministic stability are compared with the results of the frequency domain analysis in Table 2. For bridge No. 1 (bridge No. 2 in Ref. 4) a very good agreement is noted for the critical speeds obtained from the non-oscillatory time domain model and the original frequency domain model. In the case of bridge No. 2, the oscillatory time domain model yields a much closer result than that of the non-oscillatory time-domain model (about 9% higher) when compared with the frequency domain result. Further discussion of these and other results is provided in the next section.

Equations for the first and second statistical moments may be obtained if the state vector **Y**(t) governed by equation (4.11), can be approximated by a Markov vector process. Stratonovich [22] suggested that the approximation is permissible if the external random excitation $\xi(t)$ has a correlation time much shorter than the relaxation time of the system. In the frequency domain, this is equivalent to the assumption that the turbulence excitation has a broad-band frequency spectrum which extends far beyond the natural frequency of the highest significant mode. This condition may be satisfied if the turbulence field is convected at high speeds as shown by Lin and Ariaratnam [9].

The set of the first order differential equations can also be written in the following form

$$\dot{Y}_j = f_j + g_j \xi(t) \tag{4.15}$$

in which $\xi(t)$ is a stationary wide band random process that will be replaced by a Gaussian white noise with constant spectral density $D/2\pi$. This is equivalent to lumping the net area under the correlation function $R_{\xi\xi}(\tau)$ of the random process $\xi(t)$ at the origin $\tau = 0$; namely, replacing $R_{\xi\xi}(\tau)$ by a delta function at the origin. From (4.11) it is clear that f_j and g_j, are given by

$$f_j = A_{jm} Y_m \tag{4.16}$$

$$g_j = B_{jm} Y_m \tag{4.17}$$

in which a repeated index indicates a summation.

Since the random process was replaced by a white noise, equations (4.15) are the Stratonovich stochastic differential equations [22]. They can be converted to equivalent Itô stochastic differential equation [12]:

$$dY_j = m_j \, dt + \sigma_j \, dW(t) \tag{4.18}$$

in which m_j and σ_j denote the drift and the diffusion coefficients respectively and $W(t)$ is a unit Wiener process. The conversion to the form of equation (4.18) can be readily carried out by using the Wong-Zakai rule [23]. For a single white noise excitation,

$$m_j = f_j + 1/2 \, D(\partial g_j/\partial Y_m) \, g_m \tag{4.19}$$

$$\sigma_j \, \sigma_k = D \, g_j \, g_k \tag{4.20}$$

Using (4.16) and (4.17) in (4.19) yields

$$m_j = A_{jm} Y_m + 1/2 \, D \, B_{jn} B_{nm} Y_m \tag{4.21}$$

or in matrix form

$$m = (A + 1/2 \, D \, B^2) \, Y \tag{4.22}$$

Using (4.17) in (4.20)

$$[\sigma\ \sigma'] = D\ B\ Y\ Y'\ B' \tag{4.23}$$

where a prime denotes matrix transposition.

Differential equations for the first moments are obtained by inserting (4.21) into (4.18) and taking the ensemble average. The second term in (4.18) vanishes, resulting in

$$dE[Y]/dt = E[m] = F\ E[Y] \tag{4.24}$$

where matrix **F** is defined as

$$F = A + 1/2\ D\ B^2 \tag{4.25}$$

It is clear from equation (4.24) that the first moment is stable if and only if all eigenvalues of matrix **F** have negative real parts.

Differential equations for the second moments can be obtained by applying Ito's differential rule [12] to the product $Y_j Y_k$ and obtain

$$d(Y_j Y_k) = (m_j Y_k + m_k Y_j + \sigma_j\ \sigma_k)dt$$
$$+ (\sigma_j\ Y_k + \sigma_k\ Y_j)dW(t) \tag{4.26}$$

The governing equations for the second moments follow from using (4.22) and (4.23) in (4.26) and taking the ensemble average. The second term in (4.26) vanishes when taking the ensemble average. The resulting equations for the second moments can also be rearranged in matrix form,

$$\dot{C}_y = F\ C_y + C_y\ F' + D\ B\ C_y\ B' \tag{4.27}$$

in which $C_y = E[Y'Y]$ = the correlation matrix, which is a symmetric matrix.

For the purpose of stability analysis it is more convenient to convert the correlation matrix to a vector. This is done by rewriting equation (4.27) in component form as

$$\dot{C}_{jk} = (F_{ji}\delta_{mk} + F_{km}\delta_{ij} + D\, B_{ji}B_{km})\, C_{im} \tag{4.28}$$

Then letting $\nu_r = C_{jk}$ where

$$r = j(j-1)/2 + k \text{ and } k \leq j, \tag{4.29}$$

Equation (4.27) becomes

$$\dot{\mathbf{v}} = \mathbf{H}\ \mathbf{v} \tag{4.30}$$

where the elements of matrix **H** are given by

$$H_{rs} = F_{ji}\delta_{mk} + F_{km}\delta_{ij} + D\, B_{ji}B_{km}, \tag{4.31}$$

$$r = j(j - 1)/2 + k, \quad k \leq j \tag{4.32}$$

$$s = i(i-1)/2 + m, \quad m \leq i \tag{4.33}$$

Once again the stability criterion is that the eigenvalues of matrix **H** have negative real parts.

Matrix **H** is of the order of (m+1)m/2 by (m+1)m/2. In the present case of m=14, it is 105x105. The order grows rapidly as the number of state variables m increases. The calculation of the stability boundaries is carried out by fixing a spectral density level and calculating the eigenvalues of both the **F** and **H** matrices for increasing mean wind speeds. The boundary of stability is reached whenever the real part of at least one eigenvalue becomes zero.

5. Results

To illustrate the effect of oscillatory terms in the indicial functions on bridge stability, calculations were carried out for the two bridge decks given in Table 1. Figure 3 shows the variation of the complex eigenvalue of the least stable mode for bridge #1 as the flow speed increases without taking turbulence into account. The arrow indicates the direction of increasing wind speed. Instability occurs when the curve crosses the imaginary axis from left to right.

Figure 4 illustrates the stability boundaries of the mean and mean square responses for bridge 1, obtained from using the oscillatory and nonoscillatory models for the indicial function $f_{Mh}(t)$. One significant result is evident in the figure. The oscillatory model leads to a less stable bridge with zero or low level turbulence (below 0.05 sec/rad). The deterministic critical wind speed is 45 m/sec versus 51.5 m/sec for the nonoscillatory model. This trend is reversed at a higher level turbulence. From a practical point of view, the more conservative oscillatory model should be applied toward the design of this bridge, in the low turbulence range. If design is based on a second moment criterion, then turbulence has a stabilizing effect for this bridge so long as the turbulence level is not too high. It becomes destabilizing at a level above about 0.18 s/rad for the nonoscillatory model and at a much higher level for the oscillatory model.

The second moment stability always implies the first moment stability as can also be seen in Figure 4, in agreement with the Schwartz inequality.

For bridge # 2, the deterministic critical speed obtained from the oscillatory model in the time domain agrees very closely with the frequency domain result (See Table 2), suggesting that the oscillatory model is more suitable for the stochastic stability analysis. However, comparison of stability boundaries for this bridge as shown in Figure 5, indicates that the non-oscillatory model is more conservative except for very low level of turbulence. Both models also indicate a destabilizing effect of turbulence on the mean-square response when the turbulence level is sufficiently high.

6. Conclusions

The oscillatory model proposed in this paper for the self-excited indicial function is shown to be better suited for the stability investigation of certain types of bridge decks. Comparing with the traditional nonoscillatory model, the new oscillatory model is shown to yield a lower estimate for the critical mean wind speed at low levels of turbulence.

However, it also predicts a greater stabilizing effect of turbulence in some cases, or a lower destabilizing effect in other cases. Hence, the oscillatory model, where applicable, enhances the transfer of energy from the critical mode to more stable modes by means of turbulence.

A broader data base is needed for further investigation of the effect of the oscillatory behavior in self-excited loads. It was shown in Ref. 24 that the lack of sufficient data may lead to nonunique results if the parameters of the indicial functions are identified indirectly from the measured flutter derivatives. Thus, direct experimental measurement of indicial functions, including the coupled indicial functions, would be more reliable.

7. Acknowledgement

The research work reported in this paper was carried out when the author was a graduate assistant at the Center for Applied Stochastics Research at Florida Atlantic University. Sincere thanks are expressed to Professor Y. K. Lin, adviser, for his continuous and dedicated guidance and help throughout the author's studies at FAU. The project was supported by the National Science Foundation under Grant CES-8805261. Opinions, findings and conclusions are those of the writers, and do not necessarily reflect the views of NSF.

8. References

1. Theodorsen, T.: General theory of aerodynamic instability and the mechanism of flutter. NACA Report No. 496, reprinted in AIAA selected reprints on Flutter, 1969 by I. E. Garrick, 22-31.

2. Bissplinghoff, R. L.; Ashley, H.; Halfman, R. L.: Aeroelasticity. Addison - Weseley, 1957.

3. Sabzevari, A.; Scanlan, R. H.: Aerodynamic instability of suspension bridges. J. Eng. Mech. Div. ASCE 94 (1968) EM2, 489-519.

4. Scanlan, R. H.; Tomko, J. J.: Airfoil and bridge deck flutter derivatives. J. Eng. Mech. Div. ASCE 97 (1971) EM2 1717-1737.

5. Scanlan, R. H.: State of the art methods for calculating flutter, vortex-induced and buffeting response of bridge

structures," Rept. FHWA/RD-80/050, Apr. 1980, Final Report.

6. Scanlan, R. H.; Beliveau, J. G.; Budlong, K.S.: Indicial aerodynamic functions for bridge decks. J. Engng. Mech. ASCE 100 (1974) EM4 657-672.

7. Scanlan, R. H.: Role of indicial functions in the buffeting analysis of bridges. J. of Struc. Eng. Div. ASCE 110 (1984) 1433-1446.

8. Lin, Y. K.: Motion of suspension bridge in turbulent wind. J. Eng. Mech. Div. ASCE 105 (1979) 921-932.

9. Lin, Y.K.; Ariaratnam, S. T.: Stability of bridge motion in turbulent winds. J. of Struc. Mech. 8 (1980) 1-15.

10. Bucher, C. G.; Lin, Y. K.: Stochastic stability of bridges considering coupled modes. J. Eng. Mech. ASCE 114 (1988) 2055-2071.

11. Bucher, C. G.; Lin, Y. K.: Stochastic stability of bridges considering coupled modes II. J. Eng. Mech. ASCE 115 (1989) 384-400.

12. Ito, K.: On a formula concerning stochastic differentials. Nagoya Math. Journal, Japan 3 (1951) 55-65.

13. Simiu, E.; Scanlan, R. H.: Wind effects on structures. John Wiley & Sons, New York 1978.

14. Wagner, H.: Dynamicher Auftrieb fon tragflugel, Zeitschrift für angewandte Mathematik und Mechanik, Bd. 5 (1925) 17.

15. Jones, R. T.: The unsteady lift of a wing of finite aspect ratio, NACA No. 681, 1940.

16. Huston, D. R.: The effects of upstream gusting on the aero-elastic behavior of long span bridges. Ph.D. Dissertation, Dept. of Civil Eng., Princeton Univ., May 1986.

17. Shiraishi, N.; Matsumoto, M.; Mashimo, H.; Honda, A.; On vortex induced oscillation of bluff bodies. Reprint from the Memoirs of the Faculty of Eng., Kyoto Univ., Vol. XLV part 4, Oct. 1983, Kyoto Japan.

18. Shiraishi, N.; Matsumoto, M.: On classification of vortex-induced oscillation and its application for bridge structures. J. Wind Eng. and Ind. Aerod., (1983) 419-430.

19. Beliveau, J. G.; Vaicaitis, R.; Shinozuka, M.: Motion of suspension bridge subject to wind loads. J. of Struc. Div. ASCE 103 (1976) ST6 1189-1205.

20. Tsiates, G.: Random vibration of nonlinear, parametrically excited systems. Ph.D. Dissertation, Case Western Reserve University, 1984.

21. Sarkar, P. P.: Effect of wind turbulence on the stability of long span bridges. MSc Thesis, Wash. State Univ., Dec. 1986.

22. Stratonovich, R. L.: Topics in the theory of random noise", Vol. 1, Translation by R. A. Silverman, Gordon & Breach, 1963.

23. Wong, E.; Zakai, M.: On the relation between ordinary and stochastic equations. International Journal of Eng. Science (1965) 213-229.

24. Sternberg, A.: Stability and response of suspension bridges under turbulent wind excitation. Ph.D. dissertation, Florida Atlantic University, Boca Raton, FL August, 1988.

Appendix I

Elements of Matrices A and B

The nonzero elements of matrix A for stability analysis are listed as follows:

$A(1,2) = 1.0;$

$A(2,1) = -\omega_{o\alpha}^2;\ A(2,2) = -2\zeta_{o\alpha}\omega_{o\alpha};\ \ A(2,3) = P_{11}/B;$

$A(2,4) = P_{12}/U;\ A(2,5) = A(2,6) = A(2,7) = A(2,8) = A(2,9) = A(2,10) = 1.0;$

$A(3,4) = 1.0;$

$A(4,1) = P_{21};\ A(4,2) = P_{22}B/U;\ A(4,3) = -\omega_{oh}^2;$

$A(4,4) = -2\zeta_{oh}\omega_{oh};\ A(4,11) = A(4,12) = A(4,13) = A(4,14) = 1.0;$

$A(5,2) = P_3;\qquad A(5,5) = -a_{3,M\alpha}U/B;$

$A(6,2) = P_4;\qquad A(6,6) = -a_{4,M\alpha}U/B;$

$A(7,4) = P_{13}/B;\quad A(7,7) = -a_{3,Mh}U/B;$

$A(8,4) = P_{14}/B;\quad A(8,8) = -a_{4,Mh}U/B;$

$A(9,4) = P_{15}/B;\quad A(9,9) = -a_{5,Mh}U/B;$ ($a_{5,Mh}$ and P_{15} are complex)

$A(10,4) = P_{16}/B;\quad A(10,10) = -a_{6,Mh}U/B;$ ($P_{16} = P_{15}^*$ and $a_{6,Mh} = a_{5,Mh}^*$)

$A(11,2) = P_{23};\quad A(11,11) = -a_{3,L\alpha}U/B;$

$A(12,2) = P_{24};\quad A(12,12) = -a_{4,L\alpha}U/B;$

$A(13,4) = P_{33}/B;\quad A(13,13) = -a_{3,Lh}U/B;$

$A(14,4) = P_{34}/B;\quad A(14,14) = -a_{4,Lh}U/B;$

The nonzero elements of matrix B for stability analysis are listed as follows:

$B(2,1) = 2P_1;\ B(2,2) = 2P_2B/U;\ B(2,3) = 2P_{11}/B;$

$B(2,4) = 2P_{12}/U;$

$B(4,1) = 2P_{21};\ B(4,2) = 2P_{22}B/U;\ B(4,3) = 2P_{31}/B;$

$B(4,4) = 2P_{32}/U$

$B(5,2) = 2P_3;\ B(6,2) = 2P_4;\ B(7,4) = 2P_{13}/B;$

B(8,4) = $2P_{14}/B$;

B(9,4) = $2P_{15}/B$; B(10,4) = $2P_{16}/B$; (complex element and its conjugate)

B(11,2) = $2P_{23}$; B(12,2) = $2P_{24}$; B(13,4) = $2P_{33}/B$;

B(14,4) = $2P_{34}/B$;

Table 1: Structural and aerodynamic data

Bridge No.	structural Data								Source of Data
	rad/sec				kg*m	kg/m	m	m	
	ω_α	ω_h	ζ_α	ζ_h	I_s	M_s	l	B	
1.	0.74π	0.32π	0.01	0.01	8.94x10^5	1.36x10^4	490	16.4	Ref. 4 Fig. 8 Bridge #2
2.	0.74π	0.32π	0.01	0.01	1.53x10^6	2.32x10^4	1280	28.0	Ref. 4 Fig.10 Bridge #8

Table 2: Deterministic Flutter Speeds & Frequencies

Bridge No.	Frequency Domain		Time Domain			
			Non-Oscillatory		Oscillatory	
	ω_{cr} [Rad/sec]	U_{cr} [m/sec]	ω_{cr}[Rad/s]	U_{cr}[m/sec]	ω_{cr}[Rad/s]	U_{cr}[m/sec]
1.	0.712π	51.51	0.716π	51.34	0.721π	44.94
2.	0.647π	54.78	0.649π	59.26	0.636π	55.91

Table 3: Identified Parameters for Bridges 1 and 2
With and Without Oscillatory Load Model

Bridge No.		1		2	
Load	Parameter	Non-Oscillatory	Oscillatory	Non-Oscillatory	Oscillatory
M_α	$C_{1,M\alpha}$	0.17517		0.28954	
	$C_{2,M\alpha}$	-0.32363		-0.14346	
	$C_{3,M\alpha}$	0.35014		0.13153	
	$a_{3,M\alpha}$	1.29177		1.08272	
	$C_{4,M\alpha}$	0.39831		1.13159	
	$a_{4,M\alpha}$	1.58890		1.06556	
M_h	$C_{1,Mh}$	-0.0940	0.0000	3.2961	0.0000
	$C_{2,Mh}$	0.8246	0.0718	1.9674	-1.0568
	$C_{3,Mh}$	3.9922	-0.0030	-68.0211	-1.703×10^{-2}
	$a_{3,Mh}$	0.0200	3.7710	0.6224	1.7412
	$C_{4,Mh}$	-3.9830	0.0030	67.4825	-4.67×10^{-4}
	$a_{4,Mh}$	0.0204	0.0030	0.6851	5.58×10^{-4}
	$C_{5,Mh}$		0.0019		-8.65×10^{-3}
	$C_{6,Mh}$		0.4243		2.8320
	$d_{5,Mh}$		0.0865		0.1990
	k_a		0.9034		1.0368
L_α	$C_{1,L\alpha}$	-3.15835		-2.08039	
	$C_{2,L\alpha}$	-6.18546		0.14349	
	$C_{3,L\alpha}$	386.7990		-15.51302	
	$a_{3,L\alpha}$	79.2500		1.28972	
	$C_{4,L\alpha}$	3.99201		16.67732	
	$a_{4,L\alpha}$	0.70409		1.07660	
L_h	$C_{1,Lh}$	-0.00839		1.9894×10	
	$C_{2,Lh}$	-2.74372		-2.12023	
	$C_{3,Lh}$	22.7980		-0.11948	
	$a_{3,Lh}$	1.34×10^{6}		0.33203	
	$C_{4,Lh}$	1.3298		5.097×10^{-2}	
	$a_{4,Lh}$	30.8526		1.055×10^{6}	

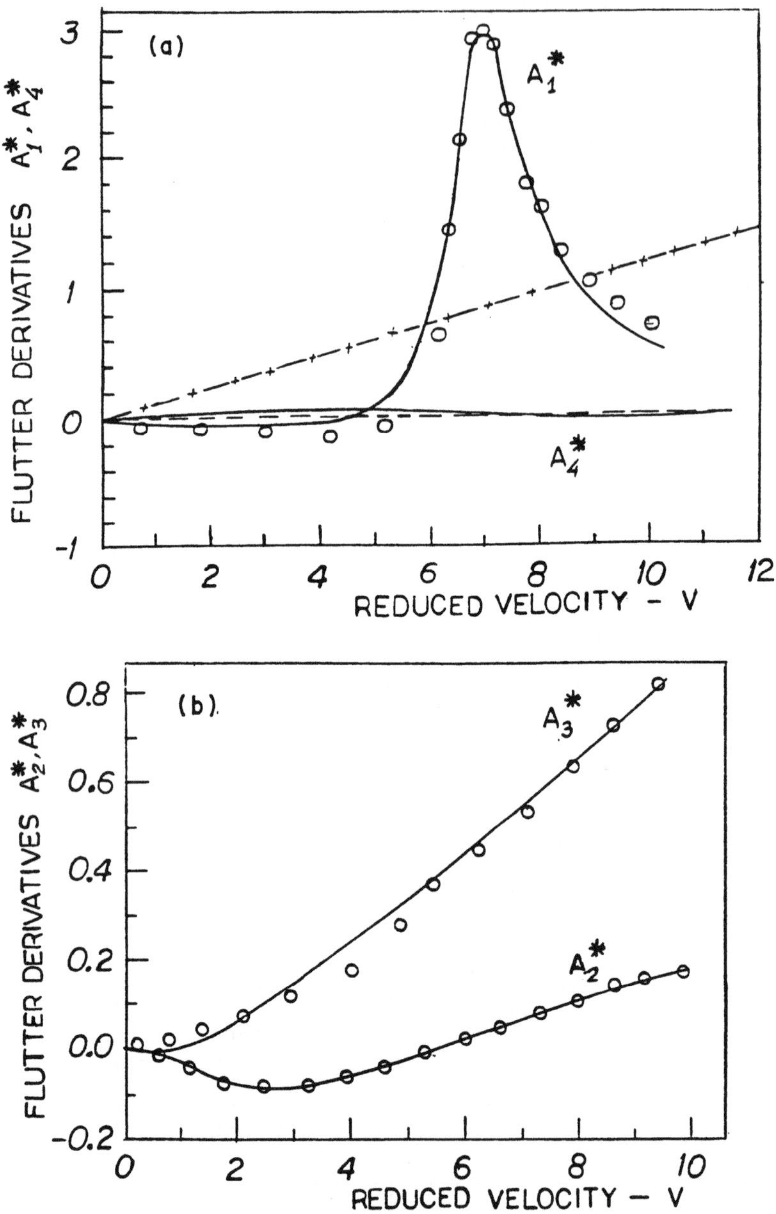

Fig. 1. Flutter derivatives for self-excited pitching moment, bridge section #1. (a) . Experimental and fitted results for M_h load, based on oscillatory indicial function model (b) Experimental and fitted results for M_α load. o o o - experimental data from Ref. 4, ____- fitted results -+-+-+ - results from Ref. 10.

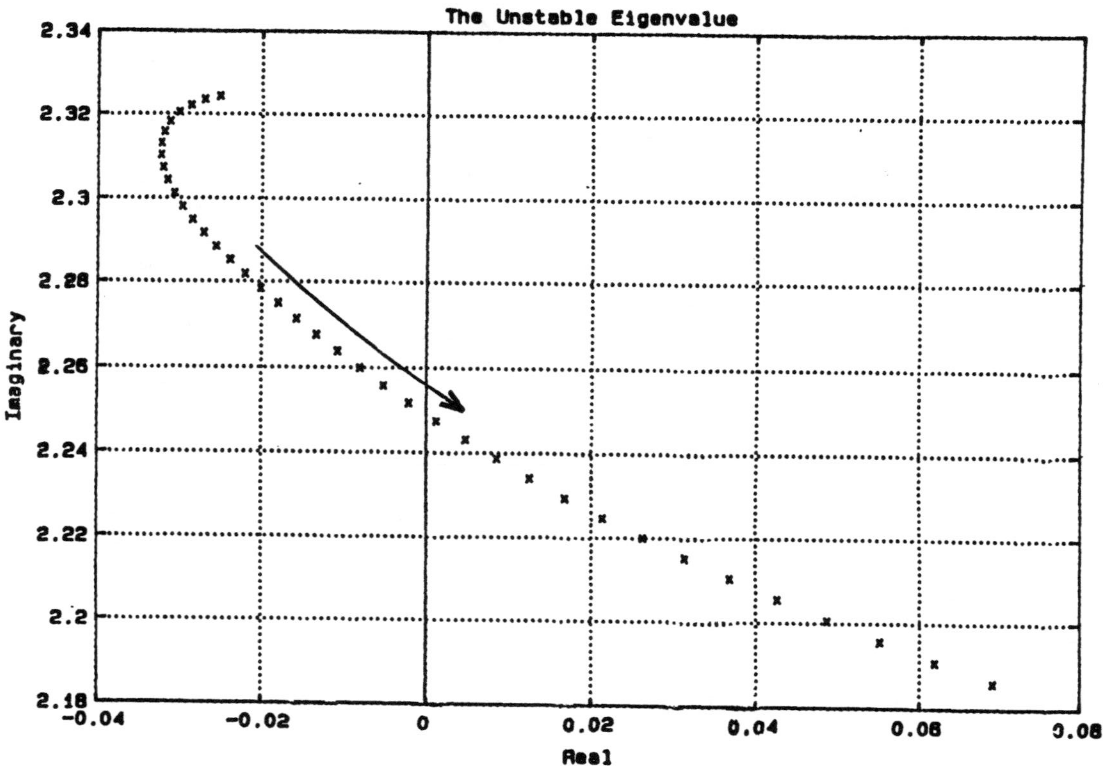

Fig. 3. Movement of the least stable eigenvalue in the complex plane due to increasing wind speed, bridge #1, without turbulence.

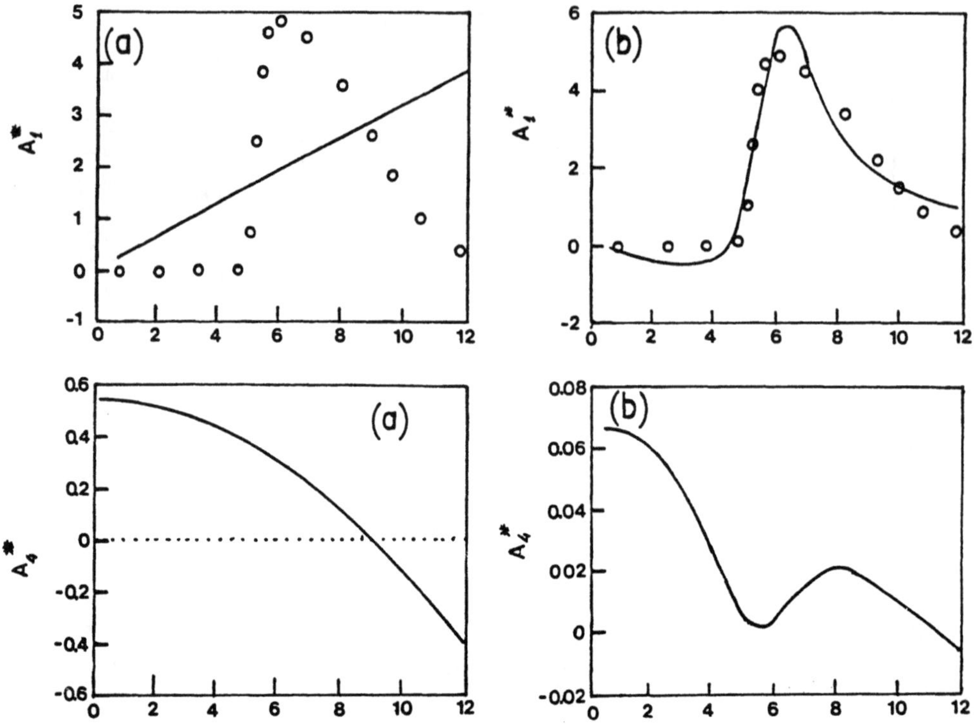

Fig. 2. Comparison of fitted results for flutter derivatives A_1^* and A_4^*, bridge section #2, (a) nonoscillatory model, (b oscillatory model. o o o - experimental data from Ref. 4, ____ - fitted results.

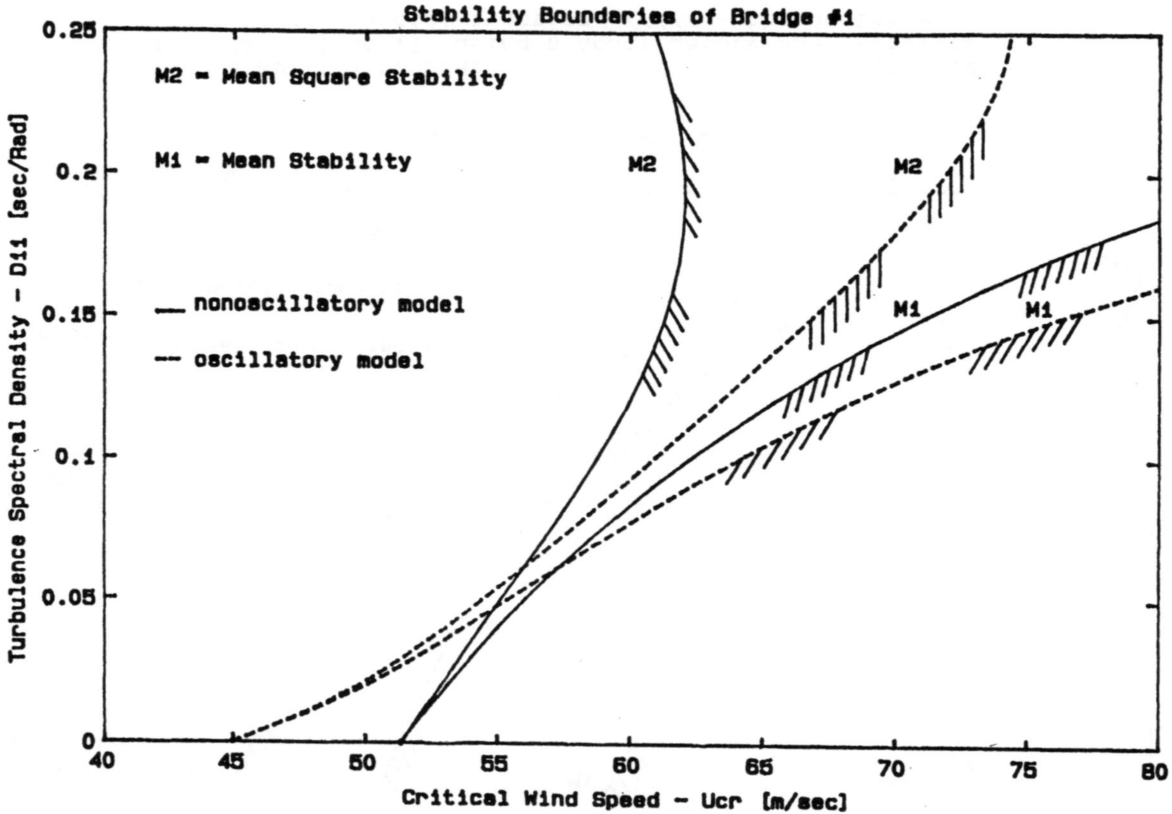

Fig. 4. Stability boundaries of the 1st and 2nd statistical moments for bridge #1, using oscillatory and nonoscillatory models for the indicial function. ///// - unstable region.

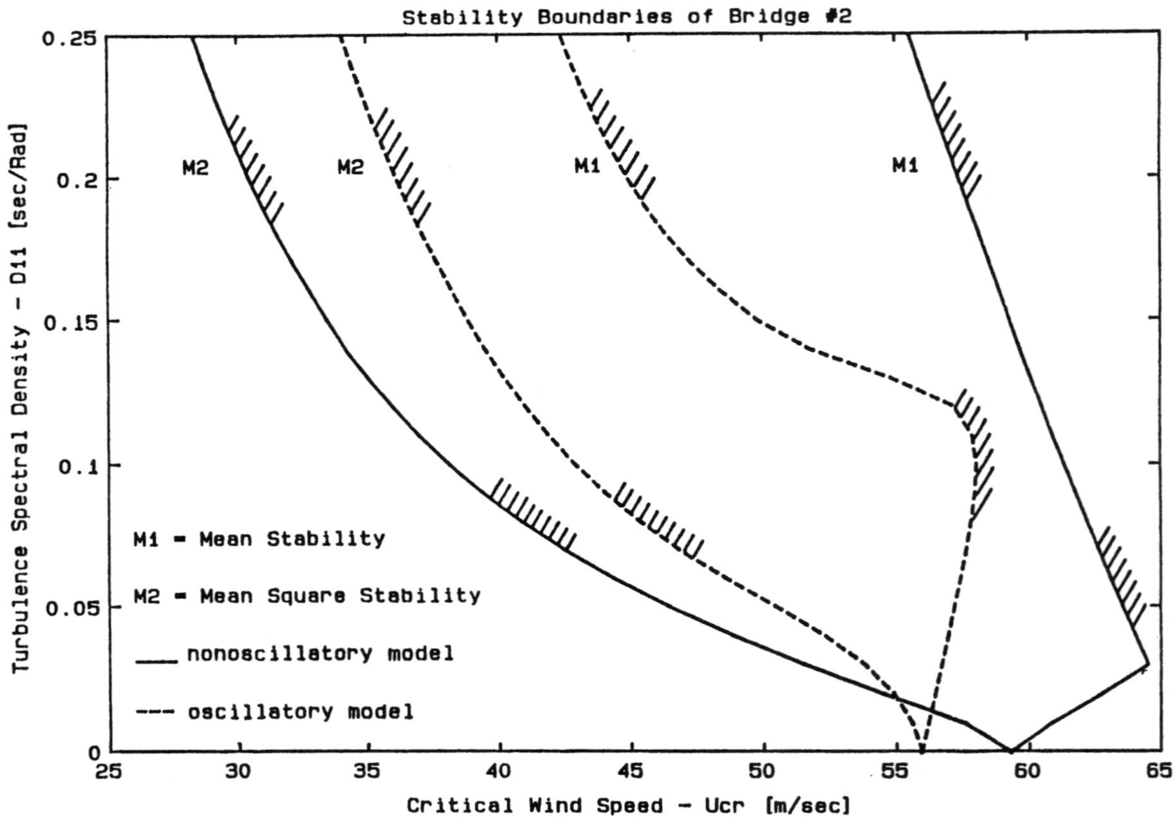

Fig. 5. Stability boundaries of the 1st and 2nd statistical moments for bridge #2, using nonoscillatory and oscillatory models for the indicial function. ///// - the unstable region

On Flow-Induced Vibration of a Circular Cylinder Placed near a Plane Boundary

T-C. SU* and Q. X. LIAN**
Center for Applied Stochastics Research
Florida Atlantic University
Boca Raton, FL 33431
USA

ABSTRACT

Experimental investigations were carried out to examine the flow features around an elastically supported circular cylinder placed near a plane boundary in a uniform stream and the associated flow-excited vibrations. The experiment was conducted in a recirculating water channel. The undisturbed uniform cross stream is parallel to the plane boundary and normal to the axis of the cylinder. The flow field was visualized using hydrogen bubble technique. Time sequential photography was used to quantify the development of the vortex pattern and the vibrational characteristics of the cylinder.

For the clearance of a quarter of the cylinder's diameter, two modes of flow-excited oscillations were observed. At lower flow speed, the longitudinal on-line oscillations were excited. Whereas, at higher flow speed, the transverse vibrations occurred. The corresponding cases of rigid cylinder were also examined. In which case, the vortex patterns were less organized and an obvious jet flow occurred within the gap between the cylinder and the plane boundary. Based on the free-stream speed and the diameter of the cylinder, the on-line oscillation occurred at a Reynolds number of around 2440, with a synchronous and periodic shedding of vortices and the reduced frequency is near to its natural frequency in still water of 1.6. While the transverse oscillations occurred at a higher Reynolds number from 3,000 ~ 4,000, with an alternate vortex shedding from cylinder and a much lower reduced frequencey of about 0.45.

INTRODUCTION

While the flow around a circular cylinder near to a plane boundary is a problem of fundamental interest, its understanding remains quite limited. The problem is of practical relevance, for example, in the dynamics of pipeline

---

* Professor of Ocean Engineering and CASR
** Senior Faculty Associate, CASR

I. Elishakoff · Y. K. Lin (Eds.)
Stochastic Structural Dynamics 2
New Practical Applications
2nd International Conference on Stochastic Structural Dynamics
May 9-11, 1990, Boca Raton, Florida

laid along a sea bed. A recent study and survey on the wake structure at various gap-to-diameter ratios can be found in Reference [1]. However, little information is currently available in the literature on the vibration of an elastically supported circular cylinder placed near a plane boundary in a uniform stream. In the present study, several limited exploratory experiments were carried out to investigate vortex shedding, flow pattern around, and the oscillation of a cylinder in a uniform steady stream near a plane boundary.

TEST FACILITIES

The experiment was performed in a water channel with recirculating water (Figure 1). Its test section is 0.3 m wide and 0.5 m long, with a water depth of 0.3 m. The flow speed is adjustable from near zero to 15 cm/s. The size of the test model and test arrangement is shown in Figure 2. The cylinder of 38 mm diameter is made of Plexiglas. It is hollow, with two small holes in both ends. When it is placed in the water channel, its interior becomes filled with water. Each end of the cylinder is mounted on a steel spring as shown in Figure 2. The other side of each steel spring is mounted on side plates. Also mounted on these side plates is a false floor. The false floor is a Plexiglas plate of 250 mm x 290 mm and is 5 mm thick, with its leading edge sharpened. The foremost point of the cylinder is placed one diameter downstream of the leading edge of the false floor. This is done to allow us to concentrate on the case when the boundary layers are thin, so that the cylinder is outside the boundary layer. When the water is at rest, the cylinder can be oscillated freely in any direction with an oscillation frequency of 2.5 Hz.

Hydrogen bubble technique was used to visualize the flow pattern and the vortex shedding around the oscillating model. The arrangement of platinum wires is shown in Figure 2. One wire is placed near the upstream edge of the cylinder, while another wire is placed near the downstream edge of the cylinder. The camera was set on one side of the channel, so that both the flow pattern and the oscillating displacement could be recorded on film. The gap between the cylinder and the false floor was a quarter of the diameter. The platinum wire for generating the hydrogen bubbles was 0.025 mm in diameter, and the frequency of the pulsating current was 8.3 Hz.

TEST RESULTS

Overview

At very low flow speeds U, the oscillation of the model is not perceptible. As speed U is increased to about 6.4 cm/s, a fore-backward oscillation (along the direction of undisturbed upstream flow) occurred. As speed U increased further, the transverse oscillation took place. The test results of each mode are presented in the following:

Fore-Backward Oscillations

For a cylinder of diameter D=38 mm placed 9.5 mm above a plane boundary, an in-line oscillation occurred at U = 6.4 cm/s. In this case, the Reynolds number based on the cylinder's diameter was $Re_D = 2440$. A typical sequence of photographs of this mode, at a frame speed of 5 frame/seconds, is shown in Figure 3. The corresponding time history of displacement of the cylinder ($\Delta x/D$, $\Delta y/D$), the locations of flow separations from the cylinder (θ_u and θ_l) and the definition sketch of the vortex system, are plotted in Figure 4.

The following flow features are noticeable:

1. The shedding of vortices V_2 and V_3 is periodic and synchronous.
2. The shedding period and the oscillating period are 0.37 seconds. Thus, the frequency f = 2.7 Hz and the reduced frequency fD/U = 1.6, which is much higher than the case of the fixed cylinder. The flow-induced oscillating frequency is, however, very close to the natural frequency in still water.
3. The location of flow separators θ_u and θ_l also changed periodically. More change is found in θ_u than θ_l, with $20° < \theta_u < 50°$ and $-45° < \theta_l < -25°$.

4. The horizontal velocity of the vortices can be obtained from analyzing time sequential photographs with the following results:

$$U_{v_1} = 1.5\ U\ ,\quad U_{v_2} = 1.4\ U\ ,\quad \text{and}\quad U_{v_3} = 2.0\ U\ . \qquad (1)$$

5. The flow speed within the gap U_s can be obtained by analyzing the space distribution of the hydrogen bubble time lines. It is found that $U_s=(1.4\sim1.5)U$. The variation is small. The predominate oscillation is along the free stream (x) direction.
6. The y component (transverse direction) oscillations are very small and are within measurement error. One could, therefore, consider no transverse oscillation occurred in this case.
7. The vortices were much better organized as compared to the corresponding case of fixed cylinder (Figure 5). In the photograph shown in Figure 5, the vortices are either weak or irregular and the jet flow in the gap is very apparent.

Transverse Oscillations

For the same cylinder D = 38 mm at S = 9.5 mm, a transverse oscillation occurred at U = 8.4 cm/s (Re_D = 3200). Typical sequential photographs showing this vibration, are given in Figure 6, and the corresponding vibration amplitude history is plotted in Figure 7, along with flow separation pattern. The frequency of transverse vibration is about 1.0 Hz and fD/U = 0.45 which is much smaller than the case of fore-backward oscillations. The shedding vortices V_2 and V_3 are periodic but alternating, such as occurred in the Karman vortex street. The relationship between the alternate vortex shedding and the direction of the cylinder's movement is indicated in Figure 7. The smaller f is partly caused by this alternate shedding, as it takes a longer time to complete a full cycle.

As the speed U increased to 10.6 cm/s, Re_D = 4030, the transverse oscillation persisted as shown in Figures 8 (photo) and 9 (plots). In this case, f = 1.25 Hz and the reduced frequency $fD/U = 0.45$, which is the same as the case when $Re_D = 3020$. Thus, the reduced frequency is independent of the Reynolds number in this mode of oscillations.

CONCLUSION

In a limited exploratory investigation, some flow features and vibration characteristics of fluid-structure interactions of a circular cylinder in the proximity of a plane surface in a uniform flow are revealed. Depending on the flow speed, distinct modes of flow-excited oscillations are observed. The vortex shedding and flow pattern are very different from the corresponding case of fixed cylinders. More experiments and additional investigations involving companion numerical simulations are underway. The detailed results are planned to be published elsewhere in the near future.

ACKNOWLEDGEMENT

This work has been supported by the National Science Foundation, Grant Number ECE-8607183. The sponsorship is gratefully acknowledged. Opinions, findings and conclusions expressed are those of the writers and do not necessarily reflect the views of NSF.

REFERENCES

1. Bearman, P. W. and Zdravkovich, M. M., Flow Around a Circular Cylinder Near a Plane Boundary, J. Fluid Mech., Volume 89, pp. 33-47.

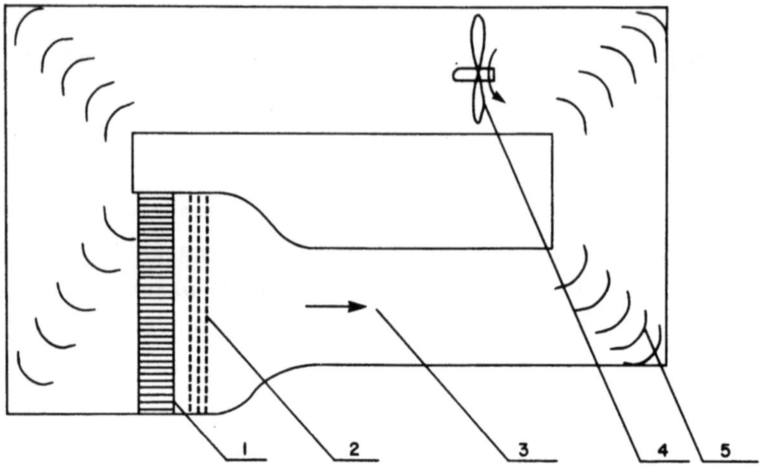

Fig. 1. The sketch of the water channel: (1) the honey-comb, (2) the screens, (3) the test section, (4) the axial flow pump, (5) the guide vanes

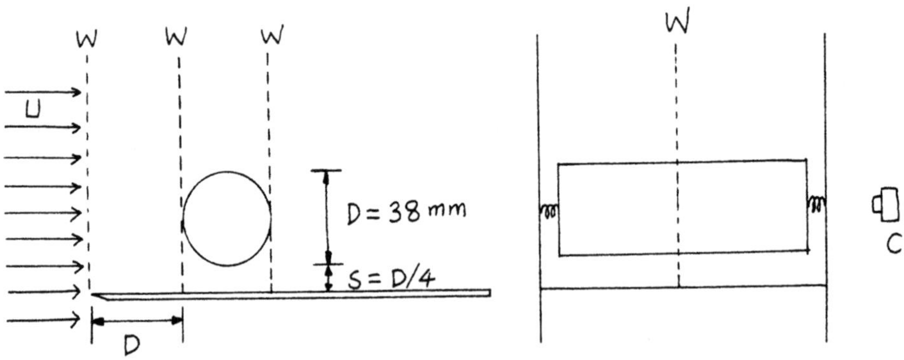

Fig. 2. The arrangement of test model: S is the gap between the cylinder and the false floor, D is the diameter of the cylinder, U is the speed of upstream blow, C is the camera

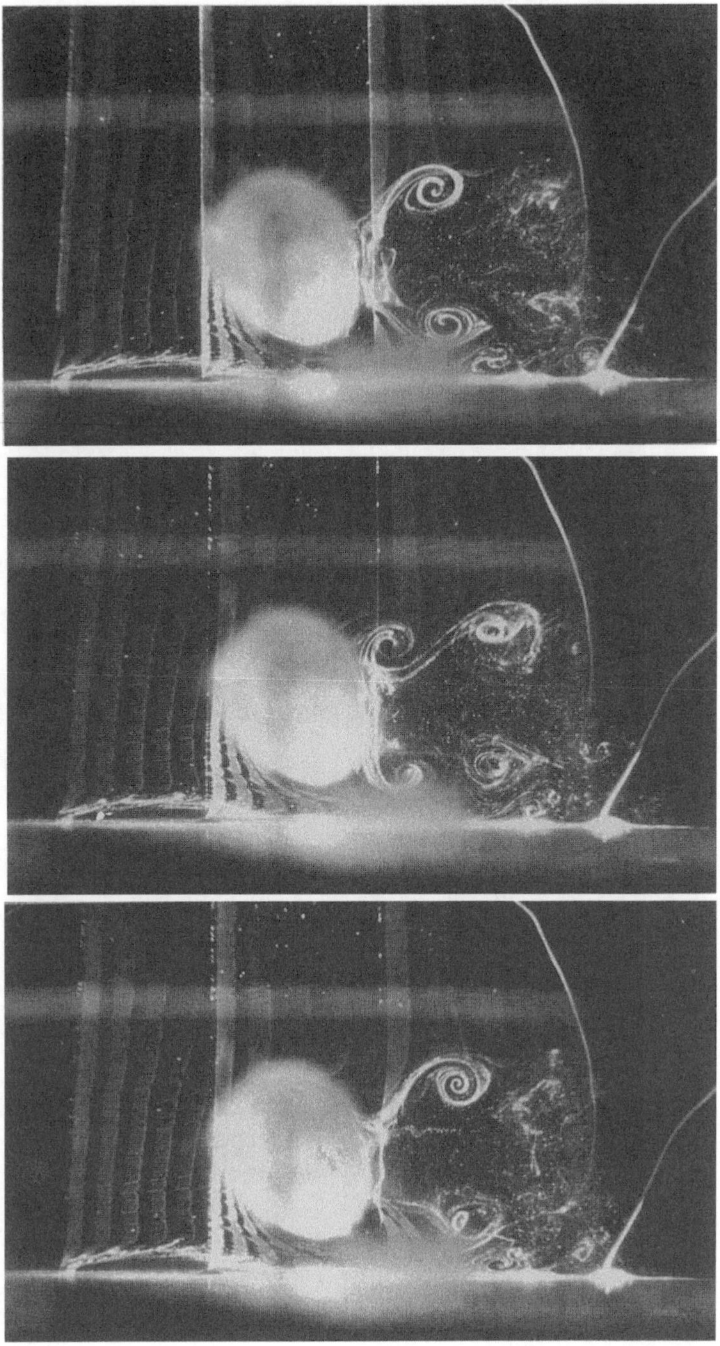

Fig. 3. Photos of typical fore-backward oscillations, $Re_D = 2440$, $S = D/4$

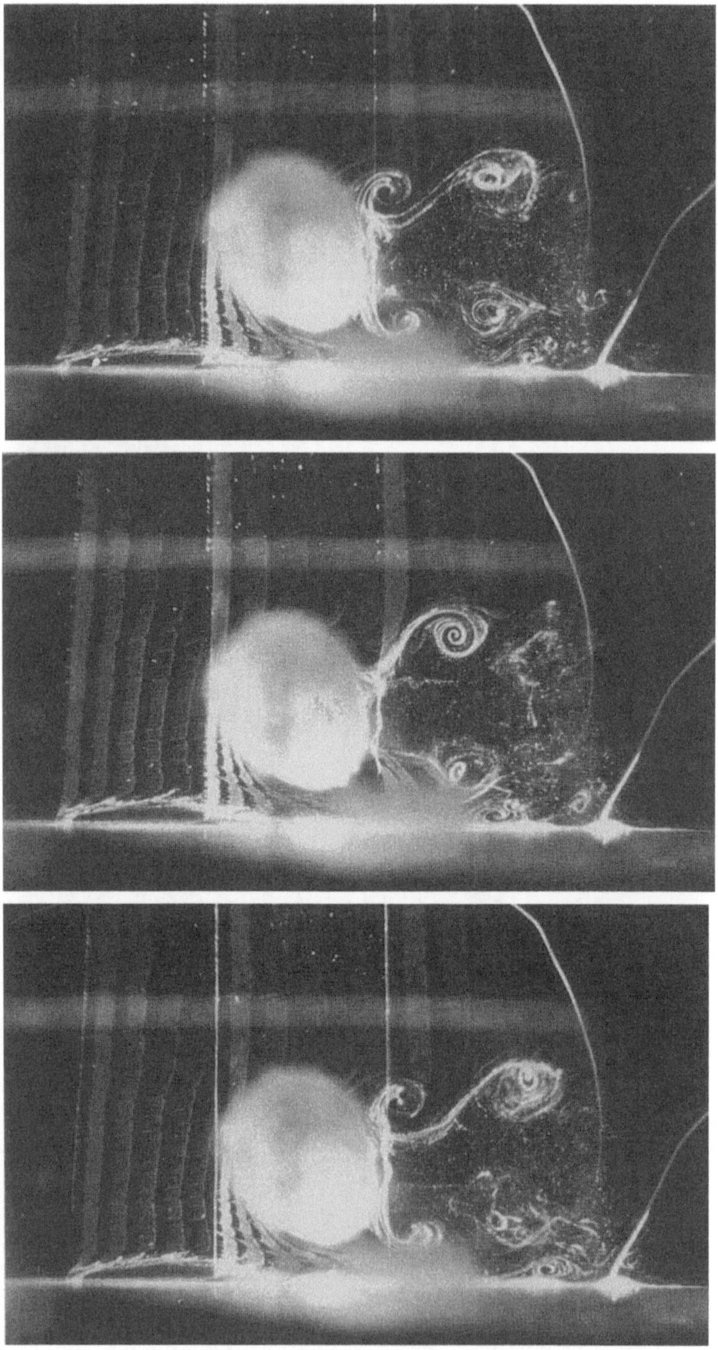

Fig. 3. (continued)

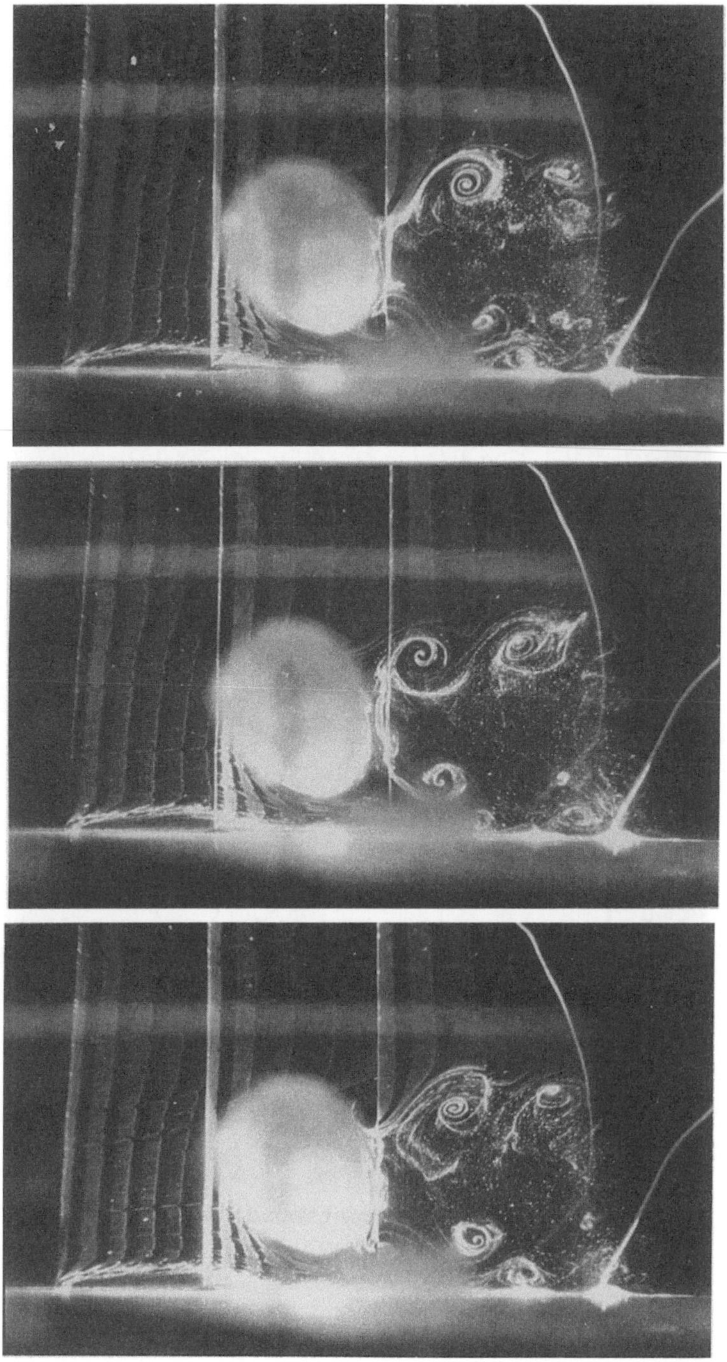

Fig. 3. (continued)

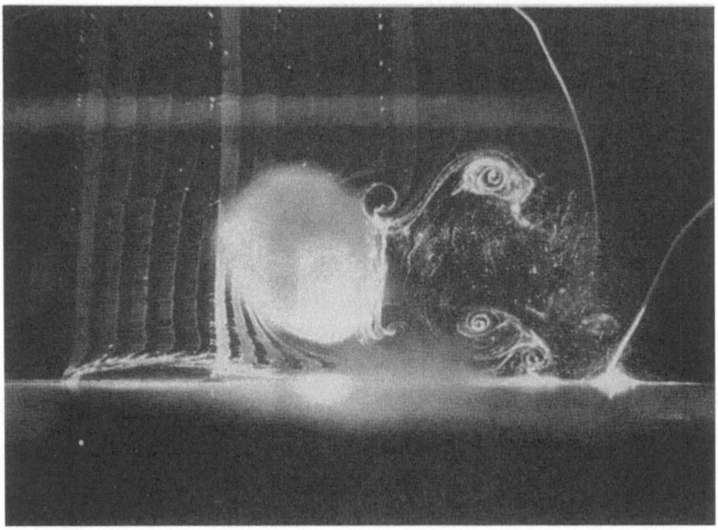

Fig. 3. (continued)

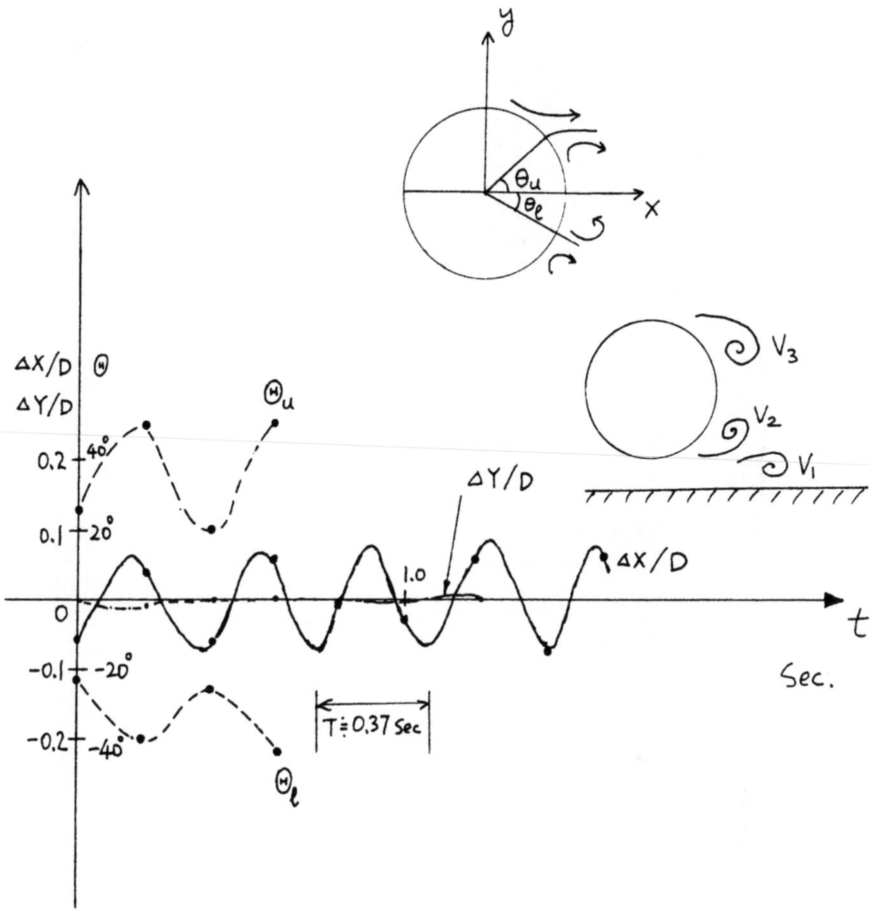

Fig. 4. The displacement of cylinder and the locations of flow separation in fore-backward oscillations

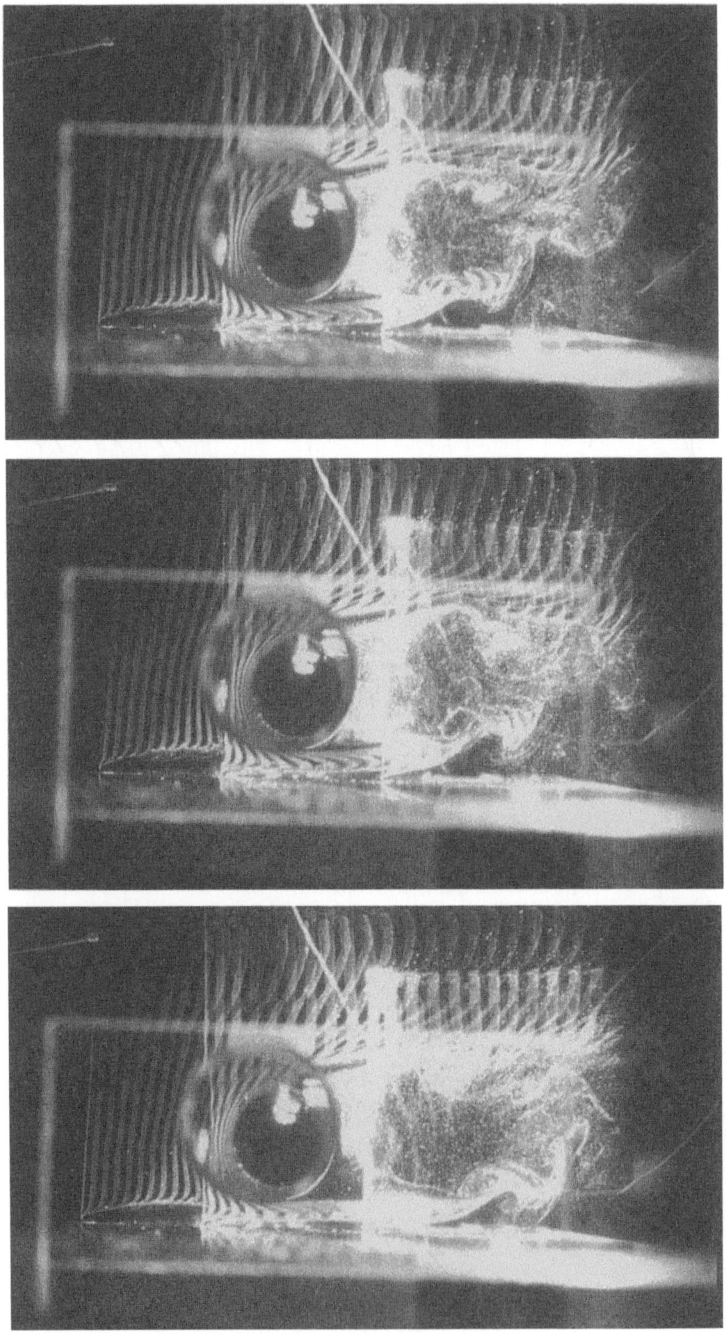

Fig. 5. Photo of flow over a fixed cylinder

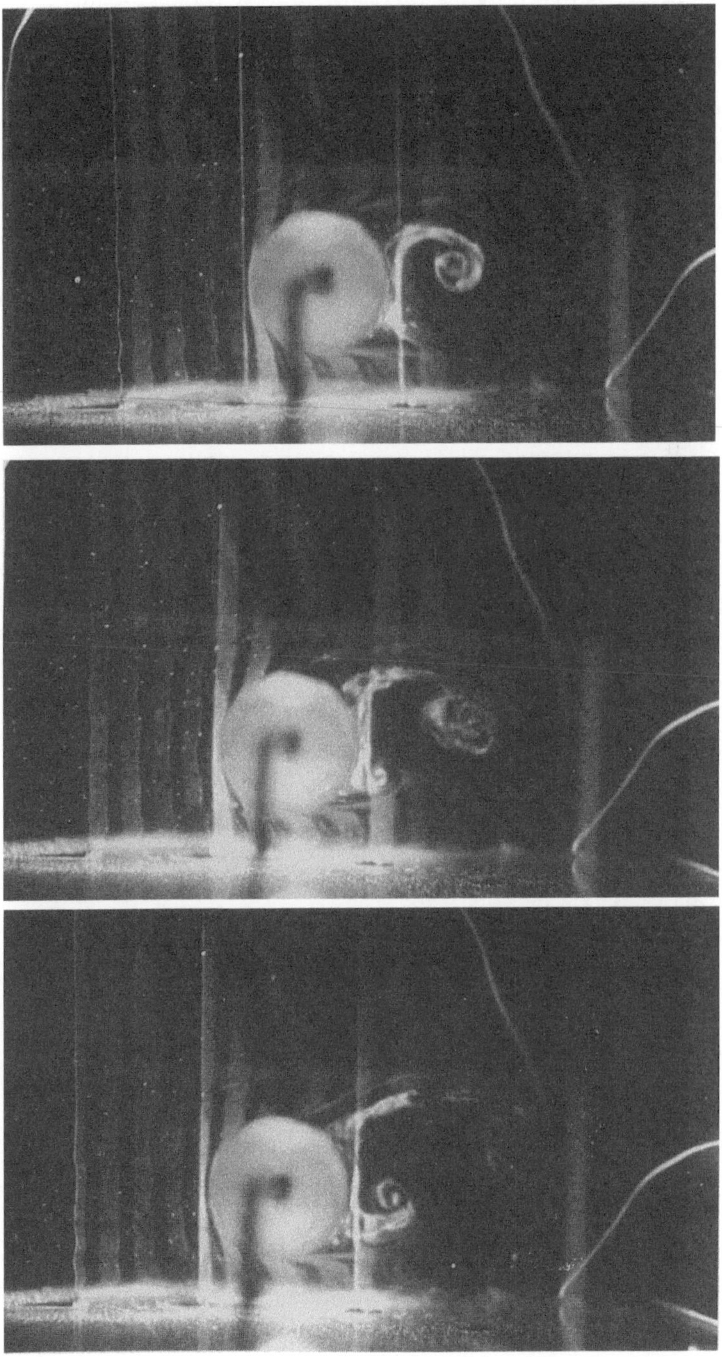

Fig. 6. Photo of a typical transverse oscillation, $Re_D = 3200$, $S = D/4$

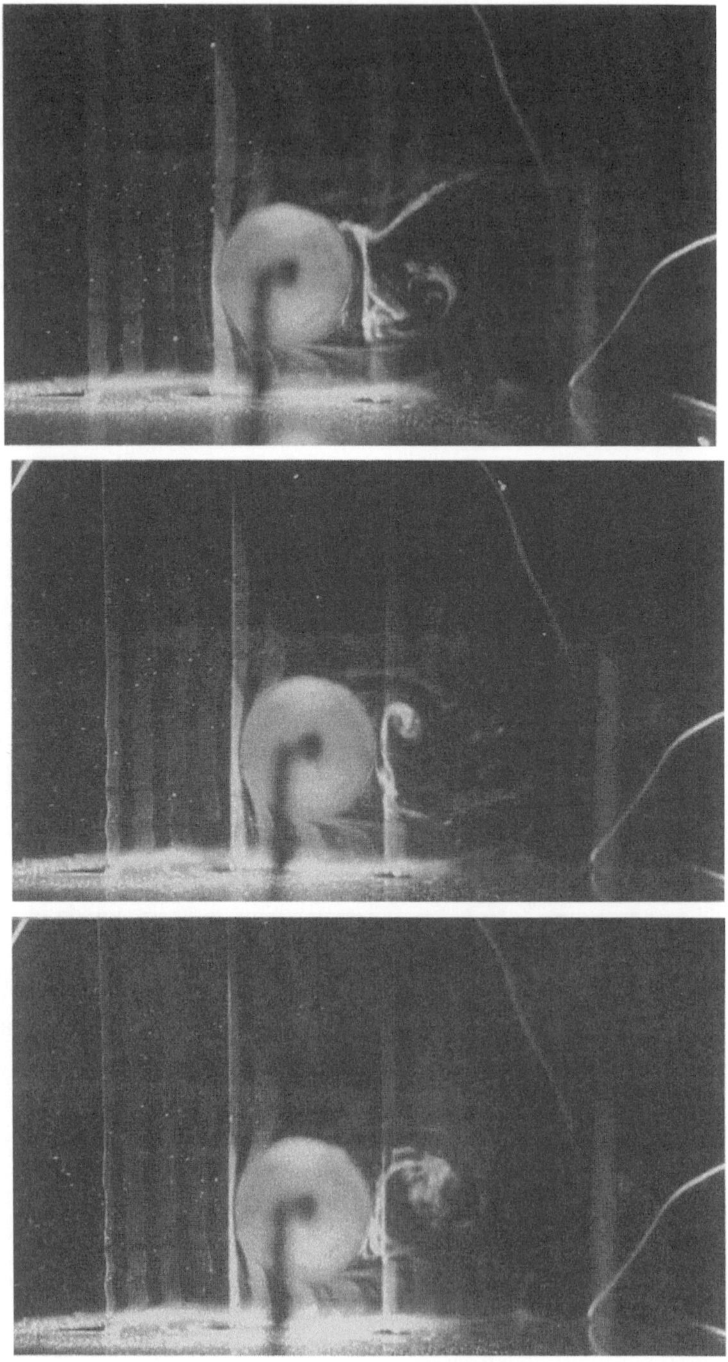

Fig. 6. (continued)

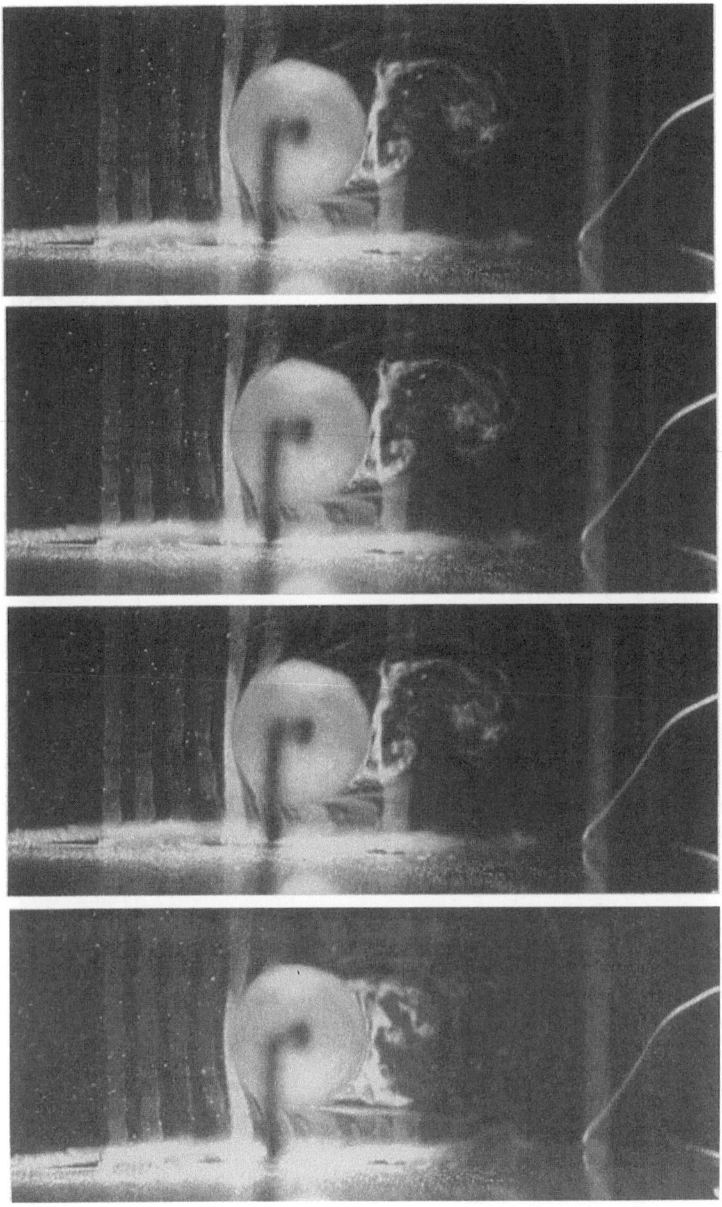

Fig. 6. (continued)

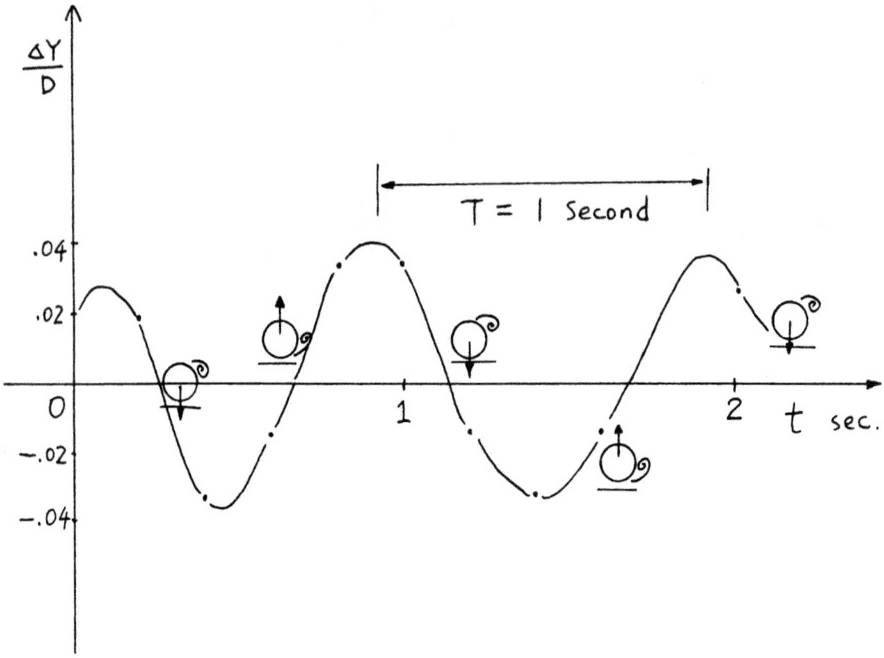

Fig. 7. The displacement of cylinder and the alternate vortex shedding in transverse oscillations, Re_D = 3200, S = D/4

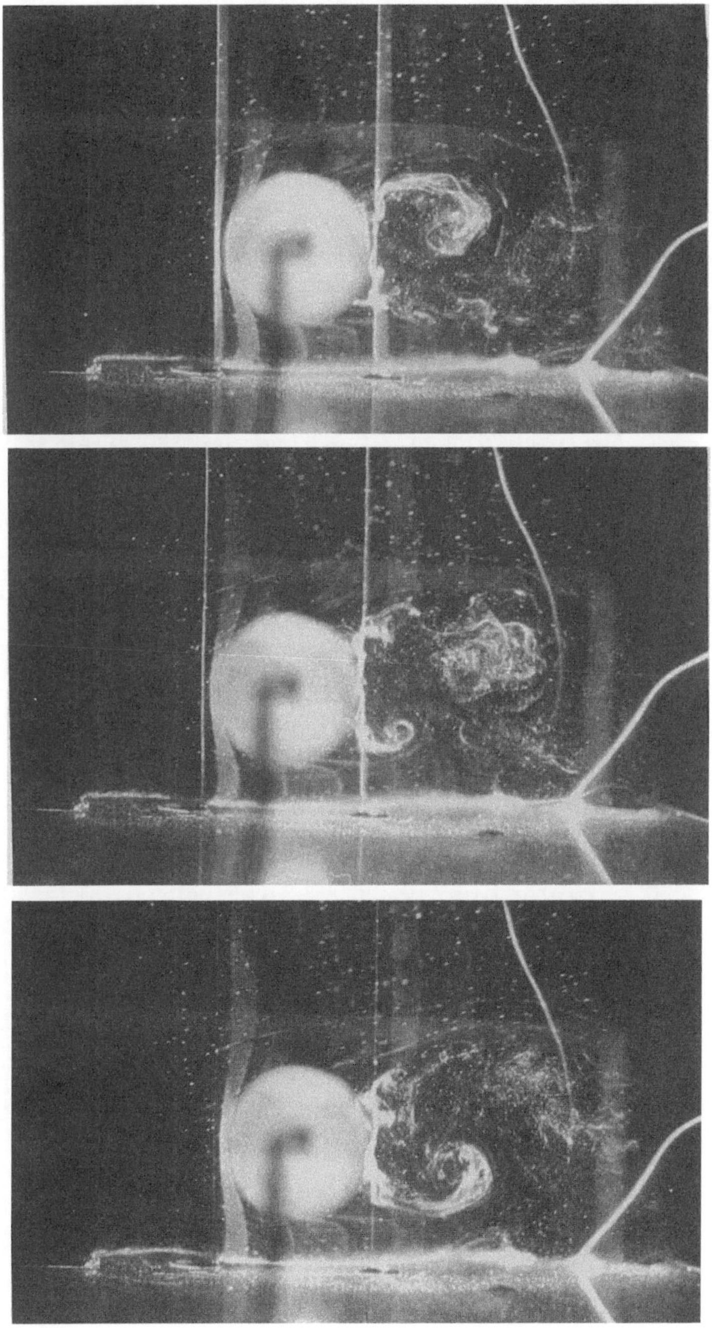

Fig. 8. Photo of a typical transverse oscillation, Re_D = 4030, S = D/4

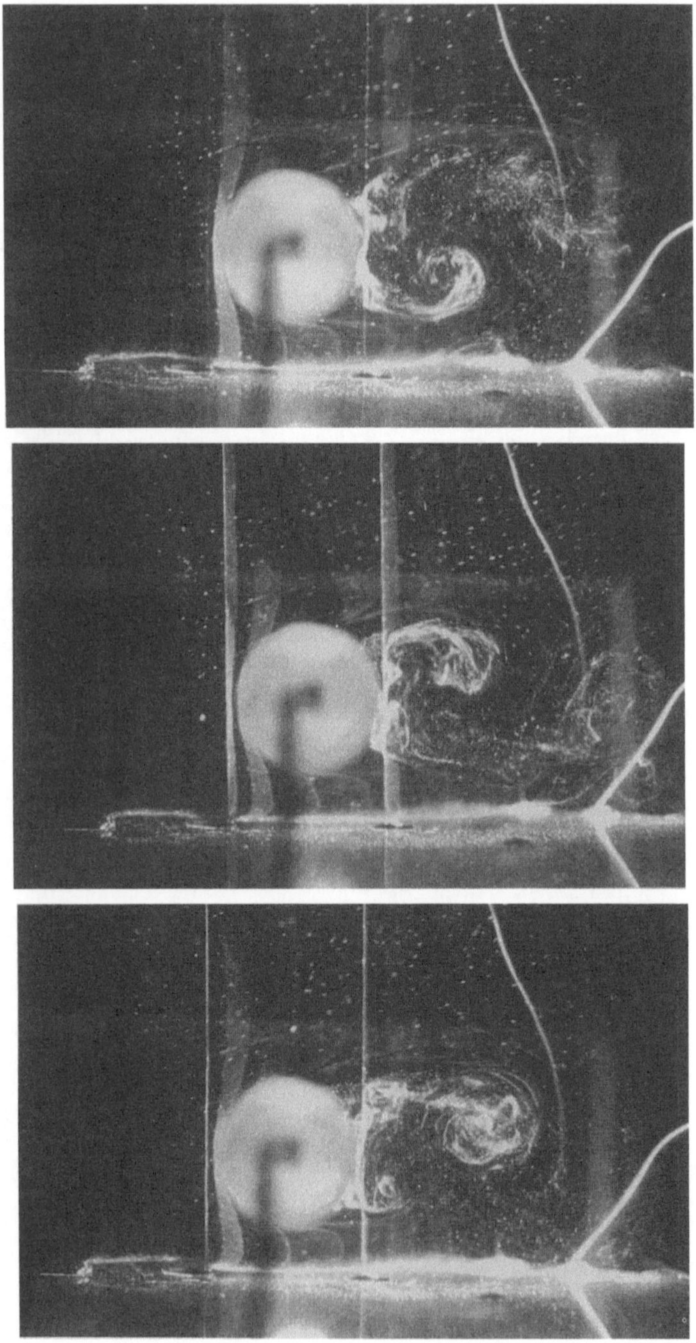

Fig. 8. (continued)

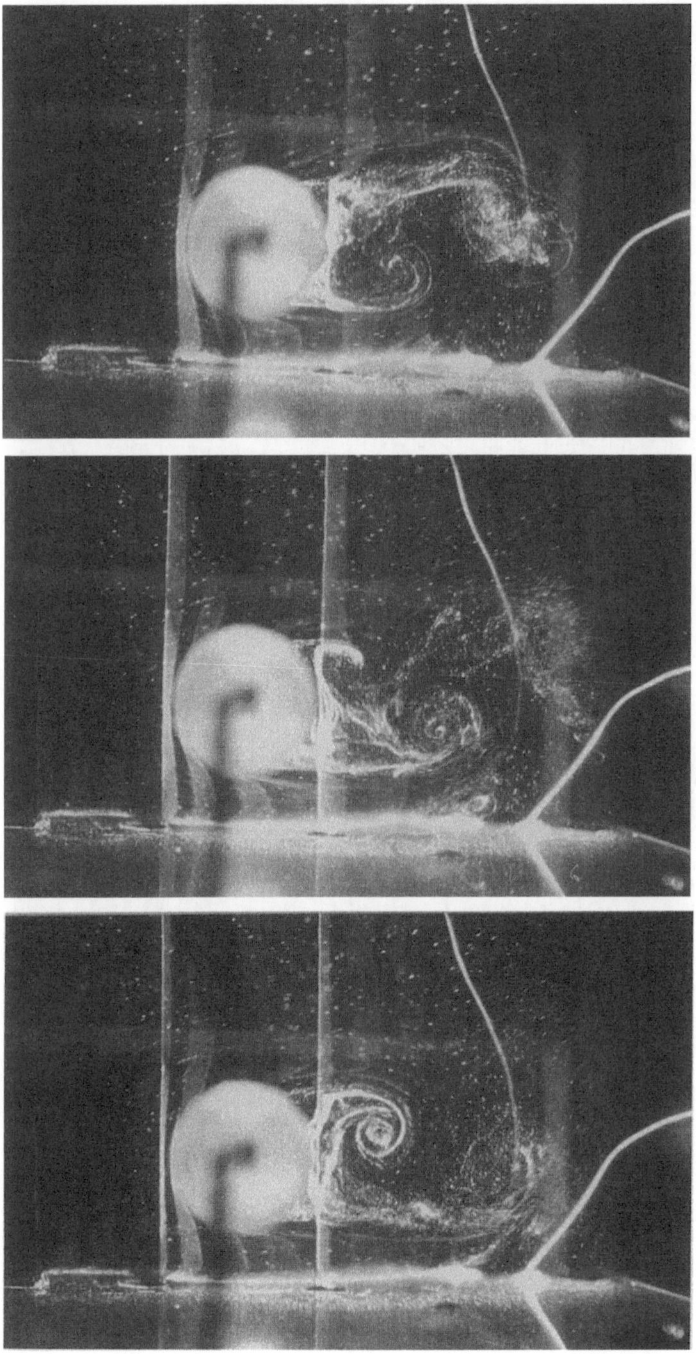

Fig. 8. (continued)

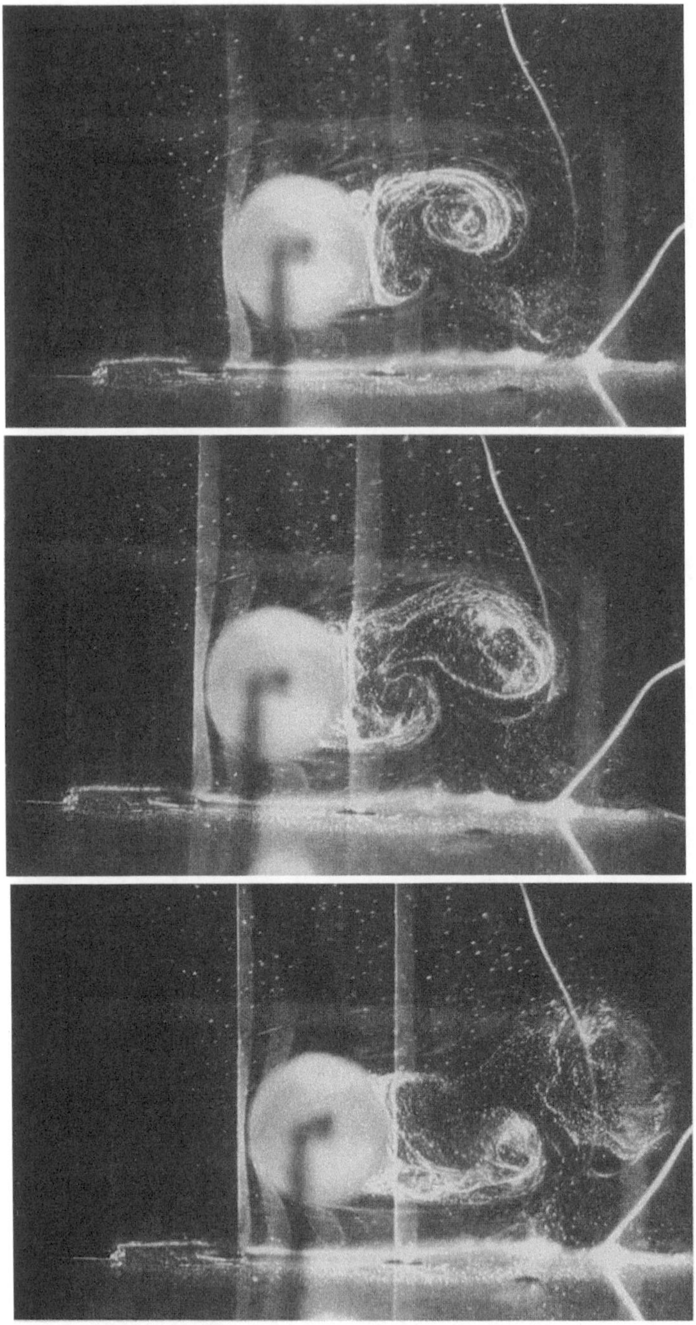

Fig. 8. (continued)

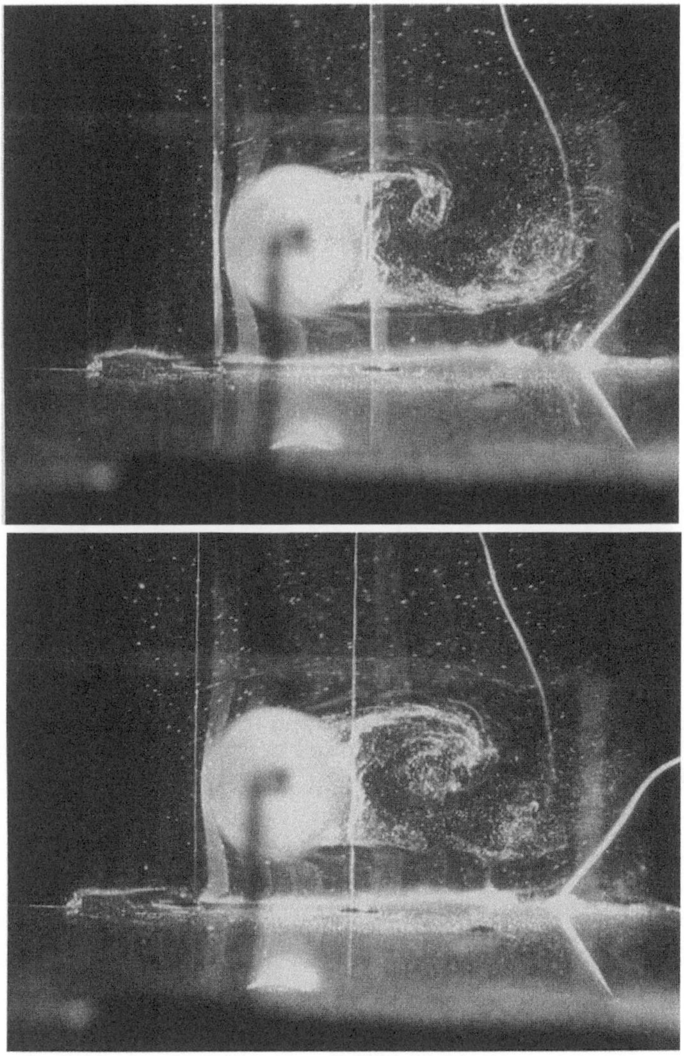

Fig. 8. (continued)

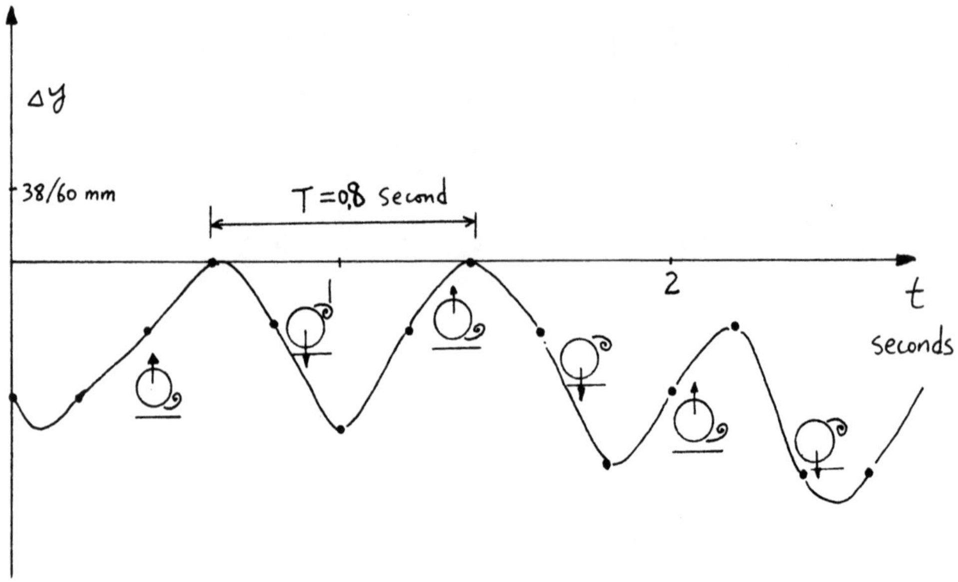

Fig. 9. The displacement of cylinder and the alternate vortex shedding in transverse oscillation, Re_D = 4030, S = D/4.

Instantaneous Optimal Control with Acceleration and Velocity Feedback

J.N. Yang, Z. Li

Dept. Civil Engr., The George Washington University
Washington, D.C. 20052;
Presently, Department of Civil Engineering,
University of California, Irvine, CA. 92117

S.C. Liu

National Science Foundation, Washington, D.C. 20550

SUMMARY

In the experimental demonstration of aseismic control systems, difficulties were encountered in the measurement of the displacement response of the structure. During earthquake ground motions, both the building and the ground are moving so that there is no absolute reference for the determination of the displacement response. An optimal control theory is proposed herein, which utilizes the measurements of acceleration and velocity responses rather than the displacement and velocity measurements. Such an optimal control law is developed based on the instantaneous optimal control theories, and it is evaluated and compared with other available optimal control laws. Numerical results indicate that the performance of the proposed optimal control law is as good as that of other optimal control laws currently available. However, the contribution of such an optimal control law to the practical implementation of active control systems for seismic hazard mitigations may be quite significant.

INTRODUCTION

Recent experimental demonstrations for the application of aseismic control systems to scaled building structures [e.g., 1-7] indicate some difficulties involved in the measurement of floor displacements. The main reason is that during earthquake ground motions, both the building and the ground are moving so that there is no absolute reference for the determination of the floor displacement. This is particularly critical for practical implementations of active control systems to full-scale buildings for earthquake hazard mitigations. Laboratory experiments [1-7] further indicate that the floor displacement response obtained by numerically integrating the velocity measurement differs significantly from the actual floor displacement due to (i) noise pollutions and (ii) error accumulations resulting from numerical integrations.

I. Elishakoff · Y. K. Lin (Eds.)
Stochastic Structural Dynamics 2
New Practical Applications
2nd International Conference on Stochastic Structural Dynamics
May 9–11, 1990, Boca Raton, Florida

Unfortunately, available optimal control theories [e.g., 8-10, 12-18] require measurements of displacement and velocity responses of the building structure. Although the instantaneous optimal open-loop control law proposed by Yang, et al. [9-10] does not require the measurements of the state vector of the structure, it is more vulnerable to a system time delay [11,17] and system uncertainties [11,16]. Since the measurements of acceleration and velocity of the structural response are much easier without involving an absolute reference, it is highly desirable to use acceleration and velocity sensors rather than displacement sensors.

The purpose of this paper is to present an optimal control law utilizing the acceleration measurements rather than the displacement measurements. This optimal control law is developed based on the instantaneous optimal control theories developed by Yang, et al. [9-10, 12-13]. For a building structure subjected to an earthquake, the performance of the proposed optimal control law is evaluated and demonstrated by comparing numerically with other available optimal control laws using both deterministic and stochastic earthquake excitations.

FORMULATION

Consider a shear-beam type building structure implemented by an active control system, such as an active mass damper or an active tendon system as shown in Fig. 1. The structure is idealized by an n degrees of freedom linear system and subjected to a one-dimensional earthquake ground acceleration $\ddot{X}_0(t)$. The matrix equation of motion can be expressed as

$$\dot{\underline{Z}}(t) = \underline{A}\,\underline{Z}(t) + \underline{B}\,\underline{U}(t) + \underline{W}_1\,\ddot{X}_0(t) \tag{1}$$

in which an under bar denotes a vector or matrix. In Eq. (1), $\underline{Z}(t)$ = a 2n state vector, $\underline{U}(t)$ = a r-dimensional control vector, $\underline{A}$ = a (2nx2n) system matrix, representing the structural characteristics of the building, $\underline{B}$ = a (2nxr) location matrix specifying the location of r controllers and $\underline{W}_1$ is an appropriate 2n vector denoting the effect of the earthquake ground acceleration $\ddot{X}_0(t)$. The state vector $\underline{Z}(t)$ consists of the displacement vector $\underline{Y}(t)$ and velocity vector $\dot{\underline{Y}}(t)$, all relative to the ground, as

$$\underline{Z}(t) = \begin{bmatrix} \underline{Y}(t) \\ \text{--------} \\ \dot{\underline{Y}}(t) \end{bmatrix} \tag{2a}$$

with the initial condition $\underline{Z}(0) = 0$. Matrices $\underline{A}$, $\underline{B}$ and $\underline{W}_1$ are given in Ref. 9.

The state vector $\underline{Z}(t)$ can also be defined in terms of the interstory deformation (drift) vector $\underline{X}(t)$ and its derivative $\dot{\underline{X}}(t)$,

$$\underline{Z}(t) = \begin{bmatrix} \underline{X}(t) \\ \text{--------} \\ \dot{\underline{X}}(t) \end{bmatrix} \tag{2b}$$

in which the jth element, $x_j(t)$, of $\underline{X}(t)$ denotes the jth interstory drift. In this situation, the matrices $\underline{A}$, $\underline{B}$ and $\underline{W}_1$ are different from those defined above; however, they are also given in Ref. 9. From the practical point of view, the use of interstory drifts as state variables is more attractive. In this paper both state vectors, Eqs. (2a) and (2b), will be used for numerical demonstrations.

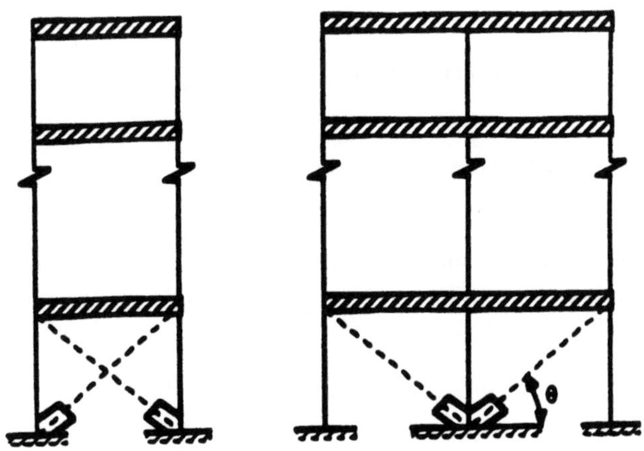

Figure 1: Structural Model With An Active Tendon Contro System: (a) Single Bay; (b) Two Bays

Following the concept of instantaneous optimal control proposed by Yang, et al. [8-9], we define a time-dependent quadratic performance index

$$J^{*}(t) = \underline{\dot{Z}}\ (t)\ \underline{Q}^{*}\ \underline{\dot{Z}}(t) + \underline{U}\ (t)\ \underline{R}\ \underline{U}(t) \tag{3}$$

in which a prime denotes the transpose of a vector or matrix. In Eq. (3), $\underline{Q}^{*}$ is a (2nx2n) positive semi-definite weighting matrix and $\underline{R}$ is a (rxr) positive definite weighting matrix.

In order to minimize the performance index given by Eq. (3), the state vector, $\underline{Z}(t)$, in the equation of motion, Eq. (1), will be expressed in terms of the finite difference as follows

$$\underline{Z}(t) = \underline{Z}(t-\Delta t) + \Delta t\ \underline{\dot{Z}}(t-\Delta t) \tag{4}$$

Substituting Eq. (4) into Eq. (1), one obtains the equation of motion in the following form

$$\underline{\dot{Z}}(t) = \underline{A}\ \underline{Z}(t-\Delta t) + \underline{A}\ \Delta t\ \underline{\dot{Z}}(t-\Delta t) + \underline{B}\ \underline{U}(t) + \underline{W}_1\ \ddot{X}_0(t) \tag{5}$$

The Hamiltonian H(t) is obtained from Eqs. (3) and (5) as

$$H(t) = \underline{\dot{Z}}'(t)\ \underline{Q}^{*}\ \underline{\dot{Z}}(t) + \underline{U}'(t)\ \underline{R}\ \underline{U}(t) + \underline{\lambda} \Big[\underline{\dot{Z}}(t) - \underline{A}\ \underline{Z}(t-\Delta t) - \underline{A}\ \underline{\dot{Z}}(t-\Delta t)\ \Delta t - \underline{B}\ \underline{U}(t) - \underline{W}_1\ \ddot{X}_0(t) \Big] \tag{6}$$

where $\underline{\lambda}$ is a Lagrangian multiplier vector.

The necessary conditions for the minimization of the performance index $J^{*}(t)$ subjected to the constraint given by Eq. (5) are as follows

$$\frac{\partial H(t)}{\partial \underline{\dot{Z}}} = 0, \qquad \frac{\partial H(t)}{\partial \underline{U}} = 0, \qquad \frac{\partial H(t)}{\partial \underline{\lambda}} = 0 \tag{7}$$

The optimal control vector $\underline{U}(t)$ can be obtained by substituting Eq. (6) into Eq. (7). Depending on the way the control vector is regulated, one obtains optimal closed-loop control (feedback), optimal open-loop control (feedforward) and optimal closed-open-loop control (feedforward and feedback) as presented in the Appendix. It is mentioned, however, that the performance of the three optimal control laws derived in the Appendix is identical under ideal control environments. For simplicity, the optimal closed-loop control vector $\underline{U}(t)$ is given in the following [see the Appendix]

$$\underline{U}(t) = -\,\underline{R}^{-1}\,\underline{B}'\,\underline{Q}^{*}\,\dot{\underline{Z}}(t) \tag{8}$$

Thus, the control vector $\underline{U}(t)$ depends on the feedback vector, $\dot{\underline{Z}}(t)$, that consists of the velocity and acceleration responses,

$$\dot{\underline{Z}}(t) = \begin{bmatrix} \dot{\underline{Y}}(t) \\ \text{------} \\ \ddot{\underline{Y}}(t) \end{bmatrix} \quad \text{or} \quad \dot{\underline{Z}}(t) = \begin{bmatrix} \dot{\underline{X}}(t) \\ \text{------} \\ \ddot{\underline{X}}(t) \end{bmatrix} \tag{9}$$

Consequently, the measurement of the displacement response $\underline{Y}(t)$ is replaced by the measurement of the acceleration response $\ddot{\underline{Y}}(t)$.

It is mentioned that the implication of minimizing the performance index, $J^{*}(t)$, given by Eq. (3) is that the quadratic function, involving the velocity response, acceleration response and control forces, is minimized at every time instant t for all $0 \le t \le t_f$, where t_f is longer than the duration of the earthquake. Although the displacement response vector $\underline{Y}(t)$ does not appear in the performance index $J^{*}(t)$, Eq. (3), it is expected that a minimization of $J^{*}(t)$ will also reduce $\underline{Y}(t)$.

Based on the instantaneous optimal closed-loop control law proposed in Refs. 9 and 10, the optimal control vector is given by

$$\underline{U}(t) = -\,\frac{\Delta t}{2}\,\underline{R}^{-1}\,\underline{B}'\,\underline{Q}\,\underline{Z}(t) \tag{10}$$

in which $\underline{Q}$ is a (2nx2n) positive semi-definite weighting matrix, and Δt is the integration step size.

The optimal control vector $\underline{U}(t)$ based on the linear quadratic optimal control law is as follows [e.g.,8,14]

$$\underline{U}(t) = -\,\frac{1}{2}\,\underline{R}^{-1}\,\underline{B}'\,\underline{P}\,\underline{Z}(t) \tag{11}$$

where $\underline{P}$ is a (2nx2n) Riccati matrix.

Finally, the matrix equation of motion for a structure implemented by an active control system is obtained by substituting Eq. (8) into Eq. (1) as

$$\left(\underline{I} + \underline{B}\,\underline{R}^{-1}\,\underline{B}'\,\underline{Q}^{*} \right) \dot{\underline{Z}}(t) = \underline{A}\,\underline{Z}(t) + \underline{W}_1\,\ddot{X}_0(t) \qquad (12)$$

in which I is a (2n x 2n) identity matrix.

Further, substitution of Eq. (10) or (11) into Eq. (1) leads to the following matrix equation of motion

$$\dot{\underline{Z}}(t) = \left(\underline{A} + \underline{B}\,\underline{G} \right) \underline{Z}(t) + \underline{W}_1\,\ddot{X}_0(t) \qquad (13)$$

in which $\underline{G} = -0.5\,\underline{R}^{-1}\,\underline{B}'\,\underline{P}$ for linear quadratic optimal control and $\underline{G} = -0.5\,\Delta t\,\underline{R}'\,\underline{B}\,\underline{Q}$ for instantaneous optimal control.

In what follows, the performance of the three optimal control laws given by Eqs. (8), (10) and (11) will be compared using a 6-story full-scale building implemented by an active tendon control system, Fig. 1(b), and subjected to earthquake ground accelerations.

STOCHASTIC EARTHQUAKE RESPONSE

The controlled structural response can be solved numerically using Eq. (12) or (13), if the time history of an earthquake ground acceleration $\ddot{X}_0(t)$ is given. The earthquake ground acceleration $\ddot{X}_0(t)$, however, varies from occurrence to occurrence and it may be characterized more appropriately by a random process. Hence, the performance of the new optimal control law will be evaluated not only using an earthquake sample time history, such as the El Centro earthquake, but also considering $\ddot{X}_0(t)$ as a random process.

The earthquake ground acceleration, $\ddot{X}_0(t)$, is modeled as a filtered shot noise. In other words, $\ddot{X}_0(t)$ is the outpout of a filter due to a shot noise exicitation,

$$\ddot{X}_g(t) = \psi(t)\ \eta(t) \tag{14}$$

in which $\psi(t)$ is a deterministic non-negative envelope function, and $\eta(t)$ is a stationary white noise with zero mean and a power spectral density S^2. Various types of envelope functions $\psi(t)$ have been used in the literature. A particular envelope function given in the following will be used: $\psi(t) = 0$ for $t < 0$, $\psi(t) = (t/t_1)^2$ for $0 \le t \le t_1$, $\psi(t) = 1$ for $t_1 \le t \le t_2$ and $\psi(t) = \exp[-c(t-t_2)]$ for $t > t_2$, where t_1, t_2 and c are parameters which should be selected appropriately to reflect the shape and the duration of the earthquake ground acceleration.

The frequency response function of the filter, denoted by $H_f(\omega)$, is given by

$$H_f(\omega) = \frac{1 + 2\zeta_g(\omega/\omega_g)\ \underline{i}}{\left[1 - (\omega/\omega_g)^2\right] + 2\zeta_g(\omega/\omega_g)\ \underline{i}} \tag{15}$$

in which $\underline{i} = (-1)^{1/2}$ and ζ_g and ω_g are parameters depending on the characteristics of the earthquake at a particular location.

Since the earthquake ground acceleration $\ddot{X}_0(t)$ has a zero mean, the mean values of the response state vector $\underline{Z}(t)$ and the active control vector $\underline{U}(t)$ are zero. The mean square values of $\underline{Z}(t)$ and $\underline{U}(t)$ are identical to the variances $\underline{\sigma}_z^2(t)$ and $\underline{\sigma}_u^2(t)$, respectively.

Let $\underline{H}_z(\omega)$ and $\underline{H}_u(\omega)$ be the frequency response vectors of $\underline{Z}(t)$ and $\underline{U}(t)$ due to a unit steady state ground acceleration, i.e., $\ddot{X}_0(t) = \exp[\underline{i}\omega t]$, $\underline{Z}(t) = \underline{H}_z(\omega)\exp[\underline{i}\omega t]$ and $\underline{U}(t) = \underline{H}_u(\omega)\exp[\underline{i}\omega t]$. The frequency response vectors, $\underline{H}_z(\omega)$ and $\underline{H}_u(\omega)$, for a controlled structure can be obtained easily from Eq. (12) or (13).

The impulse response vectors, $\underline{h}_z^*(t)$ and $\underline{h}_u^*(t)$, of $\underline{Z}(t)$ and $\underline{U}(t)$, respectively, due to the shot noise input $\ddot{X}_g(t) = \delta(t)$ are related to the frequency response vectors through the Fourier transform pair

$$\underline{h}_z^*(t) = \frac{1}{2\pi}\int_{-\infty}^{\infty} H_f(\omega)\,\underline{H}_z(\omega)\,e^{\underline{i}\omega t}\,d\omega \;;\; \underline{h}_u^*(t) = \frac{1}{2\pi}\int_{-\infty}^{\infty} H_f(\omega)\underline{H}_u(\omega)\,e^{\underline{i}\omega t}\,d\omega \quad (16)$$

in which $H_f(\omega)$ is given by Eq. (15)

The mean square response of the state vector $\underline{Z}(t)$ can be obtained easily as follows [e.g., 9].

$$\underline{\sigma}_z^2(t) = \int_{-\infty}^{\infty} |\underline{M}_z(t,\omega)|^2\,S^2\,d\omega \quad (17)$$

in which $|\underline{M}_z(t,\omega)|^2$ is a vector whose jth element is equal to the square of the absolute value of the jth element of $\underline{M}_z(t,\omega)$ given by Eq. (18)

$$\underline{M}_z(t,\omega) = \int_0^t \underline{h}_z^*(\tau)\psi(t-\tau)\,e^{-\underline{i}\omega\tau}\,d\tau \quad (18)$$

In a similar manner, the mean square vector of the active control force can be obtained as

$$\underline{\sigma}_u^2(t) = \int_{-\infty}^{\infty} |\underline{M}_u(t,\omega)|^2\,S^2\,d\omega \quad (19)$$

in which

$$\underline{M}_u(t,\omega) = \int_0^t \underline{h}_u^*(\tau)\psi(t-\tau)\,e^{-\underline{i}\omega\tau}\,d\tau \quad (20)$$

The numerical computation of the non-stationary root mean square vectors, $\underline{\sigma}_z(t)$ and $\underline{\sigma}_u(t)$, of $\underline{Z}(t)$ and $\underline{U}(t)$ can be carried out very efficiently by repeated applications of the Fast Fourier Transform (FFT) in the following manner: (i) The impulse response vectors $\underline{h}_z^*(t)$ and $\underline{h}_u^*(t)$ due to $\ddot{X}_g(t) = \delta(t)$ are computed from the corresponding frequency response vectors $H_f(\omega)\,\underline{H}_z(\omega)$ and $H_f(\omega)\underline{H}_u(\omega)$ using the FFT technique, Eq.

(16); (ii) $\underline{M}_z(t,\omega)$ and $\underline{M}_u(t,\omega)$ are computed from Eqs. (18) and (20) again using the FFT technique; and (iii) The root mean square vectors $\underline{\sigma}_z(t)$ and $\underline{\sigma}_u(t)$ are evaluated by numerically integrating Eqs. (17) and (19) and taking the square root.

NUMERICAL DEMONSTRATION

The performance of the proposed optimal control law will be demonstrated using two examples; one with the El Centro earthquake ground excitation and the other with a stochastic earthquake ground acceleration. The controlled structural response and the required active control force will be compared with those obtained using both the linear quadratic optimal control law and the instantaneous optimal control law.

A six-story full-scale building has been constructed recently in Japan by Takenaka Company in order to conduct field demonstrations of an active tendon control system and an active mass damper. The properties of the building have been provided by NCEER as follows. The (6x6) mass matrix is a diagonal matrix with each diagonal element being equal to 571.4 slugs. The (6x6) stiffness matrix K in 10^4 lb/in and the damping matrix C in lb.sec./in are given in the following

$$\underline{K} = \begin{bmatrix} 325.56 & -159.41 & 39.14 & -6.28 & 0.92 & -0.08 \\ -159.38 & 195.81 & -111.51 & 25.32 & -3.57 & 0.41 \\ 39.14 & -111.50 & 149.94 & -82.93 & 16.34 & -1.71 \\ -6.23 & 25.31 & -82.92 & 113.13 & -59.44 & 8.68 \\ 0.94 & -3.59 & 16.35 & -59.44 & 80.65 & -34.66 \\ -0.04 & 0.39 & -1.70 & 8.67 & -34.66 & 27.33 \end{bmatrix}$$

$$\underline{C} = \begin{bmatrix} 822.97 & -257.50 & 21.63 & -6.43 & -1.02 & -1.01 \\ -257.45 & 572.34 & -231.38 & 8.18 & -8.04 & -3.29 \\ 21.65 & -231.37 & 498.08 & -201.29 & -0.31 & -9.43 \\ -6.33 & 8.18 & -201.27 & 432.21 & -171.18 & -9.70 \\ -0.96 & -8.07 & -0.31 & -171.18 & 367.47 & -141.15 \\ -0.91 & 3.31 & -9.41 & -9.72 & -141.15 & 205.76 \end{bmatrix}$$

With the mass, stiffness and damping matrices above, the natural frequencies of the building are computed as 0.943, 2.765, 4.876, 7.279, 10.114 and 14.423 Hz. The damping ratio for each vibrational mode is 1%. This six-story building has two bays and an active tendon controller is installed on the first floor as shown in Fig. 1(b) where the angle of inclination θ for tendons is 51.5 degrees.

The El Centro earthquake ground acceleration scaled by a factor of 32% is shown in Fig. 2. This 32% El Centro earthquake is considered as the input excitation. Without any control system, the maximum interstory deformation x_i^* (i = 1,2,...,6) of each story unit, the maximum total acceleration of each floor $\ddot{y}_i^*$ (i = 1,2,...,6) and the maximum relative displacement y_6^* of the top floor with respect to the ground, within the entire earthquake episode, are shown in Column A of Table 1.

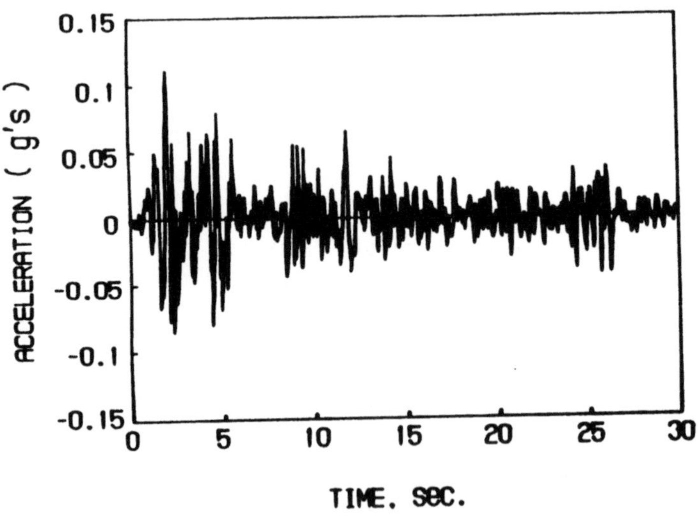

Figure 2: 32% El Centro Earthquake Ground Acceleration

With the active tendon control system, the controlled building response and the required active control force depend on the particular control law. In the present example with only one tendon controller, the $\underline{R}$ matrix consists of one element, denoted by R. For demonstrative purposes, R = 1 is used. For linear quadratic optimal control, the (12x12) weighting matrix $\underline{Q}$ is considered as a diagonal matrix in which every diagonal element is identical to q. The Riccati matrix $\underline{P}$ is computed for $q = 3 \times 10^7$ and 2×10^9, respectively. Then, within the entire earthquake episode, the maximum structural response quantities and the maximum active control force U_{max} are computed and shown in Column B of Table 1.

With the application of the instantaneous optimal control law [9,10], the weighting matrix $0.5\Delta t Q$ is partitioned as follows

$$\frac{\Delta t}{2} \underline{Q} = \alpha \left[\begin{array}{c|c} \underline{0} & \underline{0} \\ \hline \underline{Q}_{21} & \underline{Q}_{22} \end{array} \right] \tag{21}$$

in which α is a constant, and $\underline{Q}_{21}$ and $\underline{Q}_{22}$ are (6x6) matrices. Since only one controller is installed on the first floor, the i-jth element of $\underline{Q}_{21}$ and $\underline{Q}_{22}$, denoted by $Q_{21}(i,j)$ and $Q_{22}(i.j)$, respectively, can be chosen to be zero for $i = 2,3,\ldots 6$. For illustrative purpose, the following values are assigned to elements of $Q_{21}(1,j)$ and $Q_{22}(1,j)$: $Q_{21}(1,j)$ = [1660, -1080, 24.1, 73.1, -11.4, 15.1], $Q_{22}(1,j)$ = [43.5, 29.4, 7.8, 7.8, 8.0, 7.2].

The maximum building response quantities and the required maximum active control force are shown in Column C of Table 1 for $\alpha = 1.15 \times 10^5$ and 10^6, respectively.

For the proposed optimal control law, the weighting matrix $\underline{Q}^*$ is partitioned similar to Eq. (21) as

$$\underline{Q}^* = \alpha \left[\begin{array}{c|c} \underline{0} & \underline{0} \\ \hline \underline{Q}^*_{21} & \underline{Q}^*_{22} \end{array} \right] \tag{22}$$

TABLE 1: MAXIMUM RESPONSE QUANTITIES OF BUILDING: (A) WITHOUT CONTROL; (B) LINEAR QUADRATIC OPTIMAL CONTROL, EQ.(11); (C) INSTANTANEOUS OPTIMAL CONTROL, EQ.(10); AND (D) INSTANTANEOUS OPTIMAL CONTROL WITH FEEDBACK $\dot{\underline{Z}}(t)$, EQ.(8).

STORY	(A) NO CONTROL $y_6^* = 7.99$ cm		(B) $q = 3x10^7$ $U_{max} = 659$ kN $y_6^* = 5.57$ cm		(C) $\alpha = 1.15x10^5$ $U_{max} = 687.8$ kN $y_6^* = 5.29$ cm		(D) $\alpha = 2.8x10^3$ $U_{max} = 650.8$ kN $y_6^* = 5.53$ cm	
	x_i^* (cm)	$\ddot{Y}_i^*$ (cm/s^2)	x_i^* (cm)	$\ddot{Y}_i^*$ (cm/s^2)	x_i^* (cm)	$\ddot{Y}_i^*$ (cm/s^2)	x_i^* (cm)	$\ddot{Y}_i^*$ (cm/s^2)
1	0.734	120.4	0.488	113.3	0.518	107.9	0.530	123.4
2	1.453	234.5	1.024	147.6	1.011	143.5	1.036	111.7
3	1.615	283.7	1.153	203.2	1.125	197.8	1.165	157.4
4	1.753	284.5	1.171	205.7	1.148	191.5	1.143	149.8
5	1.651	331.7	1.090	203.7	1.067	199.8	0.991	193.8
6	1.135	361.2	0.790	261.9	0.784	259.0	0.718	234.6

STORY	(A) NO CONTROL $y_6^* = 7.99$ cm		(B) $q = 2x10^9$ $U_{max} = 2236$ kN $y_6^* = 2.29$ cm		(C) $\alpha = 10^6$ $U_{max} = 2129$ kN $y_6^* = 2.41$ cm		(D) $\alpha = 2x10^4$ $U_{max} = 1931$ kN $y_6^* = 2.79$ cm	
	x_i^* (cm)	$\ddot{Y}_i^*$ (cm/s^2)	x_i^* (cm)	$\ddot{Y}_i^*$ (cm/s^2)	x_i^* (cm)	$\ddot{Y}_i^*$ (cm/s^2)	x_i^* (cm)	$\ddot{Y}_i^*$ (cm/s^2)
1	0.734	120.4	0.518	256.8	0.490	260.8	0.383	185.1
2	1.453	234.5	0.488	142.0	0.513	153.3	0.591	155.4
3	1.615	283.7	0.508	158.2	0.530	157.7	0.772	172.4
4	1.753	284.5	0.635	163.8	0.629	165.3	0.723	227.5
5	1.651	331.7	0.749	154.9	0.731	152.6	0.706	198.8
6	1.135	361.2	0.648	233.9	0.637	230.8	0.670	252.2

X_i^* = MAXIMUM DEFORMATION OF iTH STORY UNIT

$\ddot{Y}_i$ = MAXIMUM TOTAL ACCELERATION OF iTH FLOOR

y_6^* = MAXIMUM DISPLACEMENT OF TOP FLOOR WITH RESPECT TO THE GROUND

U_{max} = MAXIMUM CONTROL FORCE

in which α is a constant, and $\underline{Q}^*_{21}$ and $\underline{Q}^*_{22}$ are (6x6) matrices. Again, the i-jth element of $\underline{Q}^*_{21}$ and $\underline{Q}^*_{22}$, denoted by $Q^*_{21}(i,j)$ and $Q^*_{22}(i,j)$, respectively, can be chosen to be zero for i = 2,3,...,6. For illustrative purpose, $Q^*_{21}(1,j)$ and $Q^*_{22}(1,j)$ are given as follows: $Q^*_{21}(1,j)$ = [3612, 2485, 170, 82.5, 36.8, 24.5], $Q^*_{22}(1,j)$ = [101.7, 71.1, 20.4, 20.3, 19.1, 18.1].

The maximum building response quantities, i.e., x^*_i, $\ddot{y}^*_i$ (i = 1,2...,6), y^*_6 and the required maximum control force U_{max} are presented in Column D of Table 1 for $\alpha = 2.8 \times 10^3$ and 2×10^4, respectively.

Table 1 provides a clear comparison for the performance among different optimal control laws, because the required maximum control force is approximately the same. The following observations are made from Table 1: (i) The active tendon control system is quite effective in reducing the building response quantities, whereas, the required active control force is well within the practical limit, and (ii) The difference in the performance is minimal for the three optimal control laws investigated.

We next consider that the earthquake ground acceleration $\ddot{X}_0(t)$ is a nonstationary random process with zero mean as described previously. For illustrative purpose, the parameters appearing in the envelope function, the filter and the power spectral density S^2 of the white noise are chosen as follows: t_1 = 3 sec., t_2 = 13 sec., c = 0.26 sec.$^{-1}$, ζ_g = 0.65, ω_g = 18.85 rad./sec., and S^2 = 10.824 cm^2/sec^2.

Time dependent root mean squares of the response vector, $\underline{\sigma}_z(t)$, and the control force, $\sigma_u(t)$, were computed using the three different optimal control laws described previously. The weighting matrices $\underline{Q}$ and $\underline{R}$ are identical to those used in the previous case. Within 30 seconds of the earthquake episode, the maximum root mean square values of the relative displacement of the first floor and the top floor with respect to the ground, denoted by $\bar{\sigma}_1$ and $\bar{\sigma}_6$, respectively, are presented in Table 2. Also presented in Table 2 are the corresponding maximum root mean square of the required active control force, denoted by $\bar{\sigma}_u$. As observed from Table 2, the difference in the performance for each optimal control law is minimal.

Using the proposed optimal control law, the time dependent root mean square values of the relative displacement of the top floor with respect to the ground, denoted by $\sigma_6(t)$, are plotted in Fig. 3. In Fig. 3, Curve 0 is the response without control, whereas Curves 1 and 2 are the responses with active tendon control using $\alpha = 2.8 \times 10^3$ and 2×10^4, respectively, see Table 2. The time dependent root mean squares of the

TABLE 2: MAXIMUM ROOT MEAN SQUARE OF STRUCTURAL RESPONSE AND CONTROL FORCE: (A) WITHOUT CONTROL; (B) LINEAR QUADRATIC OPTIMAL CONTROL, EQ.(11); (C) INSTANTANEOUS OPTIMAL CONTROL, EQ.(10); AND (D) INSTANTANEOUS OPTIMAL CONTROL WITH FEEDBACK $\dot{\underline{Z}}(t)$, EQ.(8).

	(A) WITHOUT CONTROL	(B) $q=3\text{x}10^7$	(C) $\alpha=1.15\text{x}10^5$	(D) $\alpha=2.8\text{x}10^3$
$\bar{\sigma}_1$ (cm)	0.136	0.077	0.082	0.080
$\bar{\sigma}_6$ (cm)	1.405	0.756	0.760	0.780
$\bar{\sigma}_u$ (kN)	0	473.3	490.6	481.5

	(A) WITHOUT CONTROL	(B) $q=2\text{x}10^9$	(C) $\alpha=10^6$	(D) $\alpha=2\text{x}10^4$
$\bar{\sigma}_1$ (cm)	0.136	0.074	0.069	0.050
$\bar{\sigma}_6$ (cm)	1.405	0.358	0.363	0.381
$\bar{\sigma}_u$ (kN)	0	1348.9	1283.2	1227.1

$\bar{\sigma}_1$ = MAXIMUM RMS OF 1ST FLOOR RELATIVE DISPLACEMENT

$\bar{\sigma}_6$ = MAXIMUM RMS OF TOP FLOOR RELATIVE DISPLACEMENT

$\bar{\sigma}_u$ = MAXIMUM RMS OF CONTROL FORCE

required active control force, $\sigma_u(t)$, are presented in Fig. 4 as Curves 1 and 2. Curve 1 in Fig. 4 is the required active control force for $\alpha = 2.8 \times 10^3$, corresponding to the case for Curve 1 of Fig. 3, whereas Curve 2 in Fig. 4 corresponds to the case for Curve 2 of Fig. 3, $\alpha = 2 \times 10^4$. Finally, the results for the time dependent root mean squares of the response and the active control force using the other two optimal control laws are very close to those shown in Figs. 3 and 4, and hence they are not presented.

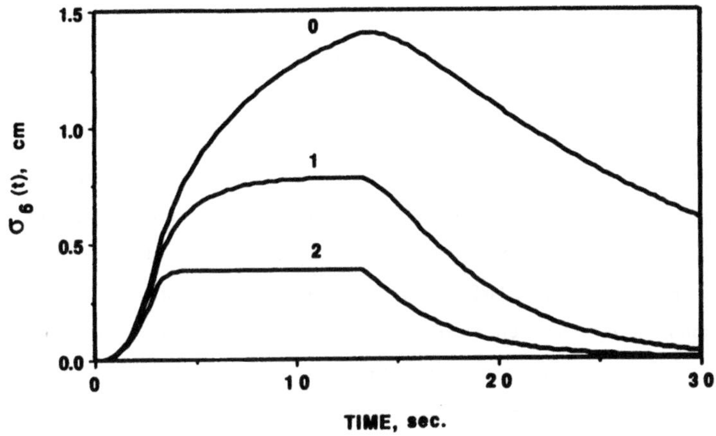

Figure 3: Root-Mean-Square of Top Floor Relative Displacement With Respect To The Ground: Curve 0 (Without Control); Curve 1 ($\alpha = 2.8 \times 10^3$); Curve 2 ($\alpha = 2 \times 10^4$).

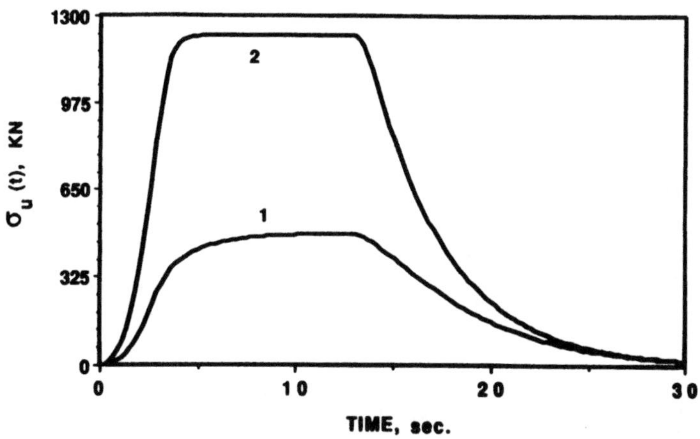

Figure 4: Root-Mean-Square of Active Control Force; Curve 1 ($\alpha = 2.8 \times 10^3$); Curve 2 ($\alpha = 2 \times 10^4$).

It is observed from Figs. 3 and 4 and Table 2 that, under stochastic earthquake ground excitations, (i) the proposed optimal control law with velocity and acceleration feedbacks performs very well, and (ii) the difference in the performance is minimal for the three optimal control laws investigated.

CONCLUSION

For practical implementations of an active control system to building structures for seismic hazard mitigations, it is important to avoid the measurement of the displacement response. A new intantaneous optimal control law is proposed, which requires the measurements of the acceleration and velocity responses rather than the displacement and velocity responses. The performance of such an optimal control law has been investigated, evaluated, and compared with other optimal control laws using both deterministic and stochastic earthquake excitations. It is demonstrated that the performance of the proposed optimal control law is as good as other optimal control laws currently available. The contribution of such an optimal control law to practical implementations of an active control system for earthquake hazard mitigation can be significant.

ACKNOWLEDGMENT

This research is partially supported by the National Science Foundation Grant No. NSF-BCS-89-04524 and NCEER Grant No. NCEER-89-2202.

REFERENCES

1. Chung, L.L., Reinhorn, A.M. and Soong, T.T., "An Experimental Study of Active Structural Control," in Dynamic Response of Structures, (G.C. Hart and R.B. Nelson, eds.), ASCE, New York, 1986, pp. 795-802.

2. Chung, L.L., Reinhorn, A.M. and Soong, T.T., "Experiments on Active Control of Seismic Structures," J. Eng. Mech., ASCE, Vol. 114, 1988, pp. 241-256.

3. Chung, L.L., Lin, R.C., Soong, T.T. and Reinhorn, A.M., "Experimental Study of Active Control of MDOF Structures," Report No. NCEER

88-0025, National Center for Earthquake Engineering Research, Buffalo, New York, 1988.

4. Chung, L.L., Lin, R.C., Soong, T.T. and Reinhorn, A.M., "Experimental Study of Active Control for MDOF Seismic Structures," J. Eng. Mech., ASCE Vol. 115, 1989, pp. 1609-1627.

5. Reinhorn, A.M., Soong, T.T., et al., "Experiments on Active Structural Control Under Seismic Loads," in Computer Utilization in Structural Engineering, J.K. Nelson, Jr. (ed.), ASCE, New York, 1989, pp. 101-110.

6. Soong, T.T., Reinhorn, A.M. and Yang, J.N., "A Standardized Model for Structural Control Experiments and Some Experimental Results," in Structural Control, H.H.E. Leipholz (ed.), Martinus Nijhoff, Amsterdam, 1987, pp. 669-693.

7. Soong, T.T., Reinhorn, A.M. and Yang, J.N., "Active Response Control of Building Structures Under Seismic Excitation," Proc. 9th World Conf. Earthquake Engineering, VIII, Tokyo/Kyoto, Japan, 1988, pp. 453-458.

8. Soong, T.T., "Active Structural Control in Civil Engineering," National Center for Earthquake Engineering Research Technical Report, NCEER-87-0023, SUNY, Nov. 1987.

9. Yang, J.N., Akbarpour, A. and Ghaemmaghami, P., "Instantaneous Optimal Control Laws For Tall Building Under Seismic Loads," National Center For Earthquake Engineering Research, Technical Report No. NCEER-TR-87-0007, June 1987.

10. Yang, J.N., Akbarpour, A. and Ghaemmaghami, P., "New Optimal Control Algorithms For Structural Control," Journal of Engineering Mechanics, ASCE, Vol. 113, No. 9, Sept. 1987, pp. 1369-1386.

11. Yang, J.N., and Akbarpour, A., "Practical Considerations For Structural Control: System Uncertainty, System Time Delay and Truncation of Small Control Forces," National Center For Earthquake Engineering Research, Technical Report No. NCEER-TR-87-0018, Sept. 1987.

12. Yang, J.N., Long, F.X. and Wong, D., "Optimal Control of Nonlinear Flexible Structures," National Center For Earthquake Engineering Research, Technical Report No. NCEER-TR-88-0002, January 1988.

13. Yang, J.N., Long, F.X., and Wong, D., "Optimal Control of Nonlinear Structures," Journal of Applied Mechanics, ASME, Vol. 55, Dec. 1988, pp. 931-938.

14. Yang, J.N., and Soong, T.T., "Recent Advancement in Active Control of Civil Engineering Structures," Journal of Probabilistic Engineering Mechanics, Vol. 3, No. 4, Dec. 1988, pp. 179-188.

15. Yang, J.N., and Wong, D., "On Aseismic Hybrid Control System," Proc. International Conference on Structural Safety and Reliability,

ICOSSAR' 89, August 1989, pp. 471-478.

16. Yang, J.N., and Akbarpour, A., "Effect of System Uncertainty on Control of Seismic-Excited Buildings," Journal of Engineering Mechanics, ASCE, Vol. 116, No. 2, Feb. 1990, pp. 462-478.

17. Yang, J.N., Akbarpour, A., and Askar, G. "Effect of Time Delay on Control of Seismic-Excited Buildings," Journal of Structural Engineering, ASCE, Vol. 116, No. 10, Oct. 1990, pp. 2801-2814.

18. Yang, J.N., Danielians, A. and Liu, S.C., "Aseismic Hybrid Control System For Building Structures Under Strong Earthquake," in Intelligent Structures, edited by K.P. Chong, S.C. Liu and J.C. Li, Elsevier Applied Science, 1990, pp. 179-195.

APPENDIX: OPTIMAL CONTROL LAWS

Substituting the Hamiltonian H(t) given by Eq. (6) into Eq. (7), one obtains the necessary conditions for minimizing the performance index $J^*(t)$ as follows

$$2\,\underline{Q}^*\ \dot{\underline{Z}}(t) + \underline{\lambda}(t) = 0 \qquad \text{(A-1)}$$

$$2\,\underline{R}\ \underline{U}(t) - \underline{B}'\ \underline{\lambda}(t) = 0 \qquad \text{(A-2)}$$

$$\dot{\underline{Z}}(t) = \underline{A}\ \underline{Z}(t-\Delta t) + \underline{A}\ \dot{\underline{Z}}(t-\Delta t)\ \Delta t + \underline{B}\ \underline{U}(t) + \underline{W}_1\ \ddot{X}_0(t) \qquad \text{(A-3)}$$

(i) Optimal Closed-Loop Control (Feedback Control)

It follows from Eq. (A-2) that the optimal control vector $\underline{U}(t)$ is proportional to the Lagrangian multiplier vector $\underline{\lambda}(t)$, i.e.,

$$\underline{U}(t) = \frac{1}{2}\ \underline{R}^{-1}\ \underline{B}'\ \underline{\lambda}(t) \qquad \text{(A-4)}$$

Let the control vector $\underline{U}(t)$ be regulated by the velocity state vector $\dot{\underline{Z}}(t)$ alone, i.e.,

$$\underline{\lambda}(t) = \underline{\Lambda}\ \dot{\underline{Z}}(t) \qquad \text{(A-5)}$$

Then, substitution of Eq. (A-5) into Eq. (A-1) yields

$$\left(2\,\underline{Q}^* + \underline{\Lambda}\right)\dot{\underline{Z}}(t) = 0 \qquad \text{(A-6)}$$

from which the unknown matrix $\underline{\Lambda}$ is obtained, for $\dot{\underline{Z}}(t) \neq 0$,

$$\underline{\Lambda} = -2\,\underline{Q}^* \qquad \text{(A-7)}$$

The optimal control vector $\underline{U}(t)$ is obtained by substituting Eq. (A-7) into Eq. (A-5) and then into Eq. (A-4) as

$$\underline{U}(t) = -\,\underline{R}^{-1}\,\underline{B}'\,\underline{Q}^*\,\dot{\underline{Z}}(t) \qquad \text{(A-8)}$$

(ii) Optimal Open-Loop Control (Feedforward Control)

Let the control vector $\underline{U}(t)$ be regulated by the measured earthquake excitation $\ddot{X}_0(t)$. Substituting Eq. (A-3) into Eq. (A-1), one eliminates $\dot{\underline{Z}}(t)$ as follows

$$\underline{\lambda}(t) = -2\,\underline{Q}^* \left[\underline{A}\,\underline{Z}(t-\Delta t) + \underline{A}\,\dot{\underline{Z}}(t-\Delta t)\,\Delta t + \underline{B}\,\underline{U}(t) + \underline{W}_1\,\ddot{\underline{X}}_0(t) \right] \qquad \text{(A-9)}$$

Further substitution of Eq. (A-9) into Eq. (A-4) yields the optimal control vector $\underline{U}(t)$ as

$$\underline{U}(t) = -\,(\underline{R} + \underline{B}'\,\underline{Q}^*\,\underline{B})^{-1}\,(\underline{B}'\,\underline{Q}^*) \left[\underline{A}\,\underline{Z}(t-\Delta t) + \underline{A}\,\dot{\underline{Z}}(t-\Delta t)\,\Delta t + \underline{W}_1\,\ddot{X}_0(t) \right] \qquad \text{(A-10)}$$

Thus, the optimal control vector is regulated by the measured external excitation $\ddot{X}_0(t)$

(iii) Optimal Closed-Open-Loop Control (Feedback and Feedforward Control)

Suppose the optimal control vector is regulated by both the feedback velocity state vector $\dot{\underline{Z}}(t)$ and the external excitation $\ddot{X}_0(t)$, i.e.,

$$\underline{\lambda}(t) = \underline{\tilde{\Lambda}}\,\dot{\underline{Z}}(t) + \underline{\tilde{q}}(t) \qquad \text{(A-11)}$$

It follows from Eq. (A-4) that

$$\underline{U}(t) = \frac{1}{2}\,\underline{R}^{-1}\,\underline{B}' \left[\underline{\tilde{\Lambda}}\,\dot{\underline{Z}}(t) + \underline{\tilde{q}}(t) \right] \qquad \text{(A-12)}$$

Now, the term 2 $\underline{Q}^* \, \dot{\underline{Z}}(t)$ appearing in Eq. (A-1) is separated into two terms such that $\underline{Q}^* \, \dot{\underline{Z}}(t) + \underline{Q}^* \, \dot{\underline{Z}}(t) + \underline{\lambda}(t) = 0$. Then, only the $\dot{\underline{Z}}(t)$ in the second term is replaced by Eq. (A-3), and $\underline{\lambda}(t)$ and $\underline{U}(t)$ are replaced by Eqs. (A-11) and (A-12), respectively; with the result

$$\left(\underline{Q}^* + \frac{1}{2} \underline{Q}^* \, \underline{B} \, \underline{R}^{-1} \, \underline{B}' \, \tilde{\underline{\Lambda}} + \tilde{\underline{\Lambda}} \right) \dot{\underline{Z}}(t) + \left(\underline{I} + \frac{1}{2} \underline{Q}^* \, \underline{B} \, \underline{R}^{-1} \, \underline{B}' \right) \tilde{\underline{q}}(t)$$

$$+ \underline{Q}^* \left[\underline{A} \, \underline{Z}(t-\Delta t) + \underline{A} \, \dot{\underline{Z}}(t-\Delta t) \, \Delta t + \underline{W}_1 \, \ddot{X}_0(t) \right] = 0 \qquad \text{(A-13)}$$

It follows from Eq. (A-13) that

$$\left[\underline{Q}^* + \left(\underline{I} + \frac{1}{2} \underline{Q}^* \, \underline{B} \, \underline{R}^{-1} \, \underline{B}' \right) \tilde{\underline{\Lambda}} \right] \dot{\underline{Z}}(t) = 0 \qquad \text{(A-14)}$$

$$\left(\underline{I} + \frac{1}{2} \underline{Q}^* \, \underline{B} \, \underline{R}^{-1} \, \underline{B}' \right) \tilde{\underline{q}}(t)$$

$$+ \underline{Q}^* \left[\underline{A} \, \underline{Z}(t-\Delta t) + \underline{A} \, \dot{\underline{Z}}(t-\Delta t) \, \Delta t + \underline{W}_1 \, \ddot{X}_0(t) \right] = 0 \qquad \text{(A-15)}$$

Thus, the unknown matrix $\tilde{\underline{\Lambda}}$ and the unknown vector $\tilde{\underline{q}}(t)$ are obtained from Eqs. (A-14) and (A-15) as follows

$$\tilde{\underline{\Lambda}} = - \left(\underline{I} + \frac{1}{2} \underline{Q}^* \, \underline{B} \, \underline{R}^{-1} \, \underline{B}' \right)^{-1} \underline{Q}^* \qquad \text{(A-16)}$$

$$\tilde{\underline{q}}(t) = \tilde{\underline{\Lambda}} \left[\underline{A} \, \underline{Z}(t-\Delta t) + \underline{A} \, \dot{\underline{Z}}(t-\Delta t) \, \Delta t + \underline{W}_1 \, \ddot{X}_0(t) \right] \qquad \text{(A-17)}$$

Effect of Spatial Variability and Propagation of Seismic Ground Motions on the Response of Multiply Supported Structures

A. Zerva

Department of Civil and Architectural Engineering
Drexel University
Philadelphia, PA 19104

Abstract

The response of lifelines, modeled as two- and three-span continuous symmetric beams of various lengths, subjected to partially and fully correlated seismic ground motions is examined. The partial correlation of the input motions consists of a term that characterizes the loss of coherence and a term that represents the apparent propagation (phase difference). The analysis suggests that the propagation effects may be neglected when the motions at the site exhibit loss of coherence; however, when the motions are coherent, phase differences may produce higher or lower response than the one induced by fully correlated motions, depending on whether the dominant modes are antisymmetric or symmetric at the location along the beam and for the response quantity (bending moment or shear force) under consideration.

Introduction

Lifelines, such as pipelines, bridges and communication transmission systems, due to their extended length, experience the effects of differential support motions during earthquakes. The spatial variability of ground motions between the supports may be attributed to the loss of coherence and the apparent propagation (phase difference) of the seismic motions. Loss of coherence may result from the variability in the rupture process, the wave transmission from an extended fault, and the various discontinuities/heterogeneities/anisotropies that the waves encounter as they propagate from the source to the site. Apparent propagation or phase difference between the motions may occur due to the delay in the arrival of the waves at the station that is further away from the source as compared to the one closer to the source. Spatial variability of ground motions may significantly influence the response of lifelines [1], and

I. Elishakoff · Y. K. Lin (Eds.)
Stochastic Structural Dynamics 2
New Practical Applications
2nd International Conference on Stochastic Structural Dynamics
May 9-11, 1990, Boca Raton, Florida

produce higher response quantities than the ones induced by the assumption of fully correlated motions. This latter assumption corresponds to equal support motions in the deterministic domain and is commonly used in engineering practice.

The effects of spatial variability and/or differential ground motions between the supports of lifelines have drawn the attention of researchers, and various structural configurations of lifelines subjected to either deterministic or stochastic, analytic or recorded, seismic ground motions have been analyzed [2-13]; the overall conclusion was that spatially varying ground motions produce different response than fully correlated motions. In a recent study [12], the writer analyzed two- and three-span beams of various lengths (short, moderate and long) subjected to input motions that exhibit loss of coherence only, with various degrees of correlation, and compared the response to the one induced by fully correlated motions. The results indicated that fully correlated motions may produce higher or lower response than partially correlated motions, depending on the dynamic characteristics of the structure, the response quantity (bending moment or shear force) that is evaluated, the position along the axis of the beam where the response quantity is evaluated, the separation between the natural frequencies of the structure and the dominant frequencies of the input motions, and on the degree of loss of coherence between the support motions. In this study, the effects of both loss of coherence and phase difference between the motions at the supports are investigated. Fully coherent and strongly incoherent motions are applied as input motions at the supports of the structures. For each case, sensitivity analyses with respect to the apparent propagation velocity are performed, and the results are compared with the response induced by fully correlated motions. In this sense, this study provides a general analysis of the effects of spatial variability on the response of long extended structures, and may be useful in the seismic safety evaluation of lifelines.

Dynamic analysis of multiply supported structures

The structural model for a N-1 span, N-support continuous beam is shown in Fig. 1a. Although the structures analyzed herein are modeled as two- and three-span beams (Figs. 1b and c), the formulation presented is valid for any number of spans and any support conditions at supports 1 and N.

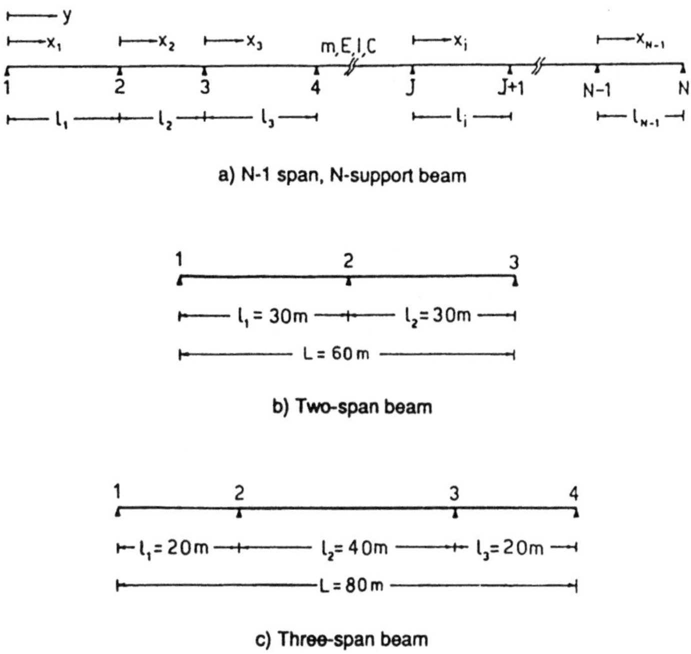

Fig.1 Structural models for continuous beams

The differential equation for the transverse motion of the structures is:

$$m\partial^2 w_t(y,t)/\partial t^2 + c\partial w_t(y,t)/\partial t + EI\partial^4 w_t(y,t)/\partial y^4 = 0 \quad (1)$$

in which, m, c, E and I indicate the mass per unit length, damping, modulus of elasticity and moment of inertia, respectively; $w_t(y,t)$ is the total displacement along the beam at location y and time t. The total displacement is composed of the quasi- (or pseudo-) static response $w_s(y,t)$ and the dynamic response $w(y,t)$ as

$$w_t(y,t) = w_s(y,t) + w(y,t) \quad (2)$$

in which,

$$w_s(y,t) = \sum_{J=1}^{N} r_J(y)\ f_J(t) \tag{3a}$$

and,

$$w(y,t) = \sum_{k=1}^{K} \Phi_k(y)\ q_k(t) \tag{3b}$$

In Eqs. (3), $r_J(y)$ is the shape function for unit support displacement at support J; $f_J(t)$ is the ground (support) displacement at support J; N is the number of supports; $\Phi_k(y)$ and $q_k(t)$ are the mode shape and the generalized coordinate for mode k under the assumption of a normal mode approach; and K is the number of modes used in the analysis. The modeshape $\Phi_k(y)$ of the continuous structure is composed of the modeshapes of the individual spans, $\phi_{kj}(x_j)$, in which x_j is the local coordinate with origin at the left support J of span j (Fig. 1a). The equations describing $\phi_{kj}(x_j)$ through the fundamental function or initial parameter approach [12,14,15] are:

$$\phi_{kj}(x_j) = [W_{kJ}\ f_1(\beta_k x_j)/2] + [\Theta_{kJ}\ f_2(\beta_k x_j)/2\beta_k]$$
$$-[M_{kJ}\ f_3(\beta_k x_j)/2\beta_k^2 EI] - [V_{kJ}\ f_4(\beta_k x_j)/2\beta_k^3 EI] \tag{5a}$$

in which, W_{kJ}, Θ_{kJ}, M_{kJ}, and V_{kJ} are the displacement, slope, bending moment and shear force, respectively, at the left support J of span j (Fig. 1a);

$$\beta_k = (m\omega_k^2/EI)^{1/4} \tag{5b}$$

ω_k is the frequency for mode k; and,

$$f_1(u) = \cosh u + \cos u;\ f_2(u) = \sinh u + \sin u;$$
$$f_3(u) = \cosh u - \cos u;\ f_4(u) = \sinh u - \sin u \tag{5c}$$

The derivation of Eqs. (5a) and (5c) is given in the Appendix. The advantage of this approach is that the constants of integration are expressed in terms of physical quantities at the supports (i.e., displacement, slope, bending moment and shear force) and the boundary conditions at the left support of each span may be directly applied in Eq. (5a). Furthermore, the approach can be easily coded for any number of spans and any boundary conditions at supports 1 and N, particularly if one notices that $df_4/du=f_3$, $df_3/du=f_2$, $df_2/du=f_1$ and $df_1/du=f_4$. The boundary conditions for the beams under consideration (Fig.1a), namely, zero displacements at the supports, continuity of slopes and equality of moments from left and right at supports 2 through N-1, and zero moments at supports 1 and N, determine the natural frequencies ω_k, and the modeshapes of the structure.

Finally, the substitution of Eqs. (2) through (5) into Eq. (1), the use of the orthogonality property of the modeshapes, and the assumption that, for low values of damping, the terms involving quasi-static velocities are smaller than the terms involving quasi-static accelerations, and may be neglected, lead to the following equations of motion in terms of the generalized coordinates:

$$\ddot{q}_k(t)+2\zeta_k\omega_k\dot{q}_k(t)+\omega_k{}^2q_k(t)=\sum_{J=1}^{N} R_{kJ}\ddot{f}_J(t) \tag{6a}$$

where,

$$R_{kJ}=-\int_0^L r_J(y)\Phi_k(y)\,dy/\int_0^L [\Phi_k(y)]^2dy \tag{6b}$$

and, $\zeta_k=\zeta_1\omega_1/\omega_k=c/2m\omega_k$ is the damping coefficient for mode k. The integrals in Eq. (6b) can be evaluated analytically, and Eq. (6a) can be solved in terms of Duhamel's integral.

<u>Response to stochastic seismic ground motions</u>

The stochastic seismic ground motions at the supports of the structures examined herein are described by the cross spectral density of ground acceleration between any two supports J and L as follows:

$$S_{\ddot{f}_J\ddot{f}_L}(\omega) = S_{\ddot{f}\ddot{f}}(\omega)\ \rho(\xi,\omega)\exp[-i\omega(y_L-y_J)/v] \tag{7}$$

in which, the first term, $S_{\ddot{f}\ddot{f}}(\omega)$, is the power spectral density of ground acceleration assumed to be the same at all locations on the ground surface; the second term, $\rho(\xi,\omega)$, denotes the spatial coherence of ground motions, and the last exponential term incorporates the effects of the apparent propagation of the motions. This formulation (Eq. (7)) for the cross spectral density of the ground motions is based on the approach of Harichandran and Vanmarcke [16]. Harichandran and Vanmarcke [16] observed that accelerograms recorded at the SMART-1 array exhibit a closure property in the lag-shift parameter, i.e., the accelerograms that are located along a straight line are shifted (aligned) relative to each other. The alignment changes only the phase (last term in Eq. (7)), but not the absolute value of the coherence (second term in Eq. (7)). Harichandran and Vanmarcke [16] also noted that, for the recorded data at the SMART-1 array, the different frequency components of the motions appear to propagate with the same apparent velocity. It has to be noted, however, that the effects of dispersion and propagation of the ground motions may be interrelated, i.e., the wave components that constitute the motions may propagate with different velocities. In this case, the formulation for the cross spectral density of the motions should be based on an alternative approach, e.g., Ref. [17].

The power spectral density of ground acceleration (Fig. 2a) is given by the Kanai-Tajimi spectrum as extended by Clough and Penzien [18]:

$$S_{\ddot{f}\ddot{f}}(\omega)= S_o\{[1+4\zeta_{g1}{}^2(\omega/\omega_{g1})^2]/[(1-(\omega/\omega_{g1})^2)^2+4\zeta_{g1}{}^2(\omega/\omega_{g1})^2]\}*$$
$$*\{(\omega/\omega_{g2})^4/[(1-(\omega/\omega_{g2})^2)^2+4\zeta_{g2}{}^2(\omega/\omega_{g2})^2]\} \qquad (8)$$

in which, S_o is the amplitude of the white-noise bedrock acceleration; ω_{g1}=5πrad/sec and ω_{g2}=0.1ω_{g1} are the resonant frequencies of two filters; and ζ_{g1}=ζ_{g2}=0.6 are the corresponding damping coefficients [19]. The advantage of the above formulation is that it allows finite values for the displacement and velocity spectra when ω->0.

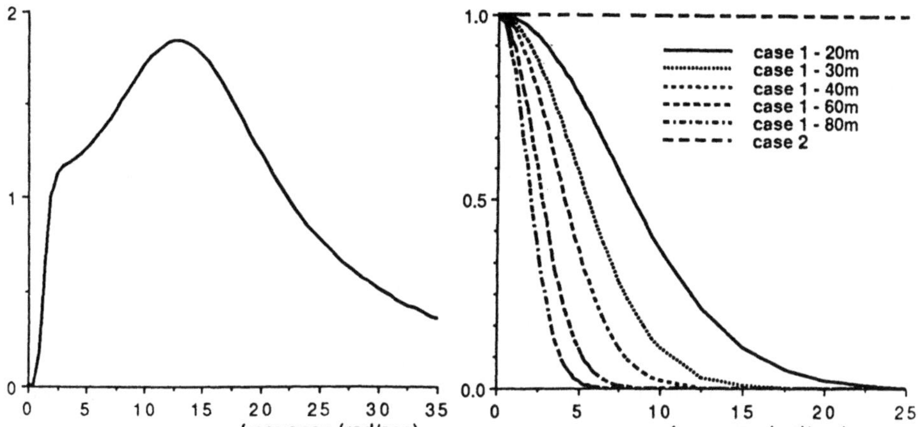

a) Normalized power spectral density of ground acceleration

b) Variation of spatial coherence with frequency at discrete separation distance

Fig.2 Stochastic input motions

The spatial correlation coefficient is of the form [20-21]:

$$\rho(\xi,\omega) = \exp\left(-(\gamma\xi\omega/v_s)^2\right) \tag{9}$$

in which, γ is a dimensionless spatial incoherence parameter; ξ is the absolute value of the separation distance between stations J and L; and v_s is the shear wave velocity in the soil. Equation (9) is based on a theoretical model [20] for scalar shear waves propagating through a random medium. Two extreme cases for ρ (Eq. (9)) are examined herein: in the first case, the ratio of the shear wave velocity over the incoherence parameter is equal to 200m/sec, and in the second case the ratio is equal to infinity. Both cases are shown in Fig. 2b for various separation distances. The fist case (case 1) produces motions with correlations that are strongly decaying with separation distance and frequency, whereas the second case (case 2), which makes ρ=1 for any separation distance and frequency, is equivalent to the assumption that the motions do not change as they propagate from one support to the other. Intermediate values of coherence between the ones presented herein have been considered by the writer in a previous analysis [12] of the effects of spatial incoherence without any phase differences on the response of multiply supported structures.

The last term in Eq. (7), $\exp[-i\omega(y_L-y_J)/v]$, in which, (y_L-y_J) is the distance between supports L and J, and v is the apparent propagation velocity, incorporates the effects of phase difference between the support motions. The values of the apparent propagation velocity used in this study are v=100, 500, 1000m/sec and infinity. The values v>500m/sec are realistic representations of the apparent propagation velocity, since the strong shaking part of the seismic motion generally propagates with the apparent velocity of the body waves at the bedrock interface, and the angle of incidence of such waves could be as high as 90° $(v\to\infty)$. The value v=100m/sec is extremely low, and is used only for illustrative purposes, i.e., in order to examine how the response of the structures varies over a wide range of propagation velocities. When $v\to\infty$ and $\rho=1$ in Eq. (7) the stochastic motions correspond to equal motions at the supports of the structure in the deterministic domain.

With the input motions of Eq. (7) and the dynamic analysis of the beams (Eqs. (2)-(6)), the power spectral density of the total displacement at location y along the beam is found to be:

$$S_{w_t(y),w_t(y)}(\omega) = \sum_{J=1}^{N}\sum_{L=1}^{N} \{r_J(y)r_L(y)/\omega^4$$

$$+ \sum_{k=1}^{K} [-\Phi_k(y)(r_J(y)R_{kL}H_k(\omega)+r_L(y)R_{kJ}H_k^{*}(\omega))/\omega^2$$

$$+ \sum_{m=1}^{K} \Phi_k(y)\Phi_m(y)R_{kJ}R_{mL}H_k^{*}(\omega)H_m(\omega)]\} S_{\ddot{f}_J\ddot{f}_L}(\omega) \qquad (10)$$

in which, ω denotes frequency; $H_k(\omega)=[\omega_k^2-\omega^2+2\zeta_k\omega_k\omega i]^{-1}$ the frequency transfer function for mode k, with $i=\sqrt{-1}$; the asterisk denotes complex conjugate; and $S_{\ddot{f}_J\ddot{f}_L}(\omega)$ is the cross spectral density of ground acceleration between supports J and L as defined in Eq. (7). The power spectral densities of bending moments and shear forces are proportional to the power spectral densities of the second and third derivative of the total displacements (Eqs. (2) and (10)) with respect to y.

Analyses of total displacements along two- and three-span beams for input motions with various degrees of coherence have been performed by the writer in a previous work [12]; it was found that the total displacements were very similar for all cases of support excitations. Analyses of total and dynamic displacements along two-span beams for earthquake excitations based on SMART-1 array data and for an apparent propagation velocity of 1000m/sec have been reported recently [11]. Mean-square values of bending moments and shear forces are determined in this study. The integration scheme for the evaluation of the mean-square response, described in detail in Ref.[12], is summarized in the following: The infinite $(-\infty,+\infty)$ ω-domain is reduced to the infinite $[0,+\infty)$ ω-domain through the use of the property that power spectral densities of stationary processes are even functions of frequency. The infinite $[0,+\infty)$ ω-domain is then transformed into the [0,1] x-domain through the transformation $\omega=(1/x)-1$. Before the transformation and in order to adequately represent the effect of the frequency transfer function of each mode at the frequency of the mode in the integration, the ω-domain is subdivided according to the following pattern:

$$[0;\omega_1-5\zeta_1\omega_1]\ldots\ldots[\omega_{k-1}+5\zeta_{k-1}\omega_{k-1};\omega_k-5\zeta_k\omega_k]$$

$$[\omega_k-5\zeta_k\omega_k;\omega_k+5\zeta_k\omega_k]\ldots\ldots[\omega_K+5\zeta_K\omega_K;+\infty)$$

and each of the above intervals is transformed to the x-domain. Gaussian integration is then used with 40 points per interval. This integration scheme has the advantage of considering the contributions of the frequency transfer functions of all modes considered and for all frequencies, the contribution of the infinite ω-domain without any truncation, in addition to being easily coded and computationally very efficient.

Numerical results

Two sets of two- and three-span beams are considered in this analysis. The following properties are used for all beams: $E=2*10^{11}N/m^2$; $m=2.45*10^3kg/m$; $I=6*10^{-2}m^4$. These values were obtained from the averages of characteristics of superstructures of actual highway bridges [22]. The first two-span beam analyzed herein (Fig. 1b) consists of spans of equal length (30m each) with total length L=60m, and the first three-span beam consists of a 40m middle span, and the end spans are 20m long, namely a total length of L=80m (Fig. 1c). The natural frequencies of these beams are as follows:

Two-span beam (L=60m):

$$\omega_k=24.27,\ 37.91,\ 97.08,\ 122.87,\ 218.43,\ 256.35,\ 388.32,\ 438.37,\ 606.74,\ 668.93,\ 873.71,\ 948.04 \ldots\ \text{rad/sec} \quad (11a)$$

Three-span beam (L=80m):

$$\omega_k=21.33,\ 54.61,\ 69.11,\ 85.31,\ 144.20,\ 218.43,\ 246.58,\ 276.45,\ 376.28,\ 491.46,\ 533.27,\ 576.79 \ldots\ \text{rad/sec} \quad (11b)$$

Continuous beams of double the length of the ones just described are also being considered: For the longer two-span beam, the length of each span is 60m with total length L=120m; for the longer three-span beam, the length of the end spans is 40m and that of the middle span 80m, giving a total length of L=160m. 8% of critical damping is used for the first mode of all structures; the damping coefficient for the higher modes is obtained from Eq. (6).

Figs. 3-7 present the results of the analyses. In the figures, case 1 represents the response induced by input motions that incorporate the effects of both loss of coherence and phase difference; case 1 at $v \to \infty$ represents the response induced by input motions with loss of coherence only. Case 2 represents the response induced by fully coherent, propagating input motions, and case 2 at $v \to \infty$ represents the response caused by fully correlated motions. The results are presented in the form of normalized square-roots of mean-square bending moments and shear forces; normalization is performed with respect to the maximum quantity (bending moment or shear force) that occurs along each beam for the case of fully correlated excitations. Both the total response quantity and the corresponding quasi-static response are shown in the figures. Generally, the first few modes (up to 4) contributed mostly to the response; however, since the differences between the responses for the various input motions were small, the number of modes K was selected when the differences between the power spectral densities for mode K and K-1 at any location along the beam was less than 2%. The evaluation of bending moments required less modes than that of shear forces; the required number of modes also depended on the degree of coherence, the value of the apparent propagation velocity, the length of the beam and the number of supports. It has to be noted, that the response induced by both partially and fully correlated motions without phase difference is symmetric about the point of symmetry of the structures, i.e., middle support for the two-span beams, and middle of middle span for the three-span beams [12]. Once the effects of propagation are incorporated into the analysis, the response seizes to be symmetric. Fully correlated motions (ρ=1 and v->∞) excite only the symmetric modes; the quasi-static response and the antisymmetric mode contribution in this case is equal to zero. All other cases excite the quasi-static response and the symmetric and antisymmetric modes.

It may be observed from all figures (Figs. 3-7) that both bending moments and shear forces are essentially constant at each location for case 1 and apparent propagation velocities higher than or equal to 500m/sec, and do not deviate from the symmetric response when v->∞. The very low apparent propagation velocity (v=100m/sec) in case 1 produces higher bending moments, particularly at the inner supports of the shorter structures, but almost insignificant differences in shear forces for the shorter two-span beam, and in bending moments and shear forces along the longer beams. This indicates that the effect of loss of coherence overshadows the propagation effects, and, actually, propagation effects may be neglected when the motions at the site are expected to be incoherent.

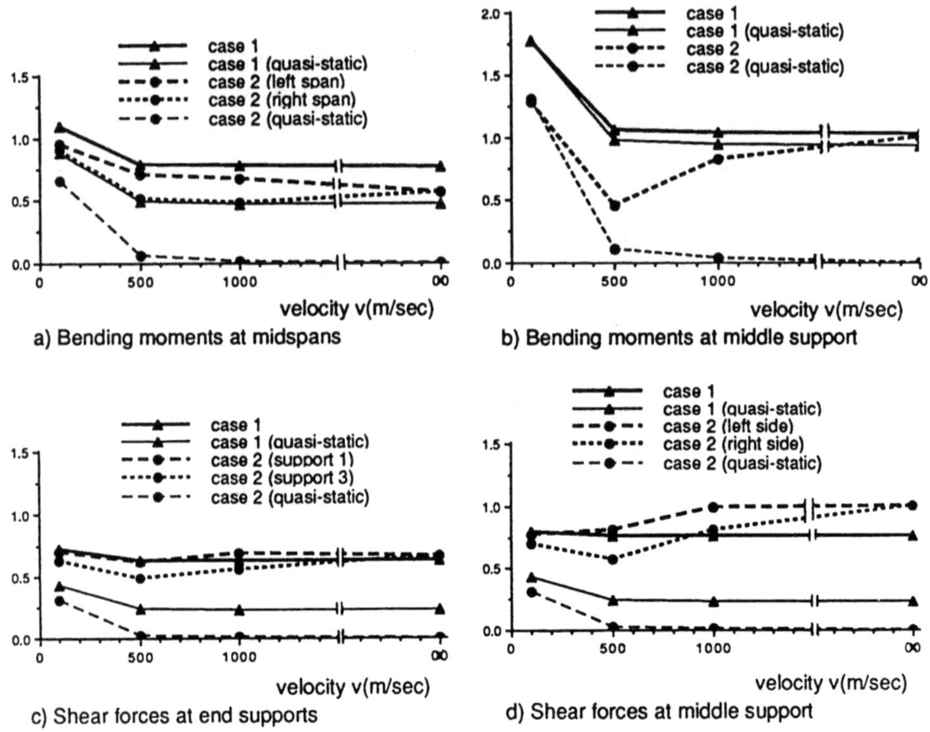

Fig. 3 Variation of response parameters along the two-span beam of total length L=60m

Bending moments at the midspans (Fig. 3a) and the middle support (Fig. 3b) of the two-span beam of L=60m for case 1 exhibit the characteristics identified in the previous paragraph, and case 2 produces lower bending moments than case 1; i.e., the effect of loss of coherence is more significant than the effect of phase difference. Low propagation velocities (v=100m/sec) produce high quasi-static terms, and, as the propagation velocity increases, the quasi-static contribution decreases. For the bending moments at the middle support of the structure (Fig. 3b), the quasi-static contribution is almost equal to the total response when v=100m/sec for both cases 1 and 2. For case 2, bending moments at the middle support (Fig. 3b) are significantly lower when v=500m/sec than for any other value of the velocity, and moments for the fully correlated motions are higher than the ones for

500<v<1000m/sec. This may be explained as follows: For the two-span beam the odd modes (k=1,3,5... in Eq. (11a)) are antisymmetric (and coincide with the modes of a simply supported beam with span length equal to 30m) and the even modes (k=2,4,6... in Eq. (11a)) are symmetric (and coincide with the modes of a propped cantilever beam with span length equal to 30m). At the middle support of the beam only the symmetric modes contribute to the dynamic response for all cases; these modes are fully excited for fully correlated motions, particularly the first symmetric mode which constitutes 98% of the total response. When v=100m/sec in case 2 the dynamic terms are almost not excited, as the velocity increases at v=500m/sec the value of the quasi-static terms drops and the dynamic terms start increasing, and continue increasing with increasing velocity. However, the drop in the quasi-static terms is much higher than the increase in the dynamic terms for v=500m/sec, since the frequency of the first symmetric mode (ω=37.91rad/sec) is much higher than the dominant frequencies of the input motions(Fig. 2a). The values of the bending moments for both cases 1 and 2 at v->∞ are almost equal: the first symmetric mode mode is fully excited in case 2 yielding large bending moments; for case 1, correlation at the frequency of this mode is already low (Fig. 2b) and its participation to the response is small, causing the sum of high quasi-static and low dynamic terms for case 1 to be equal to the high dynamic terms of case 2[12]. The excitation of the antisymmetric modes for case 2 causes differences between the values of the bending moments at the midspans of the beam (Fig. 3a): When 500<v<1000m/sec, the response at the midspan of the left span approaches from below the total response for case 1 and is higher than the symmetric response when v->∞, whereas the response at the midspan of the right span is almost equal to the quasi-static response for case 1 and lower than the symmetric response when v->∞. Consider, e.g., the response for case 2 when v=500m/sec (Fig. 3a): The response at the right span consists almost solely of the first antisymmetric mode and the quasi-static contribution, which is small. At this location the first symmetric mode is also excited, but the contribution of this mode is almost equal in magnitude to the contribution of the cross term between the first and second mode, which is negative, so that the two cancel each other. At the left span (Fig. 3a), the contribution of the first symmetric mode and that of the cross term between the first and second mode have the same sign, and add up to produce the increase.

Figs. 3c and d present the variation of shear forces at the supports of the two-span beam of total length L=60m. In this case, the highest shear force occurs at the middle support for fully correlated input motions. The dynamic terms contribute more to shear forces than to bending moments, since shear forces are one order higher derivatives with respect to y than bending moments. This is the reason that shear forces require higher number of modes than bending moments and that the quasi-static contribution is smaller for shear forces than for bending moments. At the dominant frequencies of the input motions (Fig. 2a) and at the natural frequencies of the beam, correlation for ξ=30 and 60m (Fig. 2b) is low, and the dynamic terms (particularly the first symmetric mode) for case 1 are only partially excited. Similarly, fully coherent, but propagating ground motions excite the first symmetric mode only partially. On the other hand, the fully correlated input motion excites fully the dynamic symmetric terms, and, thus, it results in higher shear forces throughout the beam. The shear forces produced by the incoherent motions (case 1) are fairly constant for any value of the propagation velocity: the quasi-static terms are slightly higher for case 1 at v=100m/sec, but as the propagation velocity increases the dynamic terms increase and the total responses for case 1 do not show any significant variation. Again, the results for case 1 are almost identical on the left and right span from the middle support of the beam, suggesting that the loss of coherency, rather than the phase difference, dominates the response. When the input motions are fully coherent but propagating, the response is sensitive to the value of the propagation velocity. Quasi-static terms appear only when $100<v<500$m/sec; when $v>500$m/sec the quasi-static term contribution is almost zero. Phase differences for case 2 produce variations in the response along the left and right side of the beam from the point of symmetry, which are caused by the antisymmetric modes and the cross terms between modes. Cross terms between modes could contribute as much as 24% to the total response (at the left side of the middle support in Fig. 3d for case 2 at v=500m/sec), suggesting that cross terms between modes should not be neglected in the analysis.

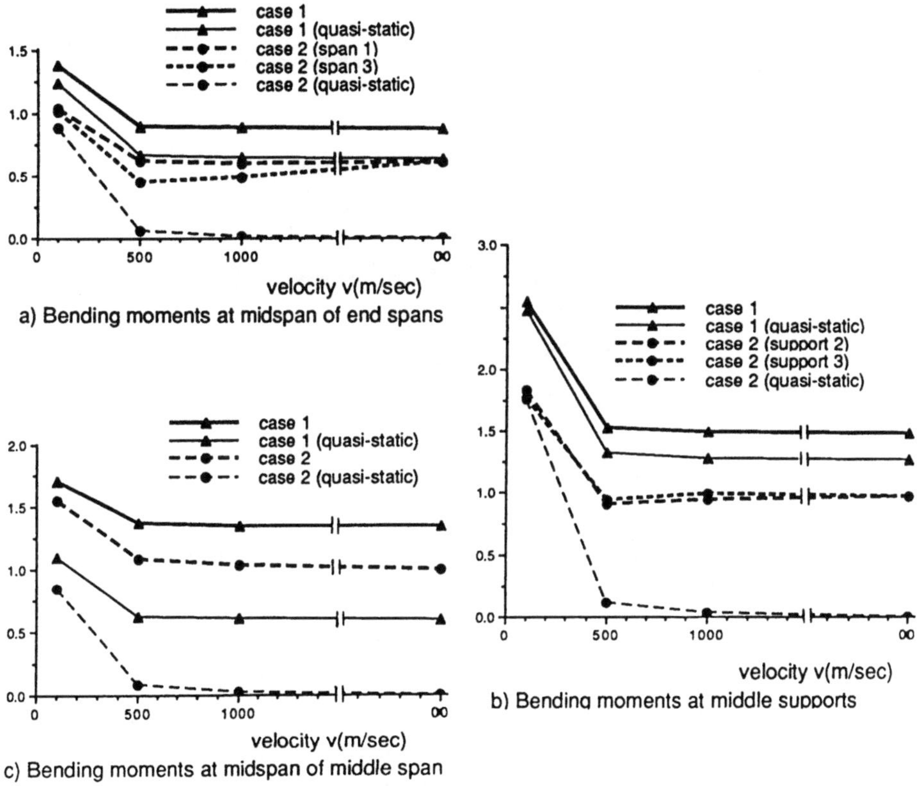

Fig. 4 Variation of bending moments along the three-span beam of total length L=80m

Figs. 4 and 5 present the variation in bending moments and shear forces for the various input motions along the three-span beam of total length L=80m. This beam has one more degree of indeterminacy than the previous one, the first mode of the structure is symmetric and its frequency (ω=21.33rad/sec in Eq. (11b)) is closer to the dominant frequencies of the input motions than the frequency of the first symmetric mode of the two-span beam.

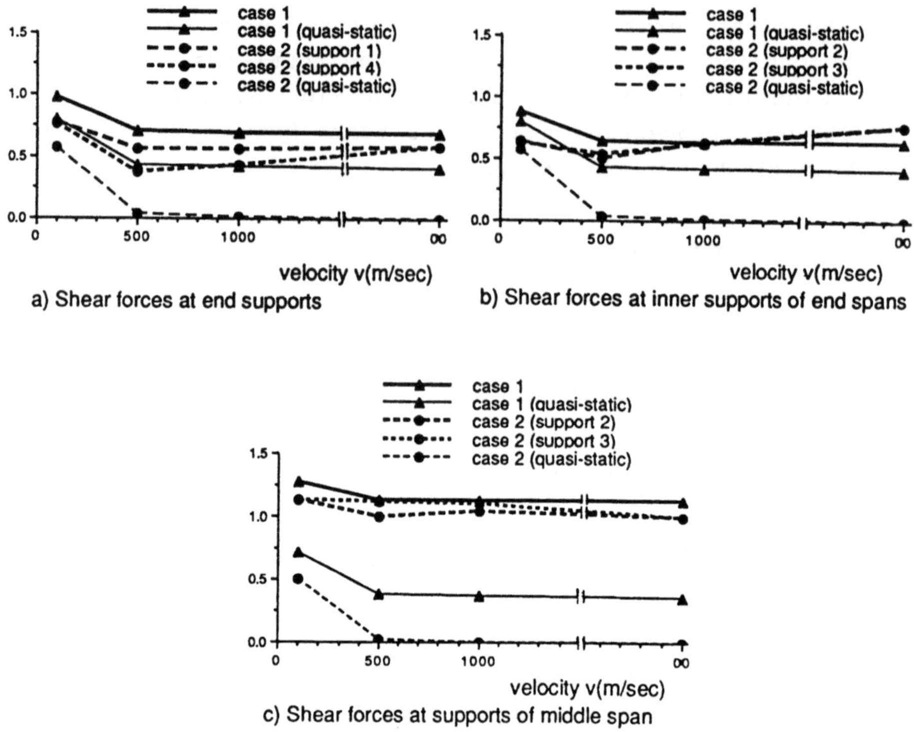

Fig. 5 Variation of shear forces along the three-span beam of total length L=80m

This fact causes the partially correlated motions (due to both loss of coherency and phase difference) to excite the first symmetric mode significantly. Furthermore, the first symmetric mode of the three-span beam is not so fully excited as was the case for the two-span beam (Fig. 3): The participation factor R_{2J} (Eq. (6b)) for the first symmetric mode of the two-span beam has the same sign for the motions of all supports, and the contributions of all motions are added. On the other hand, the participation factor R_{1J} (Eq. (6b)) for the first mode of the three-span beam is negative for the motions of supports 1 and 4, and positive for the motions of supports 2 and 3; thus, the contribution from the motions of supports 1 and 4 is subtracted from the contribution of the motions of supports 2 and 3 (Eq. (6a)). As a result of the above observations, bending moments are much higher for case 1 than for case 2; even more, bending moments for case 2 are lower than the quasi-static terms for case 1 (except at the midspan of the middle span in Fig. 4c, where the symmetric modes dominate). Both bending moments (Fig. 4) and shear forces (Fig. 5) show only slight variations for case 1 and $v>500$m/sec. Since the dynamic part of the response contributes more to shear forces than to bending moments, the differences in shear forces between cases 1 and 2 (Fig. 5) are not so significant as in bending moments (Fig. 4). Shear forces for case 2 increase with increasing propagation velocity at the inner supports of the end spans, and fully correlated motions produce the highest response (Fig. 5b). This may be attributed to the third mode of the structure, that dominates at these locations, is symmetric and fully excited by the fully correlated motions. On the other hand, fully correlated motions produce the lowest response at the supports of the middle span (Fig. 5c), whereas input motions for case 2 with $500<v<1000$m/sec produce shear forces that approach from below the response induced by case 1. This is caused mainly by the antisymmetric modes, that contribute to the response for partially correlated motions, but are not excited by the fully correlated ones. The differences between the response on the left and right side of the beam for case 2 and $500<v<1000$m/sec (Figs. 4 and 5) may be attributed to the antisymmetric modes, cross terms between modes, and cross terms between quasi-static and dynamic terms.

Figure 6 shows the variation of bending moments (Fig. 6a) and shear forces (Fig. 6b) along the longer two-span beam of total length L=120m. In this case the response induced by case 2 is higher than the one produced by the incoherent motions; the quasi-static contribution for this more flexible structure is reduced and the dynamic response dominates. The values of the natural frequencies are 1/4 the values of the L=60m beam; thus, the frequency of the first symmetric mode (ω= 37.91rad/sec/4= 9.48rad/sec) is close to the peak of the power spectral density of ground acceleration. This first symmetric mode is the one that dominates the response of the two-span beam, as discussed in the case of the shorter beam, and is fully excited by the fully correlated input motions, producing the highest response for case 2 at v->∞ both for bending moments and shear forces. The response for case 1 is essentially the same for all values of the apparent propagation velocity, since the quasi-static terms, that dominated the response when v=100m/sec along the shorter beam, contribute very little to the response of this more flexible structure. The value of the bending moment at the middle support for case 2 at v=500m/sec (Fig. 6a) is lower than all other case 2 values; a similar observation was made for the corresponding response along the shorter two-span beam (Fig. 3). The difference in this case is smaller than the one observed along the shorter beam, because the natural frequencies of the longer structure are in the same range as the dominant frequencies of the input motions. Figure 6b shows the variation of shear forces at left and right from the middle support of the longer two-span beam. Again case 1 produces lower response than case 2, mainly due to the first symmetric mode of the structure. The effect of propagation does not appear to produce significant changes in shear forces for both cases 1 and 2. The response for case 1 is essentially constant throughout the range of examined apparent propagation velocities, whereas the response for case 2 shows slight variations for the lower values of the velocities.

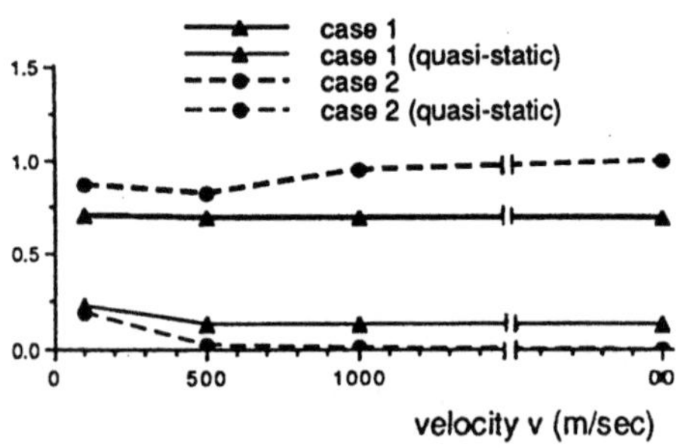

a) Bending moments at middle support

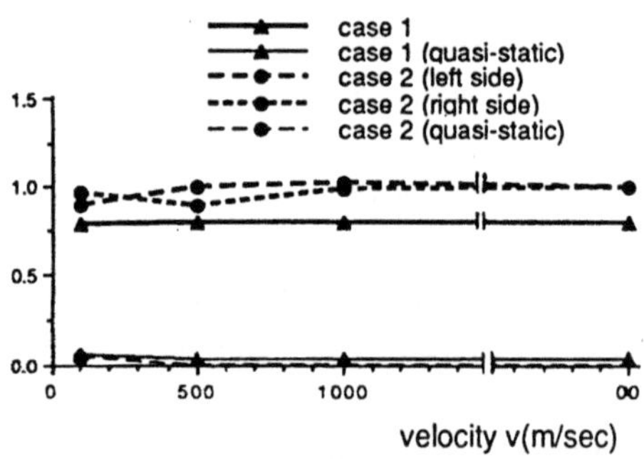

b) Shear forces at middle support

Fig. 6 Variation of response parameters along the two-span beam of total length L=120m

Figure 7 shows the variation of bending moments and shear forces along the longer three-span beam of total length L=160m. This beam is more flexible than the three-span beam of L=80m, and its natural frequencies are 1/4 the values of the shorter three-span beam. Bending moments at the midspan of the middle span (Fig. 7a) and shear forces at the supports of the end spans (Fig. 7c) are higher for fully correlated motions than for any other case. This is caused mainly by the third mode of the structure, that is symmetric and fully excited by fully correlated motions. In addition to that, the frequency of the mode (ω=69.11rad/sec/4=17.28rad/sec from Eq. (11b)) is close to the dominant frequencies of the input motions. For intermediate values of the propagation velocity for case 2, the response appears to be higher than the one induced by fully correlated motions for bending moments at the middle supports (Fig. 7b) and shear forces at the supports of the middle span (Fig. 7d).

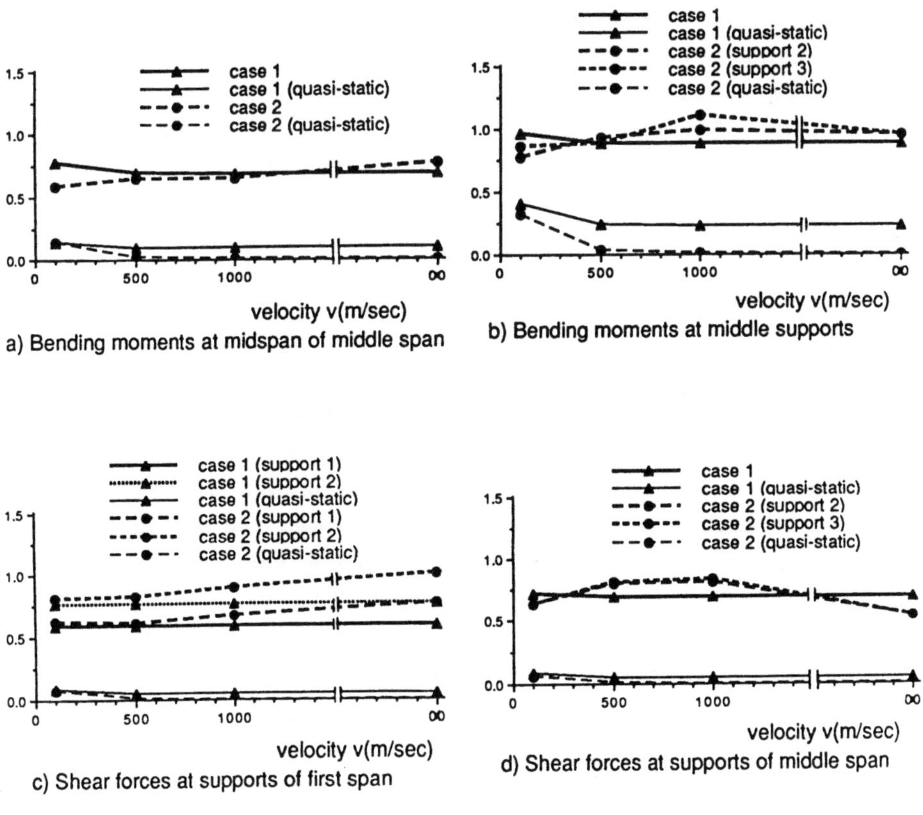

a) Bending moments at midspan of middle span

b) Bending moments at middle supports

c) Shear forces at supports of first span

d) Shear forces at supports of middle span

Fig. 7 Variation of response parameters along the three-span beam of total length L=160m

This is again attributed to the antisymmetric modes of the structure, that are not excited by the fully correlated motions. E.g., the increase in shear forces at the supports of the middle span for case 2 at v=1000m/sec is caused mainly by the fourth mode of the structure, which is antisymmetric and has a frequency of 21.33rad/sec, which is close to the dominant frequencies of the ground motions. The excitation of this mode, which dominates at the location, is also the reason that shear forces for case 1 are higher than the ones induced by full correlation. Case 1 in Fig. 7 produces essentially a constant response for all values of the propagation velocity, indicating again that the effect of loss of coherence overshadows the propagation effects.

Conclusions

The analysis presented herein examines the effect of propagating ground motions on the response of continuous structures modeled as statically indeterminate, two-and three-span beams of various lengths. Two different types of ground motions are used as input motions at the supports of the structures: in the first case (case 1) the motions exhibit a significant loss of coherence as they propagate from one support to the other; in the second case (case 2) the motions propagate along the structure without any loss of coherence. The results are compared with the response induced by fully correlated ground motions, which correspond to equal support motions in the deterministic domain; this last assumption is commonly used in engineering practice. The analysis suggested that when the motions are both incoherent and differ in phase at the supports of the structure, the response is essentially identical to the one induced by motions that are incoherent (phase difference is zero) for realistic values of apparent propagation velocities; this observation was obtained for all structures examined, and suggests that, for partially correlated motions that exhibit significant loss of coherence, the effect of loss of coherence is more important than the effect of phase difference, and the latter effect may be neglected. However, when the motions at the site become more and more coherent, the effects of apparent propagation of the motions may increase or decrease the values of bending moments and shear forces along the structures. When the symmetric modes contribute significantly to the response, fully correlated motions overestimate the response. At locations where the antisymmetric modes become significant, phase differences may either produce a higher response than full correlation, or the response at the two sides of the beams from the point of symmetry may envelop from above and below the response for fully correlated motions. Thus, for coherent motions, phase differences at the supports may produce variability in the response, suggesting that the assumption of fully correlated motions may be inadequate in modeling ground motions for the seismic safety analysis of lifeline systems.

References

1. Hall, W.J., and Newmark, N.M. Seismic design of bridges-an overview of research needs. Proc. Workshop on Earthq. Resistance of Highway Bridges, Palo-Alto, California (1979) 164-181.

2. Bogdanoff, J.L, Goldberg, J.E., and Schiff, A.J. The effect of ground transmission time on the response of long structures. Bull. Seism. Soc. Am. 55 (1965) 627-640.

3. Johnson, N.E., and Galletly, R.D. The comparison of the response of a highway bridge to uniform ground shock and moving ground excitation. The Shock and Vibration Bull. 42 (1972) 75-85.

4. Masri, S.F. Response of beams to propagating boundary excitation. Earthq. Eng. and Struc. Dyn. 4 (1976) 497-509.

5. Abdel-Ghaffar, A.M., and Rubin, L.I. Suspension bridge response to multiple-support excitations. J. Eng. Mech. Div., ASCE 108 (1982) 419-435.

6. Wilson, J. C., and Jennings, P. C. Spatial variation of ground motion determined from accelerograms recorded on a highway bridge. Bull. Seism. Soc. Am. 75 (1985) 1515-1533.

7. Zerva, A. Stochastic differential ground motion and structural response. Ph.D. thesis, Department of Civil Engineering, University of Illinois at Urbana-Champaign, Urbana, Illinois 1986.

8. Zerva, A., Ang, A. H-S., and Wen, Y.K. Lifeline response to spatially variable ground motions. Earthq. Eng. and Struc. Dyn. 16 (1988) 361-379.

9. Harichandran, R. S., and Wang, W. Response of simple beam to spatially varying earthquake excitation. J. Eng. Mech. Div., ASCE 114 (1988) 1526-1541.

10. Zerva, A. Effect of spatial variation of ground motions on bridges. Proc. 5th ASCE Specialty Conference on Probabilistic Methods (1988) 253-256.

11. Harichandran, R.S., and Wang, W. Response of indeterminate two-span beam to spatially varying seismic excitation. Earthq. Eng. and Struc. Dyn. 19 (1990) 173-187.

12. Zerva, A. Response of multi-span beams to spatially incoherent seismic ground motions. Earthq. Eng. and Struc. Dyn. 19 (1990) 819-832.

13. Lin, Y.K., Zhang, R., and Yong, Y. Multiply supported pipeline under seismic wave excitations. J. Eng. Mech. Div., ASCE 116 (1990) 1094-1108.

14. Filittov, A.T. Vibrations of Elastic Systems (in Russian). Academy of Ukrainian Sciences, Kiev, USSR 1956.

15. Robinson, A.R. personal communication; CE 474-Dynamics of Framed Structures. Department of Civil Engineering, University of Illinois at Urbana-Champaign 1982.

16. Harichandran, R.S., and Vanmarcke, E.H. Stochastic variation of earthquake ground motion in space and time. J. Eng. Mech. Div., ASCE 114 (1986) 154-174.

17. Shinozuka, M., and Deodatis, G. Stochastic wave models for seismic ground motion. Proc. of ICOSSAR 1989, San Francisco, CA (1989) 335-342.

18. Clough, R.W., and Penzien,J. Dynamics of Structures. McGraw Hill Book Co., Inc., New York, New York 1975.

19. Hindy, A., and Novak, M. Pipeline response to random ground motion. J. Eng. Mech. Div., ASCE 106 (1980) 339-360.

20. Luco, E., and Wong, H.L. Response of a rigid foundation to a spatially random ground motion. Earthq. Eng. and Struc. Dyn. 14 (1986) 891-908.

21. Luco, E., and Mita, A. Response of circular foundation to spatially random ground motion. J. Eng. Mech. Div., ASCE 113 (1987) 1-16.

22. Moses, F., Ghosn, M., and Gobieski, J. Weigh-in-motion applied to bridge evaluation. Case Western Reserve University Report No. FHWA/OH-85/12, Cleveland, Ohio 1985.

<u>Appendix</u>: Fundamental Functions or Initial Parameter Approach

The modeshape $\Phi_k(y)$ of the structure for any mode k may be expressed in terms of the modeshapes of the individual spans as follows:

$$\Phi_k(y) = \sum_{j=1}^{N-1} \phi_{kj}(y-\sum_{n=1}^{j-1} \ell_n) \; [H(y-\sum_{n=1}^{j-1} \ell_n) - H(y-\sum_{n=1}^{j} \ell_n)] \tag{A.1}$$

in which, ϕ_{kj} indicates the modeshape for span j and mode k; ℓ_n is the length of span n; and H(*) is the Heaviside function. Let

$$x_j = y - \sum_{n=1}^{j-1} \ell_n \tag{A.2}$$

be the local coordinate along span j. The modeshape for span j and mode k may then be determined from the solution of the undamped, free vibration differential equation of motion as follows:

$$\phi_{kj}(x_j) = A_{1kj}\exp(\beta_k x_j) + A_{2kj}\exp(-\beta_k x_j) + \\ + A_{3kj}\exp(i\beta_k x_j) + A_{4kj}\exp(-i\beta_k x_j) \tag{A.3}$$

where A_{1jk}, A_{2jk}, A_{3jk}, and A_{4jk} are constants, and β_k is given in Eq. (5b). Eq. (A.3) may be rewritten as:

$$\phi_{kj}(x_j) = A_{kj}\cosh(\beta_k x_j) + B_{kj}\sinh(\beta_k x_j) + \\ + C_{kj}\sin(\beta_k x_j) + D_{kj}\cos(\beta_k x_j) \tag{A.4}$$

where again A_{kj}, B_{kj}, C_{kj}, and D_{kj} are constants. The fundamental functions or initial parameter approach expresses these constants in terms of the displacement W_{kJ}, slope Θ_{kJ}, bending moment M_{kJ} and shear force V_{kJ} at the left support J of span j through the use of the following equations:

$$\phi_{kj}(x_j=0) = W_{kJ} \tag{A.5a}$$

$$\phi_{kj}'(x_j=0) = \Theta_{kJ} \tag{A.5b}$$

$$\phi_{kj}''(x_j=0) = -M_{kJ}/EI \tag{A.5c}$$

$$\phi_{kj}'''(x_j=0) = -V_{kJ}/EI \tag{A.5d}$$

where, ′, ″, and ‴ indicate the first, second and third derivative with respect to x_j. Eqs. (A.4) and (A.5) determine the constants A_{kj}, B_{kj}, C_{kj}, and D_{kj} in terms of W_{kJ}, Θ_{kJ}, M_{kJ}, and V_{kJ}, and produce Eq. (5a) in the text.

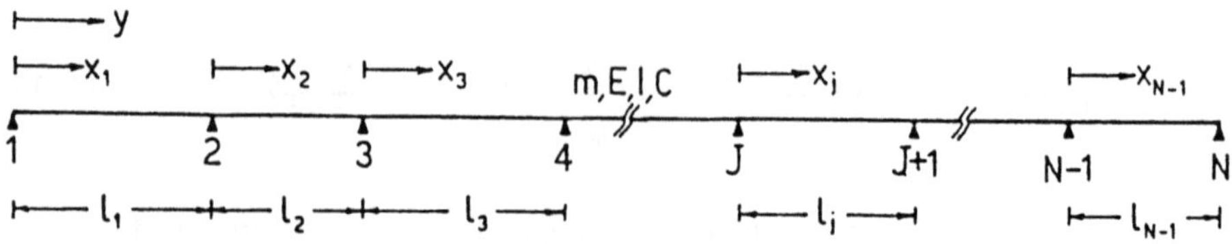

a) N-1 span, N-support beam

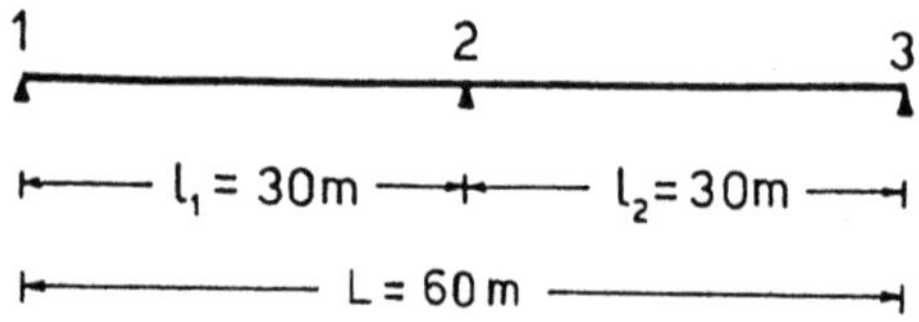

b) Two-span beam

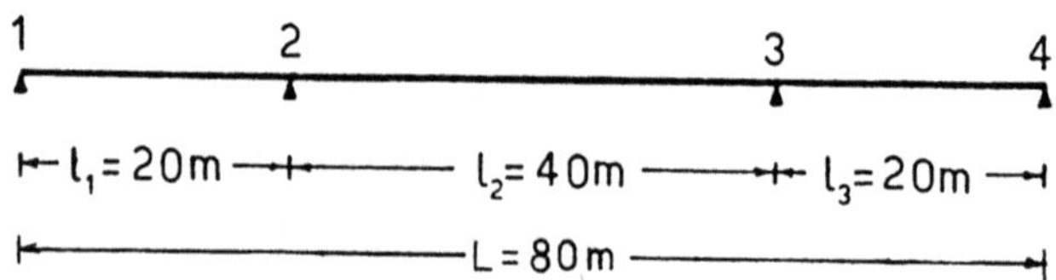

c) Three-span beam

Figure 1

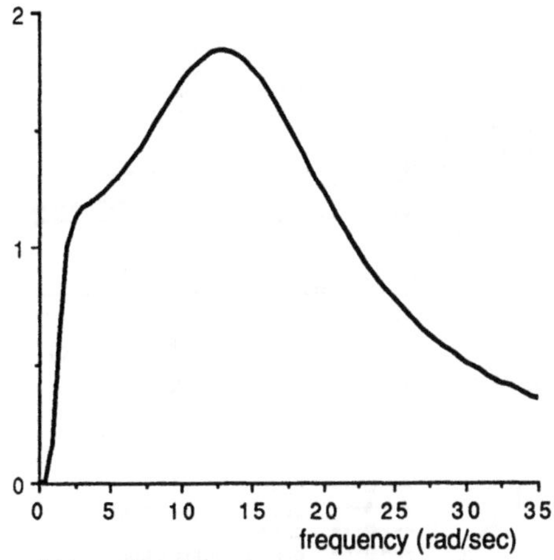

a) Normalized power spectral density of ground acceleration

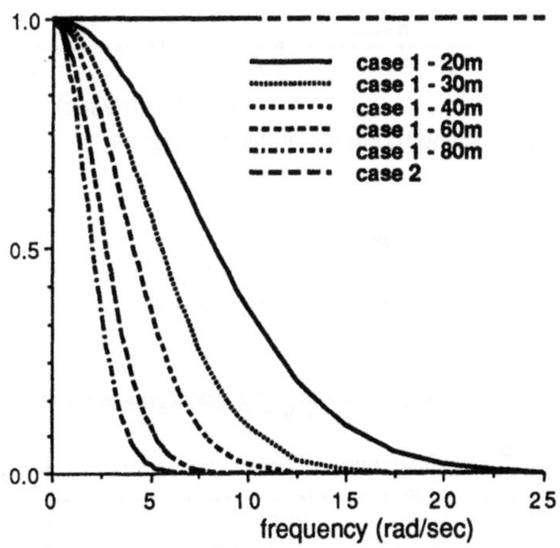

b) Variation of spatial coherence with frequency at discrete separation distance

Figure 2

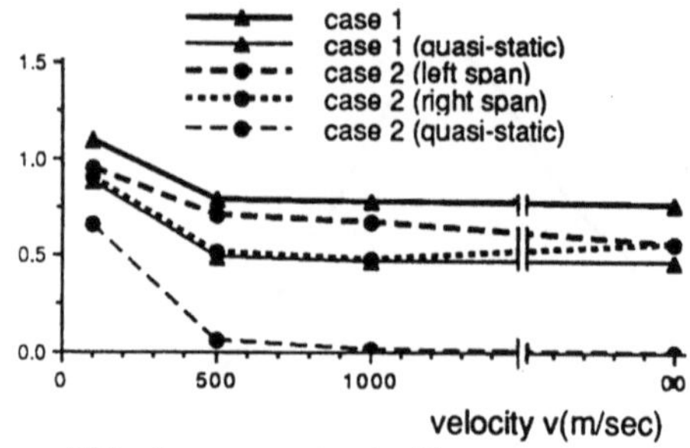

a) Bending moments at midspans

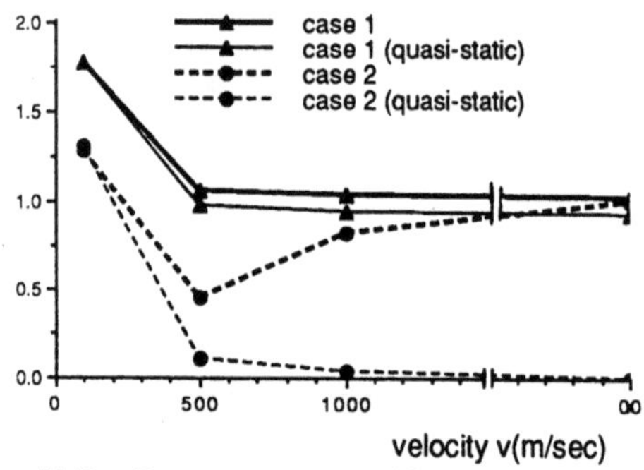

b) Bending moments at middle support

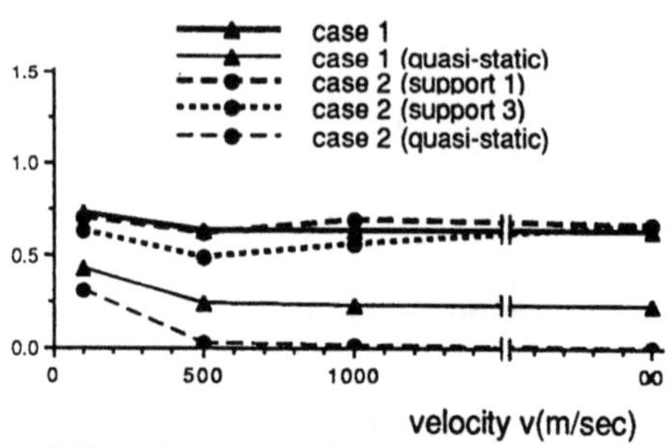

c) Shear forces at end supports

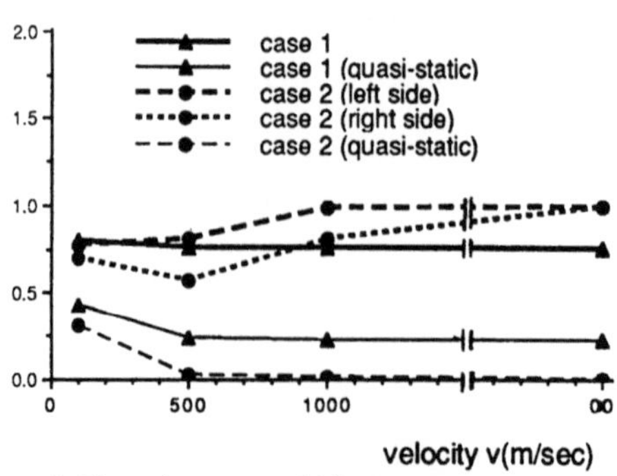

d) Shear forces at middle support

Figure 3

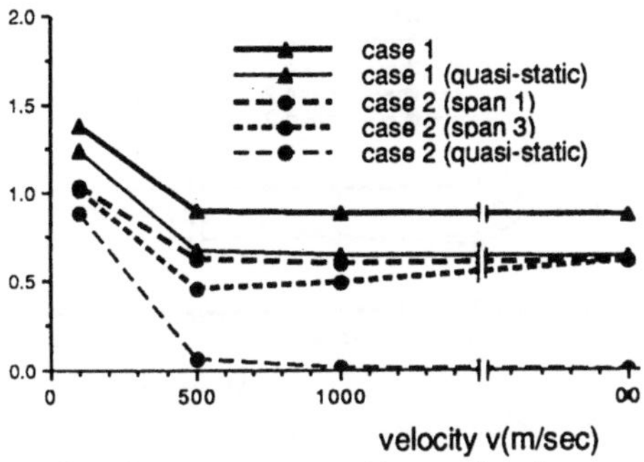

a) Bending moments at midspan of end spans

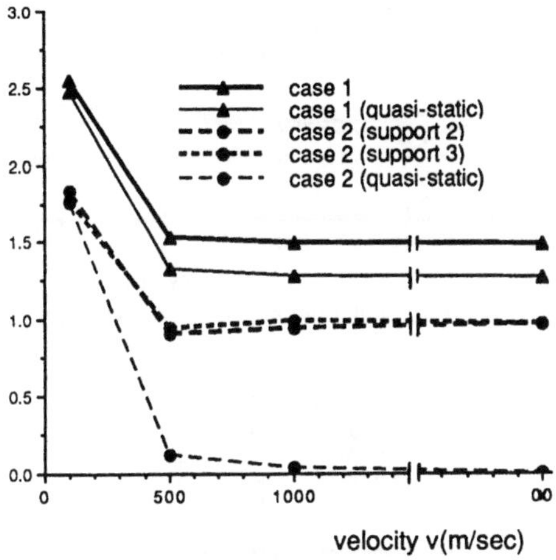

b) Bending moments at middle supports

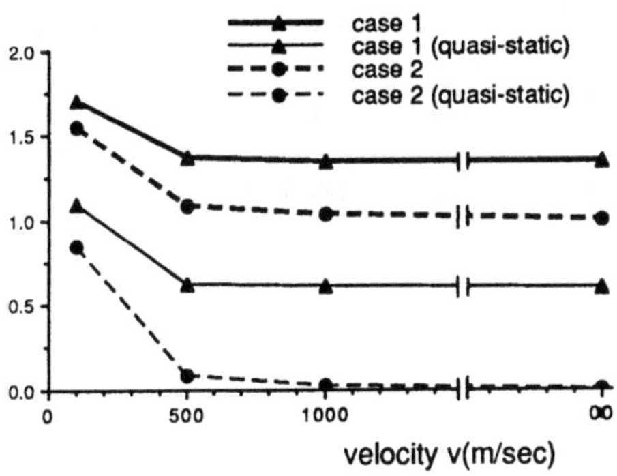

c) Bending moments at midspan of middle span

Figure 4

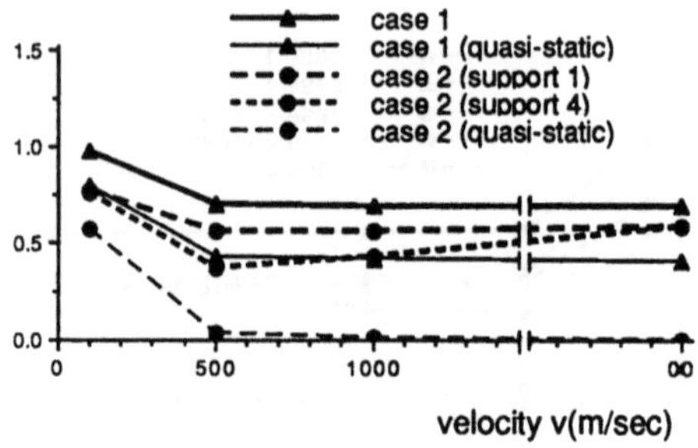

a) Shear forces at end supports

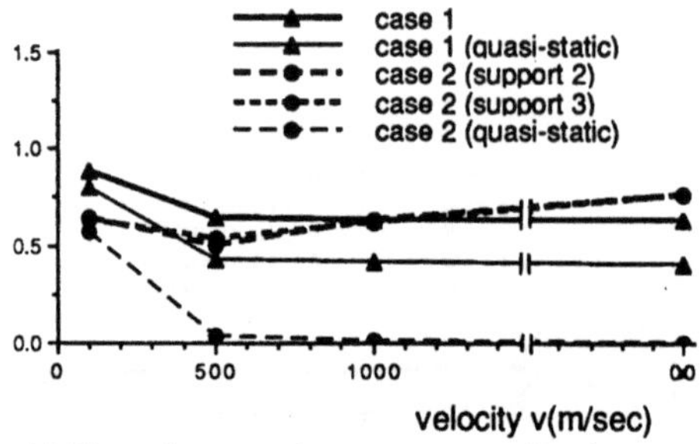

b) Shear forces at inner supports of end spans

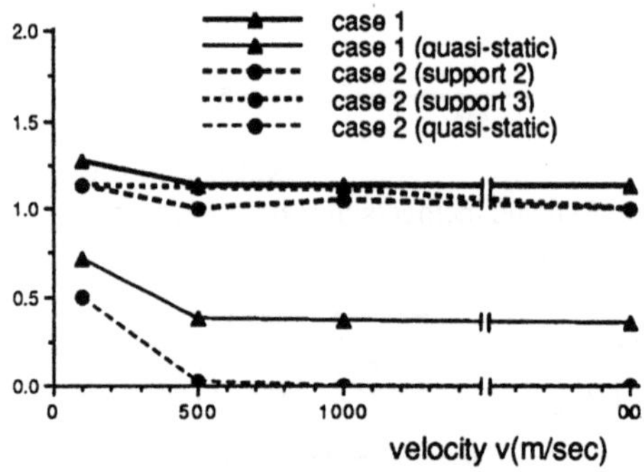

c) Shear forces at supports of middle span

Figure 5

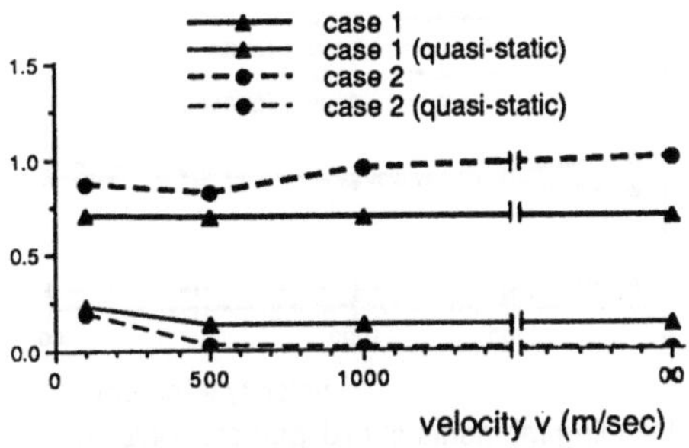

a) Bending moments at middle support

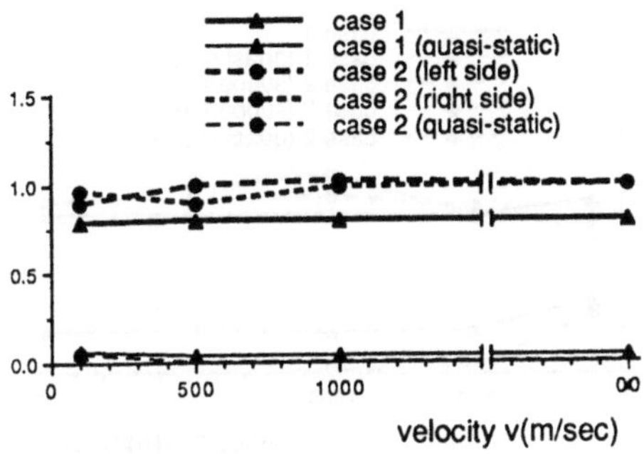

b) Shear forces at middle support

Figure 6

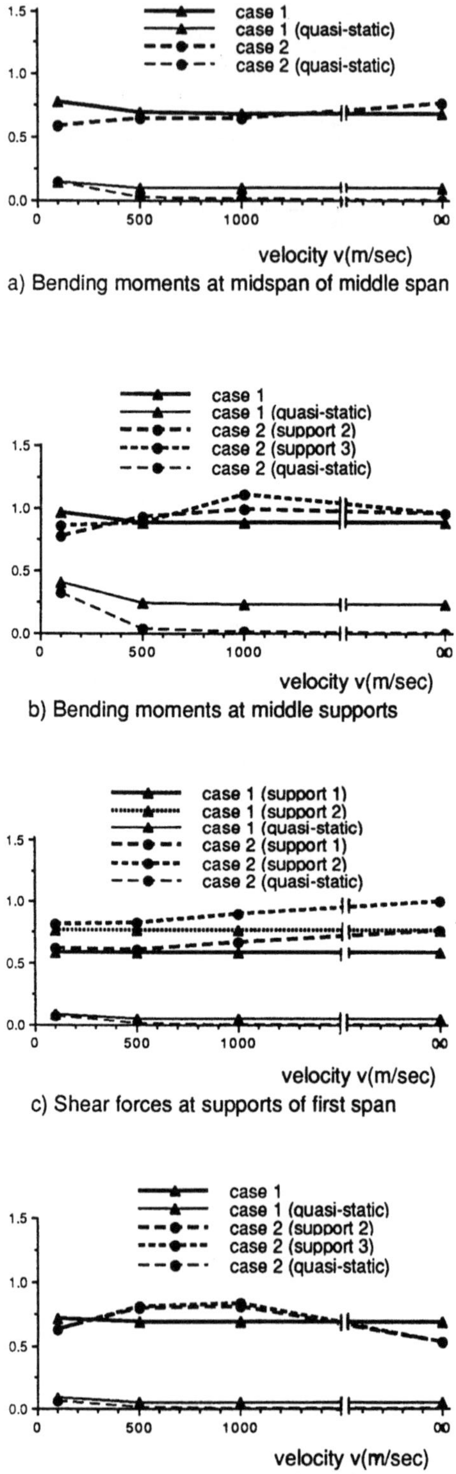

a) Bending moments at midspan of middle span

b) Bending moments at middle supports

c) Shear forces at supports of first span

d) Shear forces at supports of middle span

Figure 7

A Stochastic Finite Element Method for Real Eigenvalue Problems

W. Q. ZHU[1] and W. Q. WU[2]

[1]Department of Mechanics, Zhejiang University, Hangzhou, China

[2]Department of Civil Engineering, Zhejiang University, Hangzhou, China

Abstract

A stochastic finite element method (SFEM) based on local averages of a random vector field is developed for both distinct and repeated eigenvalues. Formulae for the variances and covariances of the eigenvalues and eigenvectors are derived. It is shown in a numerical example that, as the number of elements increases, solutions obtained from the present SFEM formulation converge much faster than those obtained from the SFEM formulation based on mid-point discretization.

1. Introduction

The eigenvalues and eigenvectors of an engineering structure are usually random due to uncertainties in its material and geometric properties. It is of considerable importance to estimate the statistics of random eigenvalues and eigenvectors for the purpose of reliability analysis and optimization of the structure. The stochastic eigenvalue problem has been studied since twenty some years ago [1-4]. To the authors' knowledge, consideration has been given only to the cases in which the eigenvalues of the mean system are all distinct. However, repeated eigenvalues corresponding to different mode shapes do occur in practice, especially when structural symmetry exists, including reflective or cyclic symmetry. In view of this, repeated eigenvalues have received attention in eigenvalue sensitivity problem in recent years [5-10], but have not in stochastic eigenvalue problem.

Uncertain material and geometric properties of a structure are usually spatially distributed. They are correlated over a

I. Elishakoff · Y. K. Lin (Eds.)
Stochastic Structural Dynamics 2
New Practical Applications
2nd International Conference on Stochastic Structural Dynamics
May 9-11, 1990, Boca Raton, Florida

certain region on a structure and should be modeled as random fields. SFEM is the only method available so far which can be used to estimate the statistics of eigenvalues and eigenvectors in this case. In the early 80's, Nakagiri and Hisada [11] developed a SFEM formulation based on mid-point discretization scheme which is capable of dealing with such eigenvalue problems. Later, Zhu and Wu [12] proposed an alternative formulation based on local averages of a random field and showed that convergence of the solution is faster than that by SFEM based on mid-point discretization as the number of finite elements increases.

In previous works [11-12], only structures with one or several types of uncorrelated random properties were considered. However, certain types of properties of a structure may be cross-correlated and they should be modeled as a random vector field. To adapt the SFEM based on the local averages of random field to the structures with several cross-correlated random properties, the formula for the covariances of the local averages of a homogeneous random scalar field over rectangular domains was generalized to that of a homogeneous random vector field with quadrant symmetry in the correlation structure [13-14]. For a nonhomogeneous random field or a non-rectangular domain, Gaussian quadrature was proposed to evaluate numerically the means and covariances for the local averages [14]. In the present paper, the theory of the local averages of homogeneous random vector field with quadrant symmetry in the correlation structure over a rectangular domain is first briefly reviewed. Then a SFEM formulation for real eigenvalue problem is developed based on local averages of random vector field. Both the cases of distinct and repeated eigenvalues are considered. Finally, an example is given to show the application and the advantage of the present SFEM formulation.

2. Local Average of Random Vector Field

The theory of local averages of a homogeneous scalar field over a rectangular domain has been well established [15].

Recently, Zhu and Ren [13] and Zhu and Wu [14] have extended the formulae for the variance and covariances to the case of homogeneous random vector fields.

For simplicity, consider a n-variate one dimensional homogeneous random vector field $Y(x) = [Y_1(x), Y_2(x), \ldots, Y_n(x)]^T$ with mean vector $m = [m_1, m_2, \ldots, m_n]^T$, variance matrix $\sigma^2 = \text{diag}\,[\sigma_1^2, \sigma_2^2, \ldots, \sigma_n^2]$ and normalized covariance matrix

$$\rho(r) = \begin{bmatrix} \rho_{11}(r) & \rho_{12}(r) & \cdots & \rho_{1n}(r) \\ \rho_{21}(r) & \rho_{22}(r) & \cdots & \rho_{2n}(r) \\ \vdots & \vdots & & \vdots \\ \rho_{n1}(r) & \rho_{n2}(r) & \cdots & \rho_{nn}(r) \end{bmatrix} \qquad (1)$$

$Y(x)$ is defined on a segment D and r is the distance between two points on D. Divide D into N elements. The local average of $Y(x)$ over element L_i is defined as

$$Y_{L_i} = Y(x_i, L_i) = \frac{1}{L_i} \int_{x_i - L_i/2}^{x_i + L_i/2} Y(x)dx$$

$$i = 1,2,\ldots,N \qquad (2)$$

where x_i is the centroid of element L_i. If the correlation structure of the field is quadrantally symmetric, i.e.,

$$\rho(-r) = \rho(r) \qquad (3)$$

then the mean vector, variance matrix and covariance matrices of the local averages can be calculated as follows [14]

$$E[Y_{L_i}] = m, \qquad i = 1,2,\ldots,N \qquad (4)$$

$$\text{Var}(Y_{L_i}) = \sigma\gamma(L_i)\sigma, \qquad i=1,2,\ldots,N \qquad (5)$$

$$\text{Cov}(Y_{L_i}, Y_{L_j}) = \sigma\frac{1}{2L_iL_j} \sum_{k=0}^{3} (-1)^k L_k^2\, \gamma(L_k)\sigma$$

$$i, j = 1, 2, \ldots N \qquad (6)$$

where $\sigma = \mathrm{diag}[\sigma_1, \sigma_2, \ldots \sigma_n]$ and $\gamma(L_i)$ is the normalized variance matrix of the local average of $Y(x)$ over L_i, which is related to $\rho(r)$ as follows

$$\gamma(L) = \frac{2}{L} \int_0^L \left(1 - \frac{r}{L}\right)\rho(r)dr \tag{7}$$

and where L_k $(k=0,1,2,3)$ are indicated in Fig. 1.

It can be shown that $\gamma(L)$ has the following properties

$$\gamma(L) = \begin{cases} I, & L \to 0 \\ \Theta/L, & L \to \infty \end{cases} \tag{8}$$

where I is unit matrix and

$$\Theta = [\theta_{ij}] = \int_{-\infty}^{\infty} \rho(r)dr \tag{9}$$

which is the matrix of correlation distance of random vector field $Y(x)$. θ_{ii} is also called the scale of fluctuation [15].

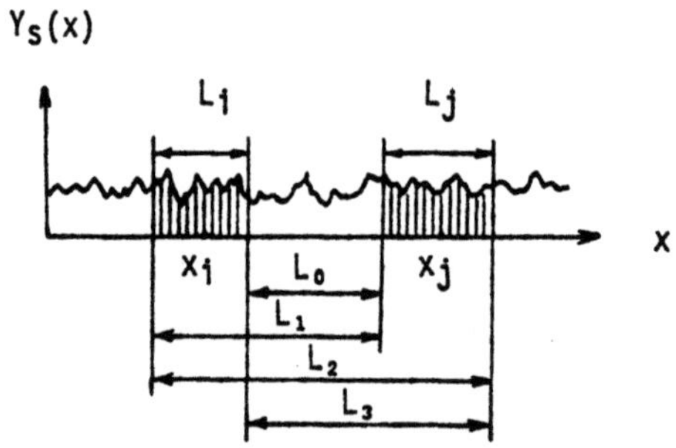

Fig. 1. $L_k(k = 0, 1, 2, 3)$ in equation (6).

Note that formulae in this section can be generalized to the case of multi-dimensional random vector fields [14]. If a random vector field is not homogeneous, or the domain over which the local average of the field is defined is not rectangular, then Gaussian quadrature can be used to calculate numerically the mean vectors and covariance matrices of the local averages of random vector fields [14].

3. Formulation of SFEM for Real Eigenvalue Problems

Suppose that n random material and geometric properties $Y_1(\mathbf{x})$, $Y_2(\mathbf{x})$, ..., $Y_n(\mathbf{x})$, of a structure are cross-correlated, where $\mathbf{x}$ is a point in a q-dimensional domain D occupied by the structure. These properties can be modeled as an n-variate q-dimensional random vector field $\mathbf{Y}(\mathbf{x})=[Y_1(\mathbf{x}), Y_2(\mathbf{x}), \ldots, Y_n(\mathbf{x})]^T$. Divide D into N finite elements and form the local averages of $\mathbf{Y}(\mathbf{x})$ over all elements

$$Y_{ve} = [Y^1_{ve}, Y^2_{ve}, \ldots, Y^n_{ve}]^T = \frac{1}{ve} \int_{ve} Y(x)dx,$$

$$e = 1, 2, \ldots, N \tag{10}$$

Let Y_{ve} represent properties in element e. Following the general procedure of deterministic FEM, we obtain the following characteristic equation for the real eigenvalue problem

$$(K - \lambda_i M)\Phi_i = 0 \tag{11}$$

where K is stiffness matrix, M mass matrix, λ_i the ith eigenvalue and Φ_i the ith eigenvector. The eigenvectors Φ_i satisfy the normalization conditions

$$\Phi_i^T M\Phi_i = 1 \qquad i = 1,2,\ldots \tag{12}$$

K, M, λ_i an Φ_i are stochastic due to the randomness in the material and geometric properties and they are functions of Y_{ve} (e=1,2,...N). We split each stochastic quantity in equation (11) into its mean and random deviation

$$K = \overline{K} + \delta K \tag{13a}$$

$$M = \overline{M} + \delta M \tag{13b}$$

$$\lambda_i = \overline{\lambda}_i + \delta\lambda_i \tag{13c}$$

$$\Phi_i = \overline{\Phi}_i + \delta\Phi_i \tag{13d}$$

Assume that the deviations of random properties from their mean values are small in some statistical sense. Then, in the first approximation, the following two sets of equations can be derived by substituting equations (13a-d) into equations (11) and (12)

$$(\overline{K} - \overline{\lambda}_i\ \overline{M})\ \overline{\Phi}_i = 0 \tag{14}$$

$$\overline{\Phi}_i^T\ \overline{M}\ \overline{\Phi}_i = 1 \tag{15}$$

and

$$(\overline{K} - \overline{\lambda}_i\ \overline{M})\ \delta\Phi_i = \delta f_i \tag{16}$$

$$\overline{\Phi}_i^T\ \overline{M}\ \delta\Phi_i = -\frac{1}{2}\ \overline{\Phi}_i^T\ \delta M\ \overline{\Phi}_i \tag{17}$$

where

$$\delta f_i = -\ \delta K\overline{\Phi}_i + \overline{\lambda}_i\ \delta M\overline{\Phi}_i + \delta\lambda_i\ \overline{M}\ \overline{\Phi}_i \tag{18}$$

and where $\overline{\lambda}_i$ and $\overline{\Phi}_i$ in equations (14) and (15) are the ith eigenvalue and eigenvector of the mean system, respectively; they can be calculated by using deterministic FEM. The calculation of the covariances of eigenvalues and eigenvectors will depend on the multiplicity of $\overline{\lambda}_i$.

3.1 All $\overline{\lambda}_i$ are distinct

If all $\overline{\lambda}_i$ are distinct, then upon pre-multiplying equation (16) by $\overline{\Phi}_i^T$, we obtain

$$\delta\lambda_i = \bar{\Phi}_i^T(\delta K - \bar{\lambda}_i \delta M)\bar{\Phi}_i \tag{19}$$

Similarly, pre-multiplying equation (17) by $\mu_i \bar{M}\,\bar{\Phi}_i$ and adding the result to equation (16), we have

$$(\bar{K} - \bar{\lambda}_i\bar{M} + \mu_i\bar{M}\,\bar{\Phi}_i\bar{\Phi}_i^T\bar{M})\ \delta\Phi_i = \delta f_i - \tfrac{1}{2}\mu_i\bar{M}\,\bar{\Phi}_i\bar{\Phi}_i^T\delta M\bar{\Phi}_i \tag{20}$$

where μ_i is a constant, which takes the value of 2 according to Fox and Kapoor[16] or the same value as λ_i according to Hu[7]. Equation (20) can be rewritten as

$$\bar{D}_i\ \delta\Phi_i = \delta f_i - \frac{1}{2}\ \mu_i\ \bar{M}\ \bar{\Phi}_i\ \bar{\Phi}_i^T\ \delta M\ \bar{\Phi}_i \tag{21}$$

where

$$\bar{D}_i = \bar{K} - \bar{\lambda}_i\ \bar{M} + \mu_i\ \bar{M}\ \bar{\Phi}_i\ \bar{\Phi}_i^T\ \bar{M} \tag{22}$$

In the first approximation, δK and δM can be expressed as

$$\delta K = \sum_{e=1}^{N} \frac{\partial K}{\partial Y_{ve}}\Big|_{\bar{Y}} \delta Y_{ve} = \sum_{e=1}^{N} [K_e^1, K_e^2, \ldots, K_e^n]\delta Y_{ve} \tag{23}$$

$$\delta M = \sum_{e=1}^{N} \frac{\partial M}{\partial Y_{ve}}\Big|_{\bar{Y}} \delta Y_{ve} = \sum_{e=1}^{N} [M_e^1, M_e^2, \ldots, M_e^n]\delta Y_{ve} \tag{24}$$

in which

$$K_e^k = \frac{\partial K}{\partial Y_{ve}^k},\ M_e^k = \frac{\partial M}{\partial Y_{ve}^k},\quad k = 1,2,\ldots n \tag{25}$$

Substituting equations (23) and (24) into equations (19) and (21) yields

$$\delta\lambda_i = \sum_{e=1}^{N} a_e^i\ \delta Y_{ve} = \sum_{e=1}^{N} [a_{1e}^i, a_{2e}^i, \ldots, a_{ne}^i]\delta Y_{ve} \tag{26}$$

$$\bar{D}_i \delta\Phi_i = \sum_{e=1}^{N} b_e^i \, \delta Y_{ve} = \sum_{e=1}^{N} [b_{ie}^i, b_{ie}^i, \ldots, b_{ne}^i] \delta Y_{re} \quad (27)$$

where

$$a_{ke}^i = \bar{\Phi}_i^T (K_e^k - \bar{\lambda}_i M_e^k) \bar{\Phi}_i, \qquad k = 1, 2, \ldots, n \quad (28)$$

$$b_{ke}^i = (- K_e^k + \bar{\lambda}_i M_e^k - \tfrac{1}{2}\mu_i \, \bar{M} \, \bar{\Phi}_i \, \bar{\Phi}_i^T \, M_e^k + a_{ke}^i \bar{M}) \bar{\Phi}_i,$$

$$k = 1, 2, \ldots, n \quad (29)$$

Then the covariances of the eigenvalues and the covariance matrices of the eigenvectors are obtained as follows:

$$\mathrm{cov}(\lambda_i, \lambda_j) = \sum_{r,s=1}^{N} a_r^i \, \mathrm{cov}(Y_{vr}, Y_{vs}) (a_s^j)^T$$

$$i, j = 1, 2, \ldots \quad (30)$$

$$\mathrm{cov}(\Phi_i, \Phi_j) = \bar{D}_i^{-1} \sum_{r,s=1}^{N} b_r^i \, \mathrm{cov}(Y_{vr}, Y_{vs}) (b_s^j)^T \, \bar{D}_j^{-T},$$

$$i, j = 1, 2, \ldots \quad (31)$$

3.2 One λ_i is a r-fold eigenvalue

Suppose that the mean system has a r-fold eigenvalue

$$\bar{\lambda}_i = \lambda_0, \qquad i = i_1, i_2, \ldots, i_r \quad (32)$$

while the other λ_i's are distinct. The sub-matrix Λ = diag $[\lambda_i, \lambda_{i_2}, \ldots, \lambda_{ir}]$ of stochastic eigenvalues and sub-matrix $\boldsymbol{\Phi} = [\phi_{i_1}, \phi_{i_2}, \ldots, \phi_{ir}]$ of eigenvectors corresponding to λ_0 can be expressed as [7]

$$\Lambda = \lambda_0 I + \delta\Lambda \quad (33)$$

$$\boldsymbol{\Phi} = \bar{\boldsymbol{\Phi}} \boldsymbol{\alpha}_0 + \delta\boldsymbol{\Phi} \quad (34)$$

where I is an rth order unit matrix, $\boldsymbol{\alpha}_0$ an rth order constant matrix, $\bar{\boldsymbol{\Phi}} = [\bar{\Phi}_{i_1}, \bar{\Phi}_{i_2}, \ldots, \bar{\Phi}_{ir}]$ the sub-matrix of orthogonal eigenvectors of the mean system corresponding to the r-fold eigenvalue λ_0, and where

$$\delta\Lambda = \text{diag}[\delta\lambda_{i_1}, \delta\lambda_{i_2}, \ldots, \delta\lambda_{ir}] \tag{35}$$

$$\delta\Phi = [\delta\phi_{i_1}, \delta\phi_{i_2}, \ldots, \delta\phi_{ir}] \tag{36}$$

The characteristic equation for these r stochastic eigenvalues and eigenvectors is

$$K\Phi - M\Phi\Lambda = 0 \tag{37}$$

while Φ satisfies the mass orthonormalization condition

$$\Phi^T M\Phi = I \tag{38}$$

It follows upon substituting equations (13a), (13b), (33) and (34) into equations (37) and (38) and proceeding in the same manner as for the case of distinct eigenvalues that

$$\delta A\alpha_0 - \alpha_0\delta\Lambda = 0 \tag{39}$$

$$\alpha_0^T\alpha_0 = I \tag{40}$$

and

$$\overline{D}_0\delta\Phi = \delta f_0 - \frac{1}{2}\mu\,\overline{M}\,\overline{\Phi}\,\overline{\Phi}^T\,\delta M\,\overline{\Phi}\,\alpha_0 \tag{41}$$

where μ is a constant and

$$\delta A = \overline{\Phi}^T(\delta K - \lambda_0\,\delta M)\overline{\Phi} \tag{42}$$

$$\overline{D}_0 = \overline{K} - \lambda_0\overline{M} + \mu\,\overline{M}\,\overline{\Phi}\,\overline{\Phi}^T\,\overline{M} \tag{43}$$

$$\delta f_0 = -\,\delta K\overline{\Phi}\alpha_0 + \lambda_0\,\delta M\overline{\Phi}\alpha_0 + \overline{M}\,\overline{\Phi}\alpha_0\,\delta\Lambda \tag{44}$$

Equations (39) and (40) describe a rth order standard eigenvalue problem. If α_0 is chosen such that no repeated root exists in equations (39) and (40), then $\delta\Lambda$ and α_0 can be uniquely determined and $\delta\Phi$ can be evaluated from equation (41). This procedure was originally proposed by Hu [7].

From equation (39) we obtain

$$\delta A\ \delta A^T\ \alpha_0 - \alpha_0\ \delta\Lambda\ \delta\Lambda^T = 0 \tag{45}$$

Taking expectation of equation (45) yields

$$\mathrm{cov}(A,A)\alpha_0 - \alpha_0\ \mathrm{cov}(\Lambda,\Lambda) = 0 \tag{46}$$

Thus, covariance matrix $\mathrm{cov}(\Lambda,\Lambda)$ and α_0 can be evaluated from equations (46) and (40).

From equations (41) and (39) we obtain

$$\bar{D}_0 \delta\Phi = \delta F \alpha_0 \tag{47}$$

where

$$\delta F = (\bar{M}\ \bar{\Phi}\ \bar{\Phi}^T - I)\ (\delta K - \lambda_0 \delta M)\bar{\Phi} - \frac{1}{2}\ \mu\ \bar{M}\ \bar{\Phi}\ \bar{\Phi}^T\ \delta M\bar{\Phi} \tag{48}$$

Then taking the covariances of both sides of equation (47) leads to

$$\mathrm{cov}(\Phi,\Phi) = \bar{D}_0^{-1} \mathrm{Cov}(F,F)\bar{D}_0^{-T} \tag{49}$$

which is independent of α_0.

Substituting equations (23) and (24) into equations (42) and (48), we obtain

$$\delta A = \sum_{e=1}^{N} A_e\ \delta Y_{ve} = \sum_{e=1}^{N}\ [A_{1e}, A_{2e}, \ldots, A_{ne}]\delta Y_{ve} \tag{50}$$

$$\delta F = \sum_{e=1}^{N} F_e\ \delta Y_{ve} = \sum_{e=1}^{N}\ [F_{1e}, F_{2e}, \ldots, F_{ne}]\delta Y_{ve} \tag{51}$$

where

$$A_{ke} = \bar{\Phi}^T(K_e^k - \lambda_0 M_e^k)\bar{\Phi} \tag{52}$$

$$F_{ke} = (\bar{M}\ \bar{\Phi}\ \bar{\Phi}^T - I)(K_e^k - \lambda_0 M_e^k)\bar{\Phi} - \frac{1}{2}\ \mu\bar{M}\ \bar{\Phi}\ \bar{\Phi}^T\ M_e^k\ \bar{\Phi} \tag{53}$$

From equations (50) and (51), we obtain

$$\mathrm{cov}(A,A) = \sum_{r,s=1}^{N} A_r\ \mathrm{Cov}(Y_{vr}, Y_{vs})\ A_s^T \tag{54}$$

and

$$\mathrm{cov}(F,F) = \sum_{r,s=1}^{N} F_r \,\mathrm{cov}(Y_{vr}, Y_{rs}) F_s^T \tag{55}$$

respectively. Equations (54) and (55) can then be used in equations (46) and (49) to evaluate the covariance matrix $\mathrm{cov}(\Lambda,\Lambda)$ of eigenvalues and the covariance matrix $\mathrm{cov}(\Phi,\Phi)$ of eigenvectors.

4. Example

Consider a cantilever beam with length ℓ = 200 mm, cross-section area s = 25 mm^2 and inertia moment I = 5 2.1 mm^4. The Young's modulus E and mass density m are assumed to be random in space with mean values E=2.058×10^5 MPa and m = 7.8 g/cm^3. They are modeled as a homogeneous random vector field $Y(x) = [Y_1(x), Y_2(x)]^T$, with the coefficients of variation $D_1 = D_2 = D = 0.1$ and a normalized covariance matrix

$$\rho(r) = \begin{bmatrix} \exp[-a|r|/\ell] & b\exp[-a|r|/\ell] \\ b\exp[-a|r|/\ell] & \exp[-a|r|/\ell] \end{bmatrix}$$

where a and b are constants defining the auto-correlation and cross-correlation, respectively, of the two component fields. The correlation distance of the field is $2\ell/a$. When a=0 the random field is reduced to a random vector. When b = 0 the two components of the field become uncorrelated.

The coefficient of variation of the primary eigenvalue obtained by using the present SFEM is shown in Fig. 2. The result obtained by using the SFEM based on mid-point discretization [11] is also shown for comparison. It is seen that the SFEM based on mid-point discretization generally over-estimates while the present SFEM under-estimates the coefficient of variation. The results of the two SFEM procedures tend to bracket the exact solution for any given mesh and approach to the exact solution as the number of elements increases. However, the present SFEM converges more rapidly compared with the SFEM based on mid-point discretization and

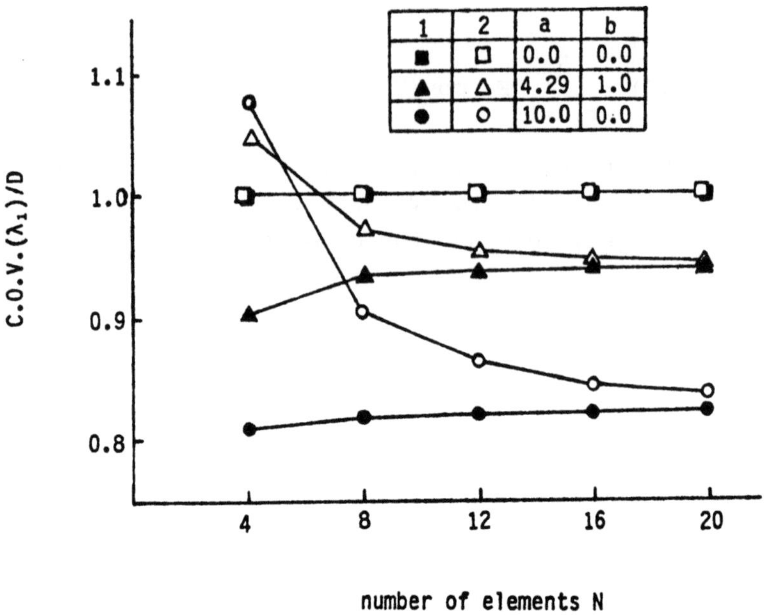

Fig. 2. The normalized coefficient of variation of primary eigenvalue c.o.v.$(\lambda_1)/D$. 1 - the presnt SFEM; 2 - the SFEM based on mid-point discretization[11].

the convergence is almost independent of the correlation distance of the underlaying random field. The convergence of the latter SFEM becomes worse when the correlation distance of the field is shorter. The better convergence of the present SFEM is due to the fact that the covariance matrix of the local averages of a random field reflects well the correlation distance of the underlaying field [14].

The foregoing discussion related to eigenvalues also applies to eigenvectors. The means and standard deviations of the first two modes of the cantilever beam calculated by using the present SFEM with 4 elements are listed in Tab. 1 for the case of a = 4.29 and b = 1.0. The results are almost the same as those obtained by using the SFEM based on mid-point discretization with 16 elements.

Order of Modes		Node No.				
		1	2	3	4	5
1st	m	1000.00	657.75	339.52	97.29	0.00
	σ	38.94	24.07	11.95	3.90	0.00
2nd	m	1000.00	-135.11	-713.75	-417.28	0.00
	σ	45.03	5.08	28.74	13.84	0.00

Table 1: The means and standard deviations of modes (mm).

5 Conclusions

A stochastic finite element method is proposed for the computation of the means and covariances of the eignevalues and eigenvectors of a structure with several spatially distributed and cross-correlated random material and geometric properties. The method is based on the local averages of the random vector field. Both the cases of distinct and repeated eigenvalues are considered. The results obtained for the numerical example indicate that the present SFEM is superior to the SFEM based on the mid-point discretization scheme in that the solution converges more rapidly as the number of elements increases.

Acknowledgement

The support of the National Science Foundation of China is gratefully acknowledged. The authors wish to express their gratitude to Professor Y. K. Lin for his reading of the manuscript and helpful comments.

References

1. Collins, J.D.; Thomson, W. T.: Eigenvalue problem for structural systems with statistical properties, AIAA Journal, 7 (1969) 642-648.

2. Hart, G. C.; Collins, J. D.: The treatment of randomness in finite element modeling, SAE paper No. 700842, 1970.

3. Hasselman, T. K.; Hart, G. C.: Modal analysis of random structural systems, Journal of Engineering Mechanics, ASCE, 98 (1972) 561-579.

4. Shinozuka, M.; Astill, C. J.: Random eigenvalue problems in structural analysis. AIAA Journal, 10 (1972) 456-462.

5. Haug, E. J.; Rousselet, B.: Design sensitivity analysis in structural mechanics, II, Journal of Structural Mechanics, 8 (1980) 161-186.

6. Chen, S. H.; Pan, H.H.: Design sensitivity analysis of vibration modes by finite element perturbation, Proceedings of the 4th International Modal Analysis Conference, 1986, 38-43.

7. Hu, H.C.: Vibration theory of multiple degree of freedom systems, Science Press, Beijiang, 1987 (in Chinese).

8. Kim, O. K.; Wallerstein, D. V.: Modal design sensitivities for multiple eigenvalues, Computer & Structures, 29 (1988) 755-762.

9. Mills-Currran, W. C.: Calculation of eigenvector derivatives for structures with repeated eigenvalues, AIAA Journal 26 (1988) 867-871.

10. Ojalvo, I.U.: Efficient computation of modal sensitivities for systems with repeated frequencies, AIAA Journal, 26 (1988) 361-366.

11. Nakagiri, S.; Hisada, T.: Stochastic finite element method applied to structural analysis with uncertain

parameters, Proceedings of International Conference on Finite Element Methods, 1982, 206-211.

12. Zhu, W. Q.; Wu, W. Q.: The application of the stochastic finite element method based on the local averages of random fields to random eigenvalue problems, Acta Aeronautica Et Astronautica Sinica, 1989, 10(2), 20-26 (in Chinese).

13. Zhu, W.Q.; Ren, Y. J.: Stochastic finite element method based on local averages of random fields, Acta Mechanica Solida Sinica, 1 (1988) 261-271.

14. Zhu, W.Q.; Wu, W.Q.: On the local average of random field in stochastic finite element analysis, Acta Mechanica Solida Sinica, 3 (1990) 1-15.

15. Vanmarcke, E.: Random fields: analysis and synthesis, The MIT Press, Cambridge, 1983.

16. Fox, R. L.; Kapoor, M. R.: Rates of change of eigenvalues and eigenvectors, AIAA Journal, 6 (1968) 2426-2429.

17. Shinozuka, M.; Yamazaki, F.: Stochastic finite element analysis: an introduction, Stochastic Structural Dynamics: progress in Theory and Applications, Eds. by S. T. Ariaratnam, G. I. Schuëller & I. Elishakoff, Elsevier Applied Science, 1988, 241-291.

MIX
Papier aus verantwortungsvollen Quellen
Paper from responsible sources
FSC® C105338

If you have any concerns about our products,
you can contact us on
ProductSafety@springernature.com

In case Publisher is established outside the EU,
the EU authorized representative is:
Springer Nature Customer Service Center GmbH
Europaplatz 3, 69115 Heidelberg, Germany

Printed by Libri Plureos GmbH
in Hamburg, Germany